MX

DNA Microarrays

METHODS EXPRESS

edited by **M. Schena**

ArrayIt Life Sciences Division, TeleChem International Inc., Sunnyvale, USA

Scion

First published 2008

A CIP catalogue record for this book is available from the British Library.

ISBN: 978 1 904842 15 6 (paperback)
ISBN: 978 1 904842 22 4 (hardback)

Scion Publishing Limited
Bloxham Mill, Barford Road, Bloxham, Oxfordshire OX15 4FF
www.scionpublishing.com

Important Note from the Publisher

The information contained within this book was obtained by Scion Publishing Limited from sources believed by us to be reliable. However, while every effort has been made to ensure its accuracy, no responsibility for loss or injury whatsoever occasioned to any person acting or refraining from action as a result of information contained herein can be accepted by the authors or publishers.

Typeset by Phoenix Photosetting, Chatham, Kent, UK
Printed by Cromwell Press, Trowbridge, Wiltshire, UK.

MX

DNA Microarrays

METHODS EXPRESS

Contents

Contributors ix
Preface xiv
Acknowledgements xv
Abbreviations xvi

Color section xix

Chapter 1. Whole human genome microarrays
Mark Schena, Milton J. Friedman, Paul Haje, Todd Martinsky, Greg Suzuki, Tony Costa, Joy Chung, Mel Ruiz, Chris Costa, Luis Machado, Rajiv Raja, and Rupal Desai
1. Introduction 1
2. Methods and approaches 4
   2.1 Recommended protocols 8
3. Troubleshooting 29
   Acknowledgements 29
4. References 30

Chapter 2. Toxicogenomics of dioxin
Craig R. Tomlinson, Saikumar Karyala, Danielle Halbleib, Mario Medvedovic, and Alvaro Puga
1. Introduction 31
   1.1 The aromatic hydrocarbon receptor signaling pathway 32
   1.2 The AHR in homeostasis and development 33
   1.3 Toxicogenomics questions 34
2. Methods and approaches 34
   2.1 Overall approach to a toxicogenomics study: three phases 34
   2.2 Recommended protocols 39
3. Troubleshooting 47
   Acknowledgements 47
4. References 47

Chapter 3. Amplified differential gene expression microarrays
Zhijian J. Chen and Kenneth D. Tew
1. Introduction 51
2. Methods and approaches 52

2.1 Recommended protocols 56
3. Troubleshooting 63
Acknowledgements 64
4. References 64

Chapter 4. DNA microarray detection and genotyping of human papillomavirus
TaeJeong Oh, SookKyung Woo, MyungSoon Kim, and Sungwhan An
1. Introduction 65
2. Methods and approaches 66
2.1 Target construction and microarray fabrication 66
2.2 Sensitivity and specificity of the HPV genotyping microarray 71
2.3 Probe labeling, hybridization, and detection 72
2.4 HPV genotyping microarrays for clinical analysis 73
2.5 Recommended protocols 75
3. Troubleshooting 79
Acknowledgements 80
4. References 80

Chapter 5. Expression profiling of transcriptional start sites
Mutsumi Kanamori-Katayama, Shintaro Katayama, Harukazu Suzuki, Yoshihide Hayashizaki
1. Introduction 81
2. Methods and approaches 83
2.1 Bioinformatic identification of promoters, target sequence design, and microarray manufacturing 83
2.2 Transcripts and tags 84
2.3 Tag cluster definition 84
2.4 Target sequence preparation and target design 84
2.5 Synthesis and amplification of first-strand cDNA 85
2.6 Fluorescent cRNA synthesis 87
2.7 Columns; detection sensitivity in oligonucleotide microarrays 88
2.8 Recommended protocols 91
3. Troubleshooting 100
Acknowledgements 100
Note added in proof 100
4. References 101

Chapter 6. Methods for increasing the utility of microarray data
Kazuro Shimokawa, Rimantas Kodzius, Yonehiro Matsumura, and Yoshihide Hayashizaki
1. Introduction 103
2. Methods and approaches 104
2.1 Spot reliability evaluation 104
2.2 Theoretical background of the method 106
2.3 Calculating absolute expression values 108

2.4 Confirming absolute expression data conversions 110
2.5 Conclusions 112
2.6 Recommended protocols 112
3. Troubleshooting 113
Acknowledgements 114
4. References 114

Chapter 7. Key features of bacterial artificial chromosome microarray production and use
Timon P.H. Buys, Ian M. Wilson, Bradley P. Coe, William W. Lockwood, Jonathan J. Davies, Raj Chari, Ronald J. DeLeeuw, Ashleen Shadeo, Calum MacAulay, and Wan I. Lam
1. Introduction 115
1.1 Surveys of DNA copy number changes 115
1.2 BAC microarrays 116
1.3 Platform choice 118
2. Methods and approaches 118
2.1 Preparation of BAC clones for printing 118
2.2 Surface chemistry 120
2.3 Microarray printing and processing 120
2.4 BAC microarray probe preparation 124
2.5 BAC microarray hybridization and washing 128
2.6 BAC microarray scanning and experimental analysis 131
2.7 Recommended protocols 135
3. Troubleshooting 139
Acknowledgements 142
4. References 143

Chapter 8. Epigenetic analysis of cellular immortalization
Aviva Levine Fridman, Scott A. Tainsky, and Michael A. Tainsky
1. Introduction 147
2. Methods and approaches 149
2.1 Methods for detecting CpG island methylation 149
2.2 Recommended protocols 152
3. Troubleshooting 158
4. References 159

Chapter 9. Microarray comparative genomic hybridization
Simon Hughes, Richard Houlston, and Jeremy A. Squire
1. Introduction 161
2. Methods and approaches 162
2.1 Recommended protocols 163
3. Troubleshooting 177
4. References 177

Chapter 10. RNA sample preparation and small-quantity RNA profiling for microarray biomarker discovery
Jianyong Shou and Lawrence M. Gelbert
1. Introduction 179
2. Methods and approaches 180
   2.1 Recommended protocols 186
3. Troubleshooting 194
4. Conclusions 198
   Acknowledgements 198
5. References 198

Chapter 11. DNA microarrays to study nonhuman primate gene expression
Stephen J. Walker
1. Introduction 201
2. Methods and approaches 203
   2.1 Human sequence-based microarrays 203
   2.2 Chapter aims 204
   2.3 Recommended protocols 205
3. Troubleshooting 223
4. References 224

Chapter 12. Enhanced microarray hybridization using surface acoustic wave mixing
Natalie Stickle, Kelly Jackson, Andreas Toegl, Frank Feist, and Neil Winegarden
1. Introduction 225
2. Methods and approaches 227
   2.1 Labeling of cDNA 227
   2.2 Surface acoustic wave mixing 227
   2.3 Recommended protocols 228
3. Troubleshooting 237
4. References 237

Appendix 1
List of suppliers 238

Index 241

Contributors

An, Sungwhan Division of Molecular Diagnostics, R&D Center of GenomicTree, Inc., 461-58 Jonmin dong Yusong Daejon, 305-811, South Korea. E-mail: info@genomictree.com

Buys, Timon P.H. British Columbia Cancer Research Centre, 675 West 10th Avenue, Vancouver, BC, Canada V5Z 1L3. E-mail: tbuys@bccrc.ca

Chari, Raj British Columbia Cancer Research Centre, 675 West 10th Avenue, Vancouver, BC, Canada V5Z 1L3. E-mail: rchari@bccrc.ca

Chen, Zhijian J. Department of Cell and Molecular Pharmacology and Experimental Therapeutics, Medical University of South Carolina, 173 Ashley Avenue, PO Box 250505, Charleston, SC 29425, USA. Present address: RheoGene, Inc., 2650 Eisenhower Avenue, Norristown, PA 19403, USA.

Chung, Joy ArrayIt Life Sciences Division, TeleChem International, Inc., 524 East Weddell Drive, Sunnyvale, CA 94089, USA. E-mail: joy@arrayit.com

Coe, Bradley P. British Columbia Cancer Research Centre, 675 West 10th Avenue, Vancouver, BC, Canada V5Z 1L3. E-mail: bcoe@bccrc.ca

Costa, Chris ArrayIt Life Sciences Division, TeleChem International, Inc., 524 East Weddell Drive, Sunnyvale, CA 94089, USA. E-mail: chris@arrayit.com

Costa, Tony ArrayIt Life Sciences Division, TeleChem International, Inc., 524 East Weddell Drive, Sunnyvale, CA 94089, USA. E-mail: tony@arrayit.com

Davies, Jonathan J. British Columbia Cancer Research Centre, 675 West 10th Avenue, Vancouver, BC, Canada V5Z 1L3. E-mail: jdavies@bccrc.ca

DeLeeuw, Ronald J. British Columbia Cancer Research Centre, 675 West 10th Avenue, Vancouver, BC, Canada V5Z 1L3.

Desai, Rupal MDS Analytical Technologies, 1311 Orleans Drive, Sunnyvale, CA 94089, USA. E-mail: rupal.desai@moldev.com

Feist, Frank Advalytix, Eugen-Sänger-Ring 4, Gewerbegebiet Brunnthal Nord, 85649 Brunnthal, Germany. E-mail: frank.feist@olympus-europa.com

Fridman, Aviva Levine Program in Molecular Biology and Human Genetics, Barbara Ann Karmanos Cancer Institute, and Department of Pathology, Wayne State University School of Medicine, 110 Warren Avenue, Detroit, MI 48201, USA. E-mail: fridmana@karmanos.org

Friedman, Milton J. ArrayIt Life Sciences Division, TeleChem International, Inc., 524 East Weddell Drive, Sunnyvale, CA 94089, USA. E-mail: milt@arrayit.com

Gelbert, Lawrence M. Lilly Research Laboratories, Eli Lilly and Company, Lilly Corporate Center DC 0424, Indianapolis, IN 46285, USA. E-mail: gelbert@lilly.com

Haje, Paul ArrayIt Life Sciences Division, TeleChem International, Inc., 524 East Weddell Drive, Sunnyvale, CA 94089, USA. E-mail: paul@arrayit.com

Halbleib, Danielle Department of Environmental Health and Center for Environmental Genetics, and Division of Environmental Genetics & Molecular Toxicology, University of Cincinnati, Cincinnati, OH 45267-0056 USA. E-mail: danielle.halbleib@uc.edu

Hayashizaki, Yoshihide Laboratory for Genome Exploration Research Group, RIKEN Genomic Sciences Center (GSC), Yokohama Institute 1-7-22 Suehiro-cho, Tsurumi-ku, Yokohama, Kanagawa 230-0045, Japan; and Genome Science Laboratory, RIKEN, 2-1 Hirosawa, Wako, Saitama 351-0198, Japan. E-mail: rgscerg@gsc.riken.go.jp

Houlston, Richard Section of Cancer Genetics, Institute of Cancer Research, 15 Cotswold Road, Sutton, Surrey SM2 5NG, UK. E-mail: Richard.Houlston@icr.ac.uk

Hughes, Simon Tumour Biology Laboratory, John Vane Science Centre, Cancer Research UK Clincial Centre, Queen Mary's School of Medicine and Dentistry, Charterhouse Square, London EC1M 6BQ, UK. E-mail: simon.hughes@cancer.org.uk

Jackson, Kelly University Health Network Microarray Centre, Toronto, Ontario, Canada. E-mail: kjackson@uhnresearch.ca

Kanamori-Katayama, Mutsumi Laboratory for Genome Exploration Research Group, RIKEN Genomic Sciences Center (GSC), Yokohama Institute 1-7-22 Suehiro-cho, Tsurumi-ku, Yokohama, Kanagawa 230-0045, Japan; and Genome Science Laboratory, RIKEN, 2-1 Hirosawa, Wako, Saitama 351-0198, Japan. E-mail: rgscerg@gsc.riken.jp

Karyala, Saikumar Department of Environmental Health and Center for Environmental Genetics, and Division of Environmental Genetics and Molecular Toxicology, University of Cincinnati, Cincinnati, OH 45267-0056, USA. E-mail: microarray@uc.edu

Katayama, Shintaro Laboratory for Genome Exploration Research Group, RIKEN Genomic Sciences Center (GSC), Yokohama Institute 1-7-22 Suehiro-cho, Tsurumi-ku, Yokohama, Kanagawa 230-0045, Japan; and Genome Science Laboratory, RIKEN, 2-1 Hirosawa, Wako, Saitama 351-0198, Japan. E-mail: rgscerg@gsc.riken.jp

Kim, MyungSoon Division of Molecular Diagnostics, R&D Center of GenomicTree, Inc., 461-58 Jonmin-dong Yusong Daejon, 305-811, South Korea. E-mail: info@genomictree.com

Kodzius, Rimantas Laboratory for Genome Exploration Research Group, RIKEN Genomic Sciences Center (GSC), Yokohama Institute 1-7-22 Suehiro-cho, Tsurumi-ku, Yokohama, Kanagawa 230-0045, Japan; and Genome Science Laboratory, RIKEN, 2-1 Hirosawa, Wako, Saitama 351-0198, Japan. E-mail: rgscerg@gsc.riken.jp

Lam, Wan L. British Columbia Cancer Research Centre, 675 West 10th Avenue, Vancouver, BC, Canada V5Z 1L3. E-mail: wanlam@bccrc.ca

Lockwood, William W. British Columbia Cancer Research Centre, 675 West 10th Avenue, Vancouver, BC, Canada V5Z 1L3. E-mail: wlockwood@bccrc.ca

MacAulay, Calum British Columbia Cancer Research Centre, 675 West 10th Avenue, Vancouver, BC, Canada V5Z 1L3. E-mail: cmacaula@bccrc.ca

Machado, Luis ArrayIt Life Sciences Division, TeleChem International, Inc., 524 East Weddell Drive, Sunnyvale, CA 94089, USA. E-mail: luis@arrayit.com

Martinsky, Todd ArrayIt Life Sciences Division, TeleChem International, Inc., 524 East Weddell Drive, Sunnyvale, CA 94089, USA. E-mail: todd@arrayit.com

Matsumura, Yonehiro Laboratory for Genome Exploration Research Group, RIKEN Genomic Sciences Center (GSC), Yokohama Institute 1-7-22 Suehiro-cho, Tsurumi-ku, Yokohama, Kanagawa 230-0045, Japan; and Division of Genomic Information Resource Exploration, Science of Biological Supramolecular Systems, Graduate School of Integrated Science, Yokohama City University, 1-7-29 Suehiro-Cho, Tsurumi-Ku, Yokohama, Kanagawa 230-0045, Japan. E-mail: rgscerg@gsc.riken.go.jp

Medvedovic, Mario University of Cincinnati, Department of Environmental Health and Center for Environmental Genetics, Division of Biostatistics and Epidemiology, and Center for Genome Information, University of Cincinnati, Cincinnati, OH 45267-0056 USA. E-mail: Mario.Medvedovic@uc.edu

Oh, TaeJeong Division of Molecular Diagnostics, R&D Center of GenomicTree, Inc., 461-58 Jonmin-dong Yusong Daejon, 305-811, South Korea. E-mail: tjoh@genomictree.com

Puga, Alvaro Department of Environmental Health and Center for Environmental Genetics, and Division of Environmental Genetics & Molecular Toxicology, University of Cincinnati, Cincinnati, OH 45267-0056 USA. E-mail: Alvaro.Puga@uc.edu

Raja, Rajiv MDS Analytical Technologies, 1311 Orleans Drive, Sunnyvale, CA 94089, USA. E-mail: rajiv.raja@moldev.com

Ruiz, Mel ArrayIt Life Sciences Division, TeleChem International, Inc., 524 East Weddell Drive, Sunnyvale, CA 94089, USA. E-mail: mel@arrayit.com

Schena, Mark ArrayIt Life Sciences Division, TeleChem International, Inc., 524 East Weddell Drive, Sunnyvale, CA 94089, USA. E-mail: mark@arrayit.com

Shadeo, Ashleen British Columbia Cancer Research Centre, 675 West 10th Avenue, Vancouver, BC, Canada V5Z 1L3. E-mail: ashadeo@bccrc.ca

Shimokawa, Kazuro Laboratory for Genome Exploration Research Group, RIKEN Genomic Sciences Center (GSC), Yokohama Institute 1-7-22 Suehiro-cho, Tsurumi-ku, Yokohama, Kanagawa 230-0045, Japan. E-mail: rgscerg@gsc.riken.go.jp

Shou, Jianyong Lilly Research Laboratories, Eli Lilly and Company, Lilly Corporate Center DC 0424, Indianapolis, IN 46285, USA. E-mail: shou@lilly.com

Squire, Jeremy A. Division of Applied Molecular Oncology, Ontario Cancer Institute, Princess Margaret, Hospital, 610 University Avenue, Toronto, Ontario, Canada M5G 2M9. E-mail: jeremy.squire@utoronto.ca

Stickle, Natalie University Health Network Microarray Centre, Toronto, Ontario, Canada. E-mail: nstickle@uhnresearch.ca

Suzuki, Greg ArrayIt Life Sciences Division, TeleChem International, Inc., 524 East Weddell Drive, Sunnyvale, CA 94089, USA. E-mail: greg@arrayit.com

Suzuki, Harukazu Laboratory for Genome Exploration Research Group, RIKEN Genomic Sciences Center (GSC), Yokohama Institute 1-7-22 Suehiro-cho, Tsurumi-ku, Yokohama, Kanagawa 230-0045, Japan; and Genome Science Laboratory, RIKEN, 2-1 Hirosawa, Wako, Saitama 351-0198, Japan. E-mail: rgscerg@gsc.riken.jp

Tainsky, Michael A. Program in Molecular Biology and Human Genetics, Barbara Ann Karmanos Cancer Institute, and Department of Pathology, Wayne State University School of Medicine, 110 Warren Avenue, Detroit, MI 48201, USA. E-mail: tainskym@karmanos.org

Tainsky, Scott A. Program in Molecular Biology and Human Genetics, Barbara Ann Karmanos Cancer Institute, and Department of Pathology, Wayne State

University School of Medicine, 110 Warren Avenue, Detroit, MI 48201, USA. E-mail: tainskys@karmanos.org

Tew, Kenneth D. Department of Cell and Molecular Pharmacology and Experimental Therapeutics, Medical University of South Carolina, 173 Ashley Avenue, PO Box 250505, Charleston, SC 29425, USA. E-mail: tewk@musc.edu

Toegl, Andreas Advalytix, Eugen-Sänger-Ring 4, Gewerbegebiet Brunnthal Nord, 85649 Brunnthal, Germany. E-mail: toegl@advalytix.de

Tomlinson, Craig R. Department of Environmental Health and Center for Environmental Genetics, and Division of Environmental Genetics and Molecular Toxicology, University of Cincinnati, Cincinnati, OH 45267-0056, USA. Present address: Department of Medicine and Department of Pharmacology and Toxicology, Dartmouth Hitchcock Medical Center, Dartmouth College, One Medical Center Drive, Lebanon, NH 03756. E-mail: Craig.R.Tomlinson@Dartmouth.edu

Walker, Stephen J. Department of Physiology, and Pharmacology and Neuroscience Program, Wake Forest University Health Sciences, Winston-Salem, NC 27101, USA. E-mail: swalker@wfubmc.edu

Wilson, Ian M. British Columbia Cancer Research Centre, 675 West 10th Avenue, Vancouver, BC, Canada V5Z 1L3. E-mail: iwilson@bccrc.ca

Winegarden, Neil University Health Network Microarray Centre, Toronto, Ontario, Canada. E-mail: winegard@uhnres.utoronto.ca

Woo, SookKyung Division of Molecular Diagnostics, R&D Center of GenomicTree, Inc., 461-58 Jonmin-dong Yusong Daejon, 305-811, South Korea. E-mail: info@genomictree.com

Preface

I must admit to an almost childish enthusiasm these days. Don't ask me why – it is just a feeling I have. Perhaps the enthusiasm flows from the realization that at no time in history have we ever had such enormous scientific acumen, and it is indeed our scientific and technological prowess that gives us an increasingly greater control of our collective destiny. And 'collective destiny' needs to be the operative term here, because obviously as the world becomes smaller we each share a greater burden not only in looking after ourselves but in taking care of each other. And how much easier it is to shoulder the needs of an expanding global population when one has the scientific wherewithal to do so. I think we can all agree that we are in a luxurious position scientifically, but how do we parlay this into advancing the human condition?

Let me be so bold as to suggest that we now have a toolkit with sufficient gain to deliver major advances in healthcare and agriculture, which in reverse order are the disciplines on which we depend for nourishment and soundness of mind and body. This new high-tech toolbox contains blueprints (genomic sequences), devices with which to decipher the blueprints (DNA microarrays), and an efficient means for making sense of the information (computers). Biochemistry's great triumvirate of genomic sequence, DNA microarray, and computer, combined with the vast knowledgebase of medicine and botany, should forthwith afford a complete understanding of genomic function in real time (gene expression), tell us how mistakes in the blueprint predispose disease (genotyping), identify individuals who bear signs of current or impending ailments (diagnostics), suggest safe and effective treatments for our maladies (drugs), quite possibly allow us to correct defective genomic segments that give rise to mental and physical disorders (gene therapy), and almost certainly supply us with a continuous and sustainable stream of new crop plants with improved nutritional attributes and growth characteristics. With DNA microarrays as the centerpiece, I believe these goals are attainable over the next several decades.

Against this backdrop, it is therefore little wonder that I jumped at the opportunity to edit a DNA microarrays methods book for Scion Publishing Ltd, not only because of the timeliness of the project and the enormous potential of DNA microarray technology, but because it is such a privilege to orchestrate a compendium of laboratory protocols provided by the world's finest microarray scientists. It has been equally flattering to work with Dr Jane Hoyle whose copy-editing skills were essential for producing highly accurate and readable content, and with Dr David Hames, Dr Jonathan Ray, and their colleagues who synergistically pilot the innovative publishing juggernaut that is Scion. Indeed, it is David's original vision for a methods book on DNA microarrays that drove the publication of the very first protocols book in the field entitled *DNA Microarrays: a Practical*

Approach published by Oxford University Press in 1999. The Scion book, *DNA Microarrays: Methods Express*, builds upon the Oxford experience and greatly extends the breadth and depth of the content by presenting detailed protocols for whole human genome microarrays and other DNA microarray applications that were but science fiction in 1999.

If the next eight years are as dynamic as the previous eight and the contents of this book are eventually rendered provincial, then I will have succeeded in my efforts to provide a compilation of DNA microarray laboratory procedures that are useful, enabling, and catalytic, and inspire the kind of scientific discovery that was hitherto little more than musing and fantasy. And while fame and fortune await those in the microarray field who are able to turn daydreams into discoveries and imagination into invention, neither fame nor fortune should blinker us to the fact that we have a moral imperative to give back more than we take, both for our own sake and for the benefit of a burgeoning global population that understands little about what we do or how we do it, but looks to us increasingly for the scientific and technological advances that we are uniquely positioned to provide.

Mark Schena
August 2007

Acknowledgements

Dedicated in memory of Dr Daniel E. Koshland, Jr.

Abbreviations

5-aza-dC	5-aza-deoxycytidine
ADGE	amplified differential gene expression
AHR	aryl hydrocarbon receptor
ANOVA	analysis of variance
aRNA	amplified RNA
ARNT	aryl hydrocarbon receptor nuclear translocator
BAC	bacterial artificial chromosome
BCP	bromochloropropane
BSA	bovine serum albumin
CAGE	cap analysis of gene expression
CCD	charge-coupled device
CGH	comparative genomic hybridization
Ct	threshold cycle
CTAB	cetyl triammonium bromide
DEPC	diethylpyrocarbonate
DMSO	dimethylsulfoxide
DOP-PCR	degenerate-oligonucleotide-primer PCR
DTT	dithiothreitol
EST	expressed sequence tag
GAPDH	glyceraldehyde 3-phosphate dehydrogenase
GIS	gene identification signature
GSC	genome signature cloning
HPV	human papillomavirus
IVT	*in vitro* transcription
KEGG	Kyoto Encyclopedia of Genes and Genomes
LB	Luria–Bertani
LM-PCR	linker-mediated PCR
MIAME	Minimum Information About a Microarray Experiment
MPG	magnetic porous glass
NHP	nonhuman primate
PBMC	peripheral blood mononuclear cell
PBS	phosphate-buffered saline
PCR	polymerase chain reaction
QRT-PCR	quantitative reverse transcriptase polymerase chain reaction
READ	Riken Expression Microarray Database
RIN	RNA integrity number

RT	reverse transcriptase
RT-PCR	reverse transcriptase polymerase chain reaction
RTS	representative transcript set
SAGE	serial analysis of gene expression
SAW	surface acoustic wave
SDS	sodium dodecyl sulfate
SMRT	Submegabase-resolution tiling set
SNP	single-nucleotide polymorphism
SRED	spot reliability evaluation score for DNA microarrays
SSC	saline sodium citrate
TCDD	2,3,7,8-tetrachlorodibenzo-*p*-dioxin
TGF-β	transforming growth factor-β
t.p.m.	transcripts per million
TSS	transcription start site
TU	transcriptional unit
X-gal	5-bromo-4-chloro-3-indolyl β-D-galactopyranoside

Color section

Chapter 1. Whole human genome microarrays

Figure 5. A gene expression readout for all of the genes in the human genome (see page 7). The ArrayIt H25K whole human genome microarray was hybridized with a fluorescent probe mixture derived from human mRNA. Amplification and labeling of the mRNA was performed using the SenseAmp and cDNA synthesis kits (Genisphere). Aminoallyl-modified cDNA products were labeled with Cy3 and Cy5 dyes (Amersham) and hybridized to H25K according to the instructions of the manufacturers. H25K was scanned for fluorescence emission using an Axon 4000B microarray scanner (Molecular Devices). TIFF data corresponding to expression values for 25 509 human genes and 795 controls were coded to a color palette with IMAGEJ software (NIH). The scanned image corresponds to an 18 × 54 mm printed microarray.

Chapter 2. Toxicogenomics of dioxin

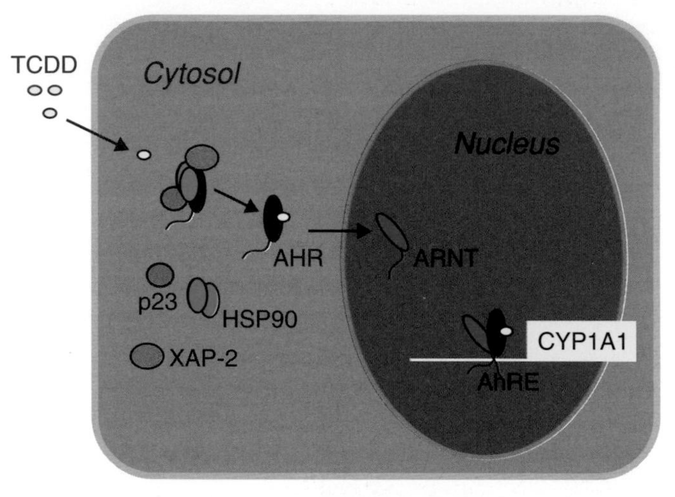

Figure 1. AHR signaling pathway (see page 32).
The chemical toxin TCDD moves across the plasma membrane into the cytosol where it binds to a high-affinity AHR that exists in the unbound state as a complex with other cellular proteins. TCDD-bound AHR moves into the nucleus, binds specifically to high-affinity chromosomal sites, and regulates the expression of cellular genes.

Chapter 3. Amplified differential gene expression microarrays

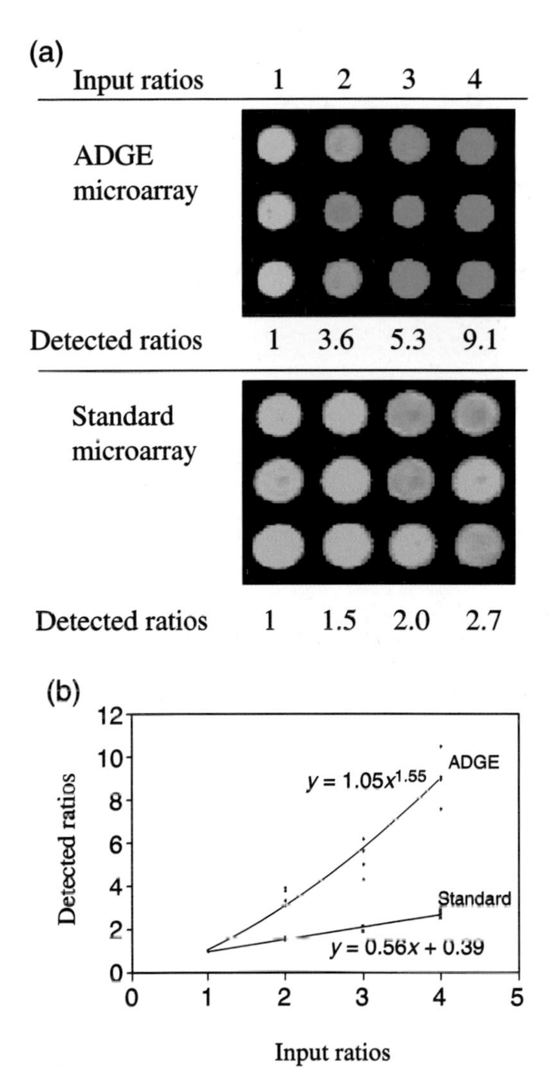

Figure 2. Comparative results of ratio magnification with an ADGE microarray versus a traditional microarray (**see page 54****).**
The clones corresponding to the 12 spots on a chip were amplified by PCR using primers bearing a *TaqI* site at the end. After digesting with *TaqI*, an equivalent amount of DNA from each clone was ligated to the CT or TT adapters. The CT and TT adapter-linked DNA fragments were mixed in ratios of 1 : 1 for the three clones of the first column, 1 : 2 for clones of the second column, 1 : 3 for clones of the third column, and 1 : 4 for clones of the fourth column. The mixed probes were hybridized to the microarray chip. The detected ratios are averages of three spots in each column normalized to the value of the first column (*a*). The relationship between detected ratios (y) and input ratios (x) was $y=1.05x^{1.55}$ with $R^2=0.97$ for ADGE microarray and $y=0.56x+0.39$ with $R^2=0.96$ for standard microarray (*b*).

Chapter 4. DNA microarray detection and genotyping of human papillomavirus

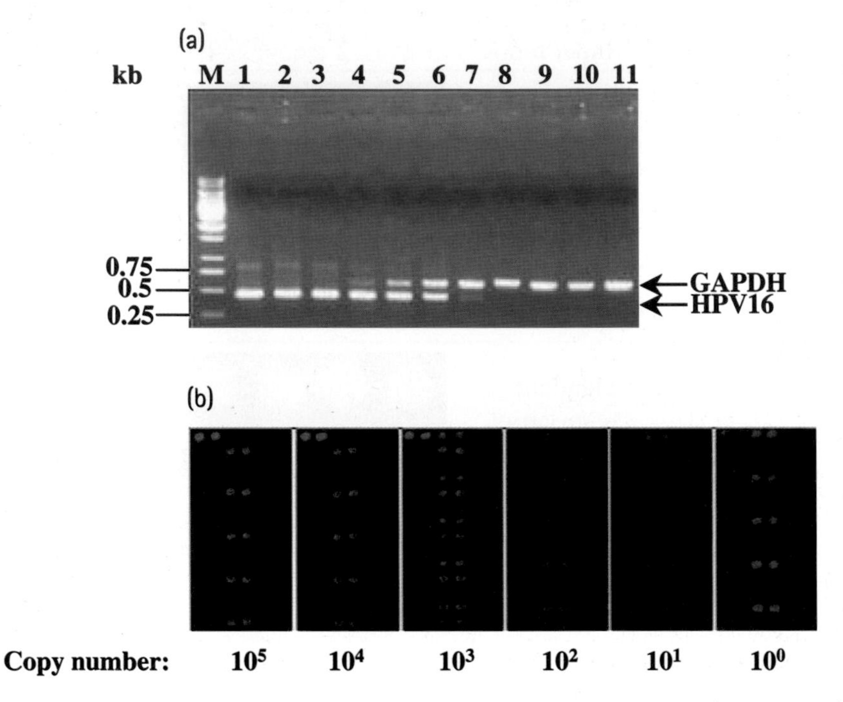

Figure 3. Comparison of the sensitivity of PCR versus an HPV DNA microarray (see page 71).
(*a*) Agarose gel electrophoresis of MYH PCR products with a serial dilution of the HPV16 plasmid. Lanes: M, 1 kb ladder; 1, 10^{10} copies; 2, 10^9 copies; 3, 10^8 copies; 4, 10^7 copies; 5, 10^6 copies; 6, 10^5 copies; 7, 10^4 copies; 8, 10^3 copies; 9, 10^2 copies; 10, 10^1 copies; 11, 10^0 copy. (*b*) HPV DNA microarray hybridization results using amplicons generated by MYH PCR starting with the designated number of copies of the HPV16 plasmid.

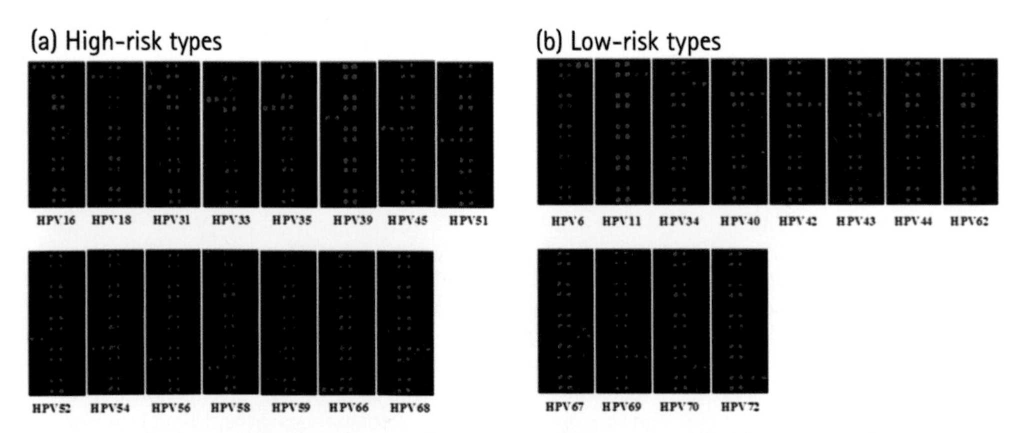

Figure 4. Microarray hybridization specificity of 27 HPV types amplified from plasmids (see page 72).
(*a*) Hybridization results with high-risk HPV types. (*b*) Hybridization results with low-risk HPV types. The HPV type is indicated below each microarray scan.

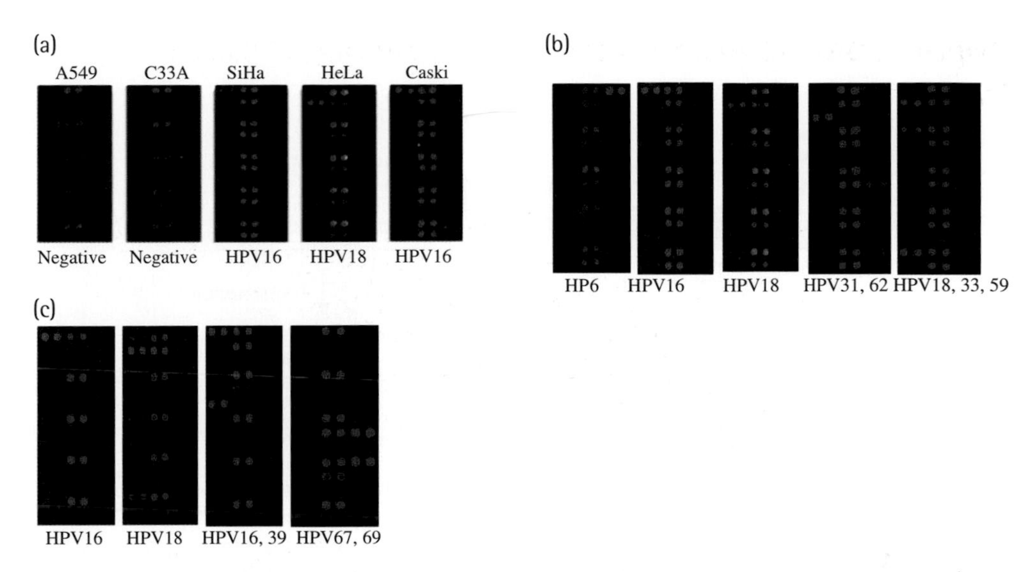

Figure 5. HPV genotyping assay using cell lines and cervical scrapes as the starting samples (see page 74).
(*a*) The HPV-negative cell lines A549 and C33A showed hybridization signals with PCR control targets only, whereas the HPV-positive cell lines SiHa, HeLa, and Caski showed positive signals on the PCR controls, HPV-positive controls and each cognate type-specific target. (*b*) HPV genotyping results using HPV DNA microarrays and cervical scrapes as the starting samples. (*c*) HPV genotyping results using 30 nt type-specific oligonucleotide targets with cervical scrapes as the starting samples. Each HPV type is designated below each microarray scan.

Chapter 5. Expression profiling of transcriptional start sites

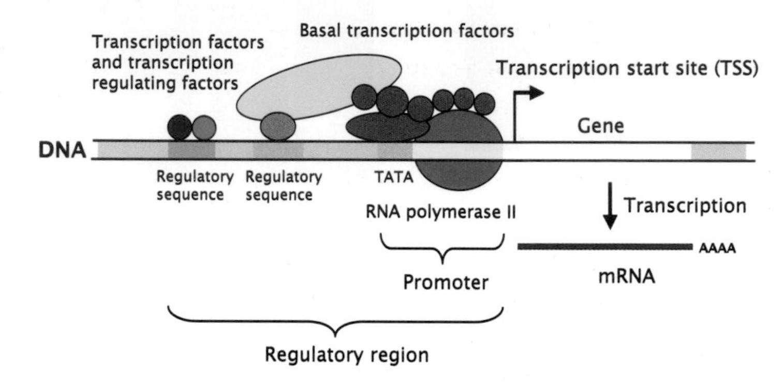

Figure 1. Schematic illustration of the regulatory region of a typical eukaryotic gene (see page 82). The promoter is the *cis*-acting DNA sequence where the general transcription factors and the polymerase assemble. Promoters typically contain one or more canonical core promoter sequences such as the TATA box. Gene regulatory proteins including transcription factors and transcription regulating factors bind to distal regulatory or enhancer sites and modulate the rate of transcription initiation. The transcription start site (TSS) defines the nucleotide position wherein mRNA synthesis initiates. Cellular mRNAs are often polyadenylated (AAAA) at the 3′ end of the nascent transcript.

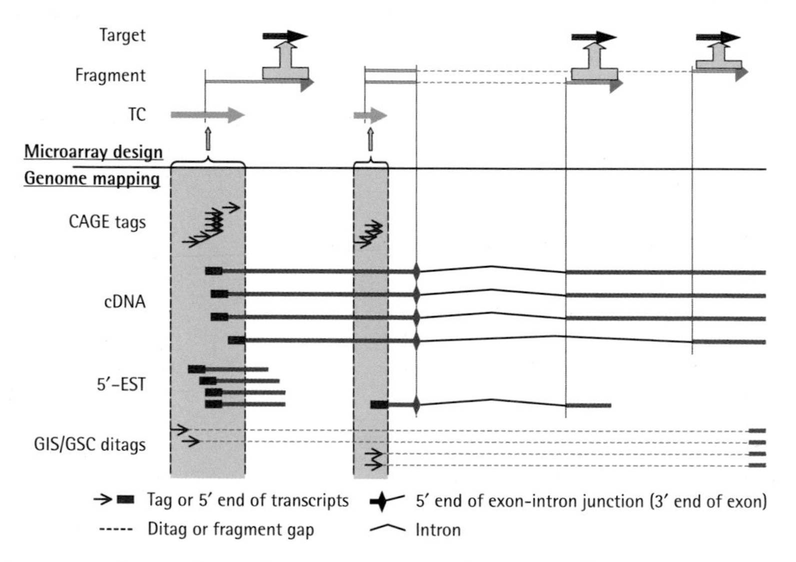

Figure 2. Concepts of target design for TSS microarrays (see page 85).
Lines in the bottom half of the figure depict full-length transcripts or cDNAs (orange) and the 5′ ends (CAGE, 5′-EST or GIS/GSC) aligned to the genomic sequence. The upper portion of the figure shows the regions used for microarray design including the actual microarray target sequences (dark blue arrows). The tag cluster represents overlapping regions of transcripts and 5′-end tags. The fragment for target design (turquoise arrow) begins from the major transcription start site and extends 120 bp from that site. If the fragment overlaps the 5′ end of an exon–intron junction, the fragment 'jumps' the intron to the closest 3′ exon. In the target design phase, double-stranded molecules are chosen using primer specificity and amplification efficiency as criteria. According to the Agilent Probe Design Service, appropriate 60 bp regions in each fragment would be suggested for use as microarray targets.

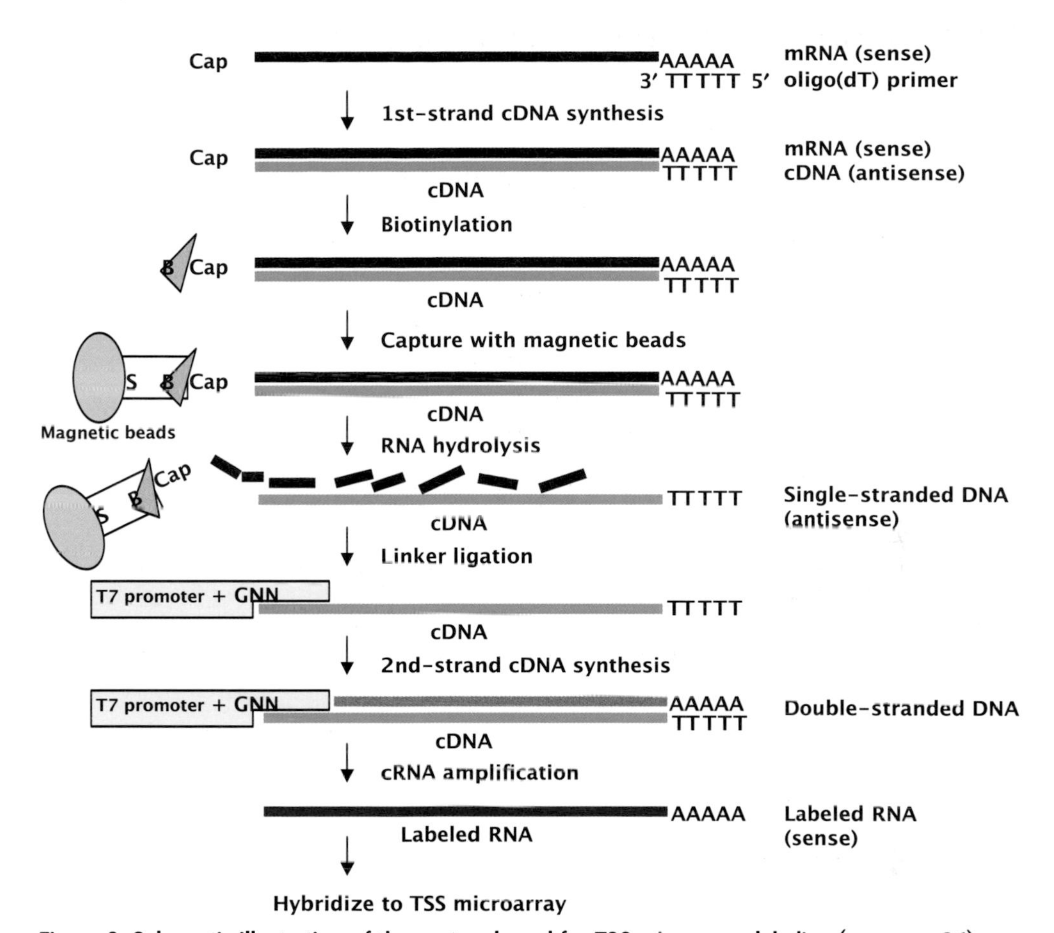

Figure 3. Schematic illustration of the protocol used for TSS microarray labeling (see page 86). To avoid amplification biases and to ensure labeling from bona fide 5′ ends, 'cap-trapper' full-length cDNA selection was combined with cRNA amplification as a new labeling technique suitable for TSS microarray experiments. Cellular mRNAs (blue lines) are primed with oligo(dT) (TTTTT) and extended with RT to yield nascent cDNA (green lines). 'Cap trapping' with biotin (B)- and streptavidin (S)-conjugated magnetic beads highly enriches the number of full-length cDNAs. A primer sequence containing a T7 promoter sequence is annealed to the 5′ ends of the cDNAs, followed by second-strand (pink line) cDNA synthesis. The cRNA (red line) for hybridization to TSS microarrays is synthesized from the double-stranded cDNA templates using T7 RNA polymerase (see protocols for details).

Chapter 7. Key features of bacterial artificial chromosome microarray production and use

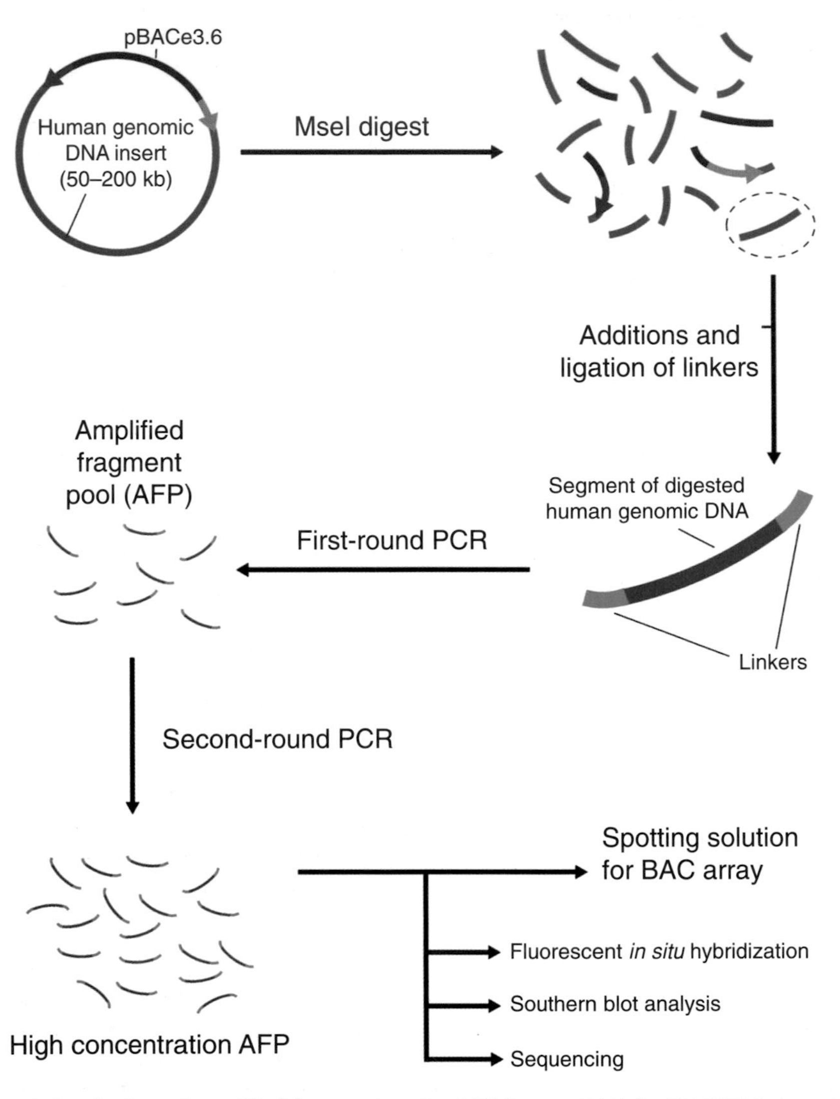

Figure 1. Production of amplified fragment pools of BAC clone DNA by LM-PCR for use in microarray printing (see page 119).
BAC clone DNA is isolated and digested with *Mse*I to release the human genomic insert from the vector. Linkers are then ligated to the *Mse*I-digested ends of the genomic insert and amplified by LM-PCR. The remaining product is further amplified using another round of LM-PCR, which is used to verify the clone identity and the creation of a printing sample for microarray manufacturing.

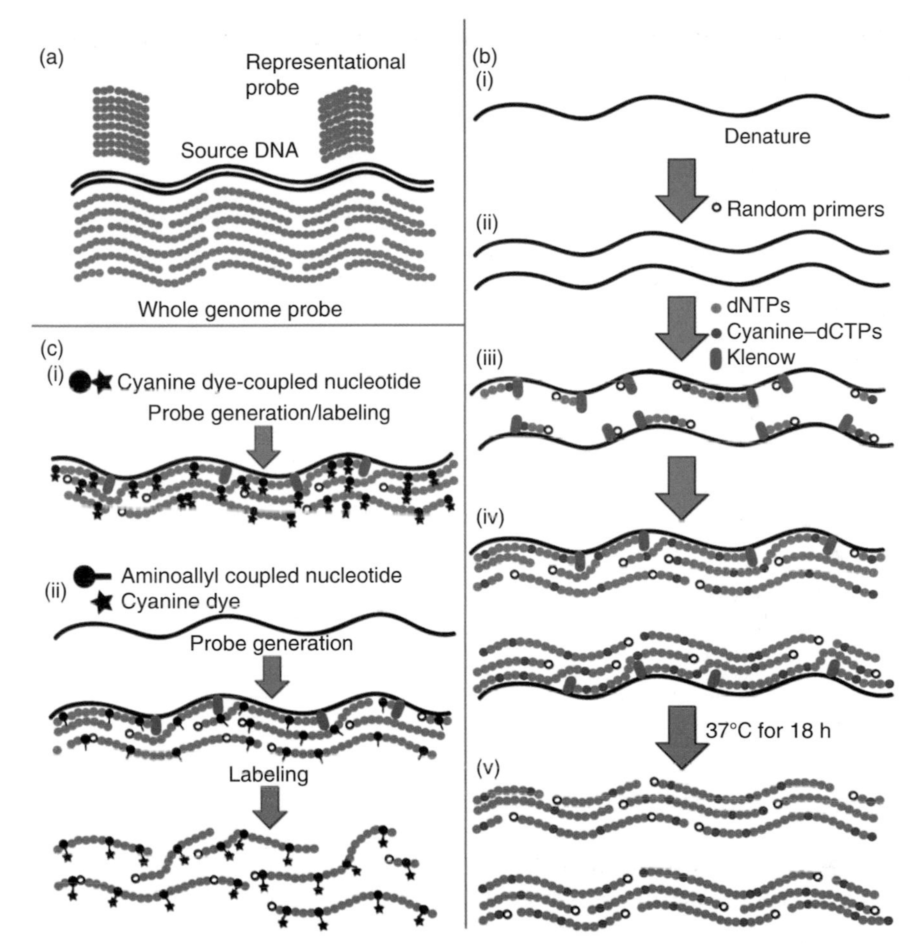

Figure 3. Whole genome and representational DNA labeling (see page 126).
(a) Genome sampling vs whole genome labeling. Labeled DNA probes are shown as grey circles demonstrating the differential genome coverage of both methods. (b) Random priming method of whole genome labeling for use in BAC microarray CGH experiments. (i) Genomic DNA from both test and reference samples are added individually to a mixture containing random primers and buffer and heated to 100°C to denature the double strands. (ii) The solution is then snap-cooled on ice to prevent strand reannealing. A mixture of dNTPs, Klenow enzyme, and either Cy3–dCTP or Cy5–dCTP are added to the DNA and the resulting mixture is placed at 37°C. (iii) The random primers anneal to the ssDNA template leaving a 3′-OH that allows Klenow to initiate DNA synthesis by incorporating both dNTPs and cyanine-labeled dCTP. (iv) Klenow continues moving along the DNA template, displacing previously synthesized stands, resulting in linear amplification of the starting material. (v) The resulting probe contains incorporated cyanine dye, allowing its use in microarray CGH experiments. (c) Differences between direct and indirect labeling. (i) In direct labeling, cyanine-coupled nucleotides are incorporated directly into generated probe. (ii) In indirect labeling, aminoallyl-coupled nucleotides are incorporated into the probe, and fluorophores are bound to the nucleotides in a second step.

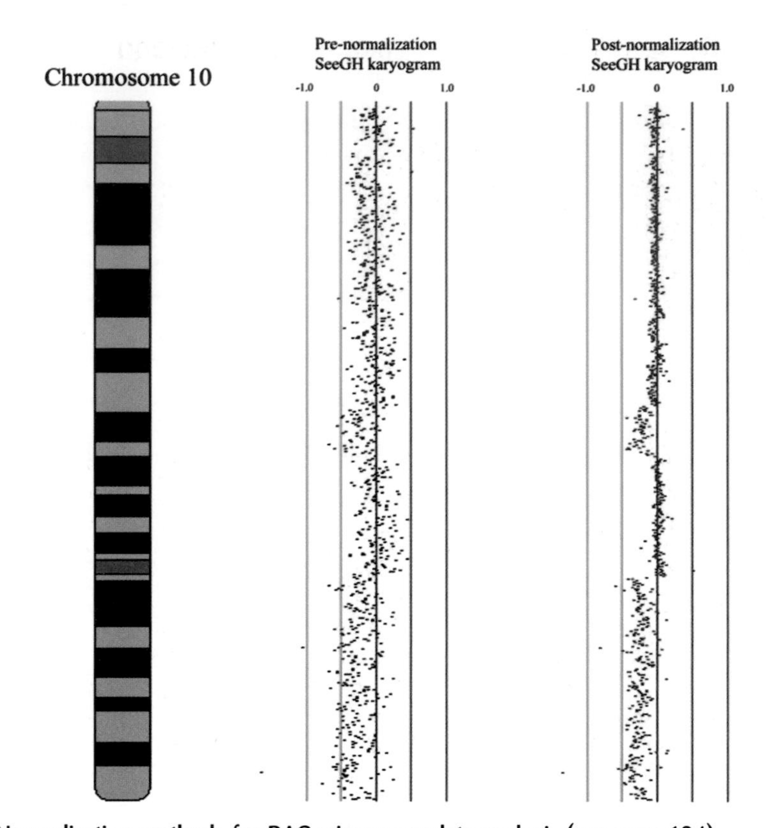

Figure 5. Normalization methods for BAC microarray data analysis (see page 134).
The effects of microarray CGH data normalization for a competitive hybridization of a cancer cell line against its drug-resistant derivative. Normalized and non-normalized $\log_2$ signal intensity ratios were plotted using SeeGH software. Clones with standard deviations among the triplicate spots of >0.09 or a signal-to-noise ratio of >3 were filtered from the analysis. Chromosome arm 10q is represented on the left. Vertical lines denote $\log_2$ signal ratios from −1 to 1 with copy number increases to the right (red lines) and decreases to the left (green lines) of zero (purple line). A $\log_2$ signal ratio of zero represents equivalent copy number between the hybridized samples. Each black dot represents a single BAC clone. Normalization was performed using a custom normalization program with the parameters set as a default (89).

Chapter 9. Microarray comparative genomic hybridization

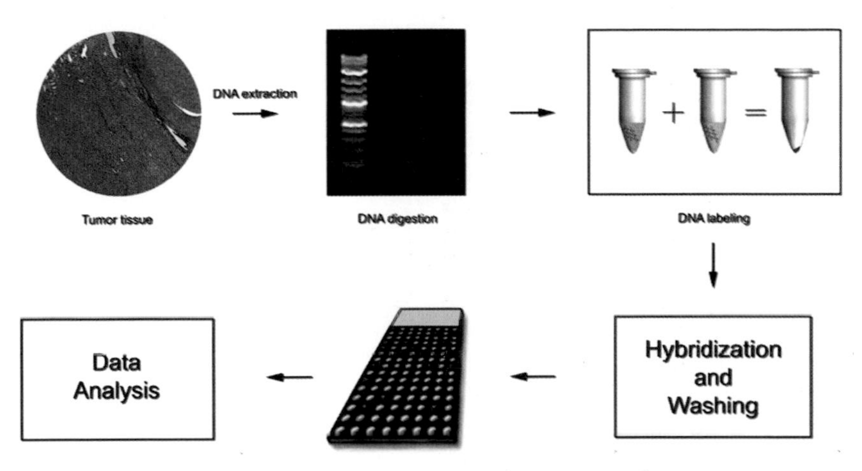

Figure 1. Graphical representation of the CGH technique (see page 162).
Test genomic DNA is first extracted from tumor tissue and control DNA is obtained from either commercial sources or from normal tissue. The DNA is digested using a restriction enzyme prior to labeling either directly or indirectly with different fluorescent dyes. The differentially labeled DNAs are combined and hybridized to a microarray of DNA sequences. Labeled tumor DNA competes with labeled normal reference DNA and the ratio of fluorescence for the two dyes can then be used to determine chromosomal loss or gain.

Chapter 11. DNA microarrays to study nonhuman primate gene expression

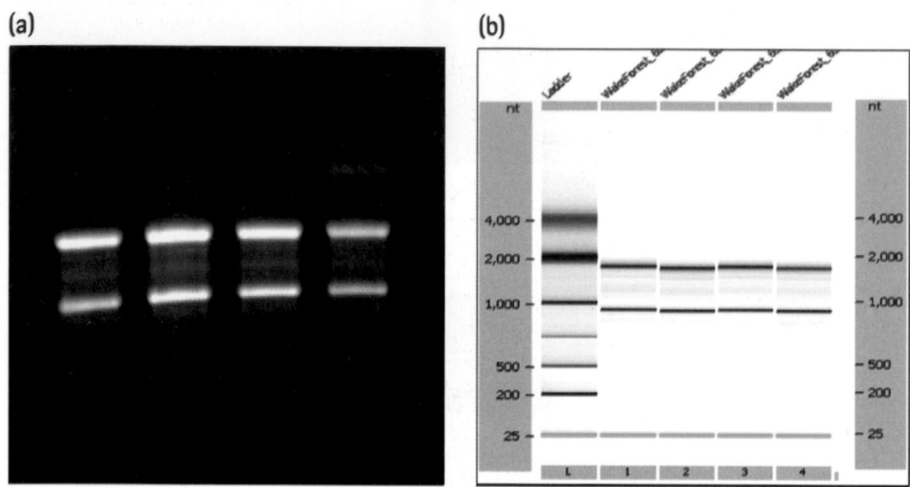

Figure 1. Analysis of RNA preparations from NHP blood (see page 209).
(*a*) Total RNA was isolated from 2 ml of whole blood drawn from four individual animals and purified using TRI Reagent BD (Molecular Research Inc.). Following purification clean-up with an RNeasy kit (Qiagen), the RNA was eluted with 10 μl of nuclease-free dH_2O. A volume of 1 μl of RNA was used to determine concentration. Approximately 1 μg of RNA, based on Nanodrop analysis, was fractionated by 1% formamide gel electrophoresis, and the stained gel was photographed as shown. (*b*) A Bioanalyzer Instrument (Agilent) was used to analyze four NHP RNA samples prior to labeling for microarray analysis: lane 1, sample 89, 1.2 ratio and 9.2 RNA integrity number (RIN); lane 2, sample 92, 1.2 ratio and 9.3 RIN; lane 3, sample 94, 1.2 ratio and 9.3 RIN; and lane 4, sample 96, 1.0 ratio and 9.4 RIN.

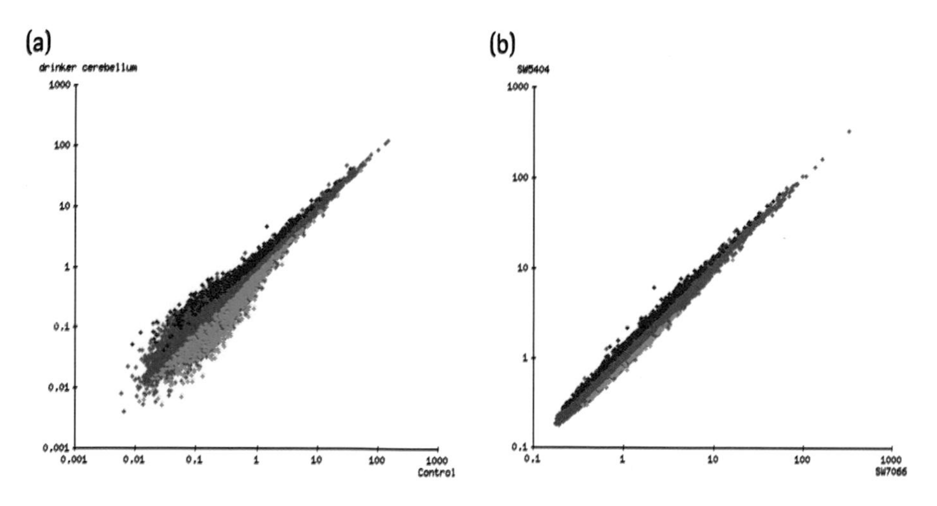

Figure 2. Scatter plots showing NHP RNA-generated microarray datasets normalized using MAS 5.0 software (*a*) or rma (*b*) (see page 215).

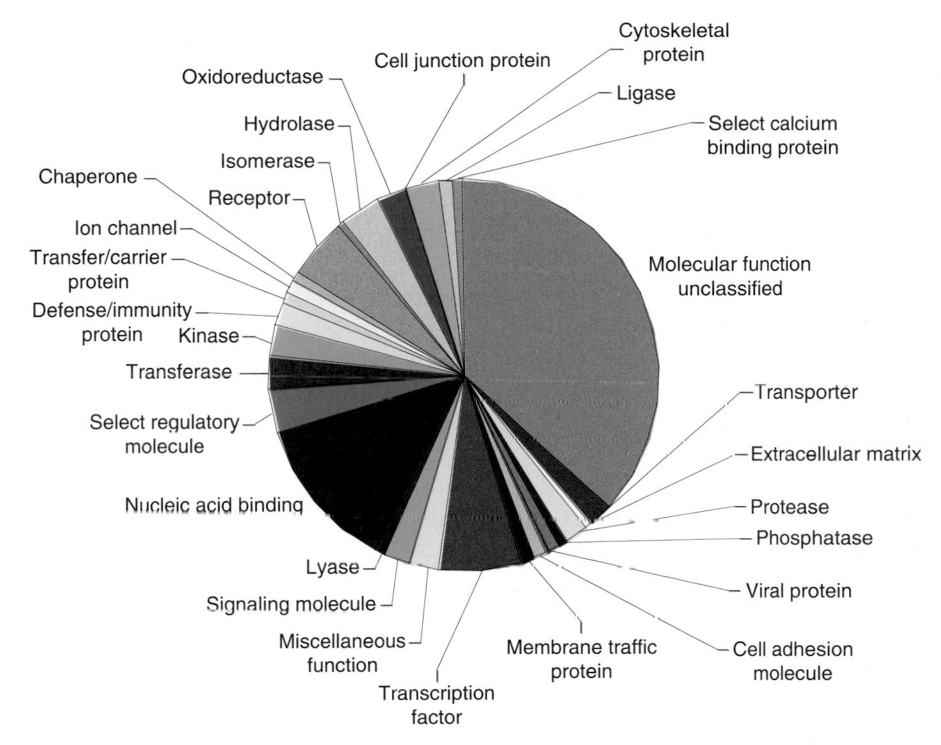

Figure 4. Molecular function diagram from analysis using PANTHER software and microarray results from four cynomolgus monkey whole blood RNA samples (see page 223).

CHAPTER 1

Whole human genome microarrays

Mark Schena, Milton J. Friedman, Paul Haje, Todd Martinsky, Greg Suzuki, Tony Costa, Joy Chung, Mel Ruiz, Chris Costa, Luis Machado, Rajiv Raja, and Rupal Desai

1. INTRODUCTION

A dozen years ago, my colleagues and I published the first paper on microarrays (1). The 1995 *Science* paper utilized DNA microarrays containing 45 cDNAs to measure the expression of 45 genes in *Arabidopsis thaliana*, a small mustard plant used widely as a model organism for studying higher plant biology. Since the 1995 *Science* publication, more than 25 000 papers utilizing microarray technology have appeared in the scientific literature (see *Fig. 1*), dozens of books have been written on the subject (see for example 2–4), the technology has been featured on television, and thousands of commercial products supporting a billion dollar industry have been developed. Recently, the United States Food and Drug Administration (FDA) spearheaded the ambitious MicroArray Quality Control program to streamline the use of microarrays for expression profiling and diagnostics, suggesting an important role going forward for DNA and protein microarrays in human health care (5). Why have the scientific and medical communities embraced microarray technology with such enthusiasm?

The answer to this question is, in part, because microarrays are extremely tractable devices: ordered arrays of microscopic elements on planar substrates that allow specific binding of genes and gene products. These simple criteria are met by an increasing number of important microarray devices. The arrangement of microarray elements in rows and columns allows high-speed three-dimensional motion control technologies (robots) to be used for microarray manufacturing. Microscopic elements can be printed by *in situ* synthesis of biochemical building blocks or by *ex situ* deposition of samples using sophisticated and highly miniaturized semi-conductor, contact printing and ink-jet approaches. Planar substrates

DNA Microarrays: *Methods Express* (M. Schena, ed.)

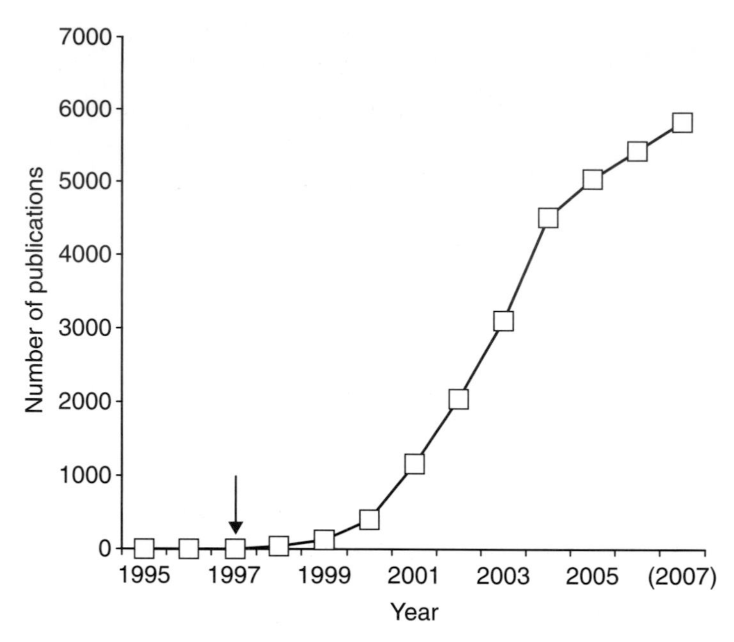

Figure 1. Microarray publications in the scientific literature over time. The number of scientific papers (squares) was plotted as a function of the year of publication. The search term 'microarray or microarrays or microarrayed or microarraying' was used to search manuscript titles and abstracts in the PubMed (MEDLINE) database of the National Center for Biotechnology Information (www.ncbi.nlm.nih.gov/sites/entrez?db=PubMed). The year 2007 is shown in parentheses to denote the fact that it reflects an approximate number of publications, and the arrow corresponds to the appearance of commercial products on the microarray market.

composed of glass and silicon materials permit low reaction volumes and fluorescent, chemiluminescent, and colorimetric labeling and detection of important biochemical reactions. Specific binding events between target and probe molecules are accurately detected using scanning devices that utilize lasers, light-emitting diodes, photomultiplier tubes, charged-coupled devices, and other advanced detection components. Large quantities of microarray data are quantified and modeled using precocious algorithms and computational tools. Microarray technology is highly interdisciplinary, tapping into and uniting major academic fields including biology, biochemistry, chemistry, computer science, engineering, mathematics, and physics (see *Fig. 2*).

Microarrays have also been widely embraced because microarray technology is rather like Churchill's democracy, being 'the worst technology except for all the others'. Whilst microarrays are not without their shortcomings, at present these devices represent the best way to perform scores of extremely powerful biological assays spanning the gamut from

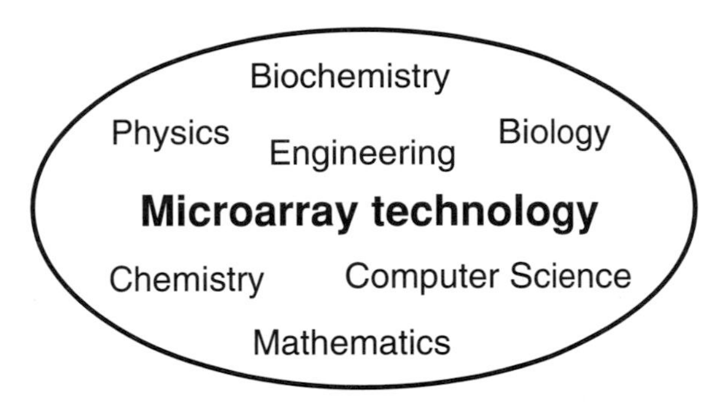

Figure 2. DNA microarray technology in major academic disciplines.
Biology, biochemistry, chemistry, computer science, engineering, mathematics, and physics are utilized extensively in the microarray field.

gene expression and genotyping to protein-binding studies and serum profiling. Microarrays of cDNAs, oligonucleotides, patient amplicons, small molecules, peptides, proteins, patient serum, antibodies, carbohydrates, artificial chromosomes, live cells, and many other sample types are manufactured on a daily basis in laboratories worldwide. Every organism in the biosphere is amenable to microarray analysis, with human, mouse, rat, fruit fly, worm, mustard plant, yeast, and bacteria being among the most commonly studied. The capacity to print either oligonucleotides or patient DNAs permits the use of both single- and multi-patient genotyping formats (6). Microarray assays exploit parallelism, miniaturization, and automation, the same elegant methodological cornerstones that have driven the proliferation of the massive microelectronics industry.

There are also 'cultural' advantages to working in the microarray field. Practitioners of this exciting new science tend to be bright, enthusiastic, creative, and cooperative. Microarray technology represents an expanding opportunity, which maintains a high level of optimism in the field and minimizes conflicts. The high impact of microarrays on many different research areas affords grant applicants a strong chance of federal funding. The commercial success of the microarray industry has generated significant product revenues and has captivated the investment community, providing ample research budgets in the private sector. The burgeoning intellectual property landscape has not escaped the attention of patent and trial attorneys, who preside over the increasingly important transactions of initial public offerings, mergers and acquisitions, patent prosecution, and patent infringement cases in the microarray field.

For the scientist, microarray 'chips' possess a host of practical advantages over earlier filter and membrane assays. The use of fluorescence instead of radioactivity increases user safety, eliminates disposal costs, and allows the deployment of multi-color fluorescent methods. Chips are also much easier to handle than filter membranes, allow the use of barcodes, and can be stored as permanent archives. Diverse silane attachment chemistries render glass a superior support material compared with porous filter papers. Atomically flat, planar glass substrates can be coated with gels, nitrocellulose, nylon, polylysine, and other materials when traditional surface chemistries are required for specialized microarray applications.

Myriad theoretical and practical advantages of DNA microarrays have driven a steady increase in microarray complexity (the number of genes represented per microarray). The initial 1995 *Science* paper reported on 45 plant genes (1). In 1996, microarrays representing 1000 human genes were manufactured (7). By 1997, several groups had constructed microarrays representing the entire 6000-gene genome of the yeast *Saccharomyces cerevisiae* (8, 9), and a year later genome-wide expression analysis of the yeast cell cycle was reported (10). The logical extension of the DNA microarray, predicted in 1996 (7), was to manufacture single chips representing all 25 000 genes in the human genome. The successful implementation of whole human genome microarrays, a 'holy grail' of microarray technology, has been achieved in dramatic fashion recently and is, appropriately, the subject of this chapter.

2. METHODS AND APPROACHES

The sequencing of the human genome (11) has afforded a cohesive strategy for whole human genome microarray manufacture (see *Fig. 3*). Oligonucleotides representing unique gene targets are selected from the human genome database, 'crunched' against the genomic sequence to ensure uniqueness, and manufactured into microarrays using *in situ* synthesis or *ex situ* deposition. Samples of mRNA are isolated from human cells, labeled with either fluorescent or nonfluorescent tags, hybridized to the microarrays, and scanned for the emitted signals. With whole-genome microarrays, single strengths provide a quantitative readout for the expression levels of all of the genes in the human genome. Changes in gene expression are detected using either two-color fluorescent labeling and detection approaches on single microarrays, or by comparing the signals obtained in single-color mode across multiple microarrays. Expression data are then assembled into databases and mined and modeled using the tools of computational biology. Correlations between gene expression data and other information types including genotyping, drug toxicity, pathology, and so forth permit prognostic and diagnostic medical decisions to be made (see *Fig. 3*).

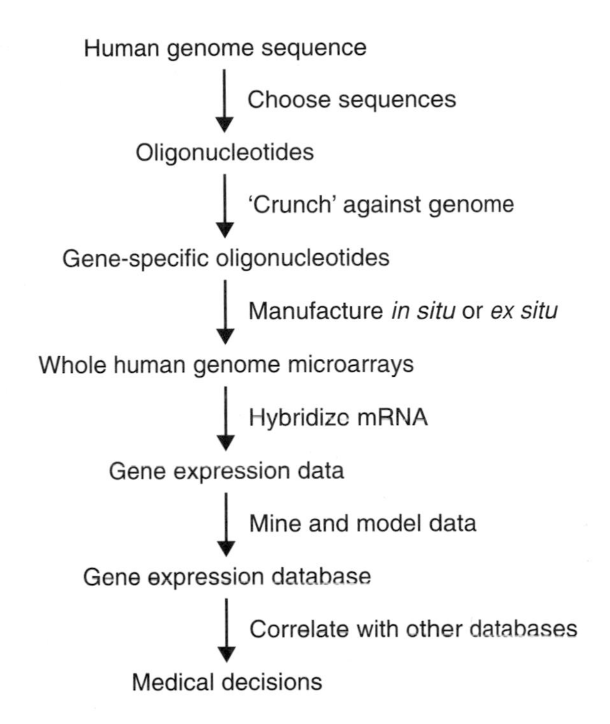

Figure 3. Flow-chart for whole human genome microarray manufacture and use. The finished human genome sequence is used as a blueprint from which oligonucleotides of 25–60 nucleotides in length are chosen as gene identifiers. An analysis or 'crunch' against the genome affords gene-specific oligonucleotides that provide unique hybridization sequences free of redundant sequence motifs and other negative attributes. Whole human genome oligonucleotide microarrays are manufactured using *in situ* or *ex situ* approaches and the resultant microarrays are reacted with labeled mixtures derived from human mRNA. Microarray signals provide a quantitative expression readout for all of the genes in the human genome. Mining and modeling gene expression data allows archiving in a gene expression database. Correlations with other types of biological information including genotyping, drug toxicity, and oncology provide the basis for making medical decisions.

The complexity and cost of manufacturing whole human genome microarrays has largely but not exclusively restricted this activity to the private sector (see *Table 1*). Microarray technology companies utilize state-of-the-art clean-room manufacturing facilities equipped with the most advanced robotics, printing technologies, surface chemistries, oligonucleotide reagents, and processing equipment. Tight control of every step in the manufacturing process ensures whole human genome microarrays of outstanding quality, and quality control and assurance is key to producing data of the highest precision and accuracy.

Table 1. Whole human genome microarray products from the leading commercial vendors

Company	Product	Manufacturing method	Description
Affymetrix, Inc., Santa Clara, CA, USA	U133A 2.0 Human Genome Microarray	GeneChip *in situ* synthesis technology using photomasks on a 12.7 × 12.7 mm glass chip	A total of 500 000 25mers representing 18 400 human genes and transcripts
Agilent Technologies, Inc., Palo Alto, CA, USA	44K Whole Human Genome Microarray	SurePrint *in situ* synthesis ink-jet technology on a 25 × 75 mm glass slide	A total of 44 000 60mers representing 41 000 human genes and transcripts
Applied Biosystems, Foster City, CA, USA	Applied Biosystems Human Genome Survey Microarray v2.0	Proprietary *ex situ* contact printing technology on a 40 × 50 mm glass chip	A total of 32 878 60mers representing 29 098 human genes and transcripts
ArrayIt Division, TeleChem International, Inc., Sunnyvale, CA, USA	H25K Whole Human Genome Microarray	ArrayIt patented *ex situ* contact printing technology on a 25 × 76 mm glass substrate	A total of 26 304 60mers representing 25 509 fully annotated human genes and 795 controls
Illumina Inc., San Diego, CA, USA	Sentrix Human-6 Expression Beadchip	Sentrix *ex situ* Beadchip technology on a 25 × 75 mm glass slide	A total of 46 000 79mers (29 nt reference address + 50 nt gene) representing genome-wide transcripts
NimbleGen Systems, Inc., Madison, WI, USA	Human Whole Genome Expression Microarray	MAS *in situ* synthesis based on digital light processing technology on a 25 × 76 mm glass substrate	A total of 385 000 60mers representing 47 633 gene targets

Whole human genome microarrays manufactured by *in situ* synthesis combine the four DNA bases in a stepwise manner to produce gene-specific oligonucleotides at specific locations on the glass substrate. Affymetrix, NimbleGen, and Agilent carry out high-density *in situ* oligonucleotide synthesis using photomasks, digital light processing, and ink-jet dispensers, respectively (see *Table 1*). Applied Biosystems, ArrayIt, and Illumina manufacture the gene-specific oligonucleotides in an *ex situ* manner and then deposit the oligonucleotides onto the substrate using contact printing (Applied Biosystems and ArrayIt) or microspheric beads (Illumina) (see *Table 1*).

This chapter presents protocols for mRNA purification, amplification, labeling, and hybridization developed for use with the 44K Human Genome Microarray from Agilent and the H25K Human Genome Microarray from

ArrayIt. By virtue of being designed from the human genome sequence rather than from expressed sequence tags, H25K offers a number of advantages over some other whole-genome microarrays including very high gene representation in AceView and GenBank (see *Fig. 4*, also available in the color section) and strong signal intensities across the entire human genome (see *Fig. 5*, available in the color section). Although the protocols described herein were optimized using the Agilent 44K and ArrayIt H25K microarrays, it is likely that these protocols will be applicable to all whole human genome microarray platforms. Validation of the amplification and labeling protocols on a gene-by-gene basis suggests that the protocols presented here are likely to provide accurate and reliable gene expression measurements for all of the genes in the human genome (12).

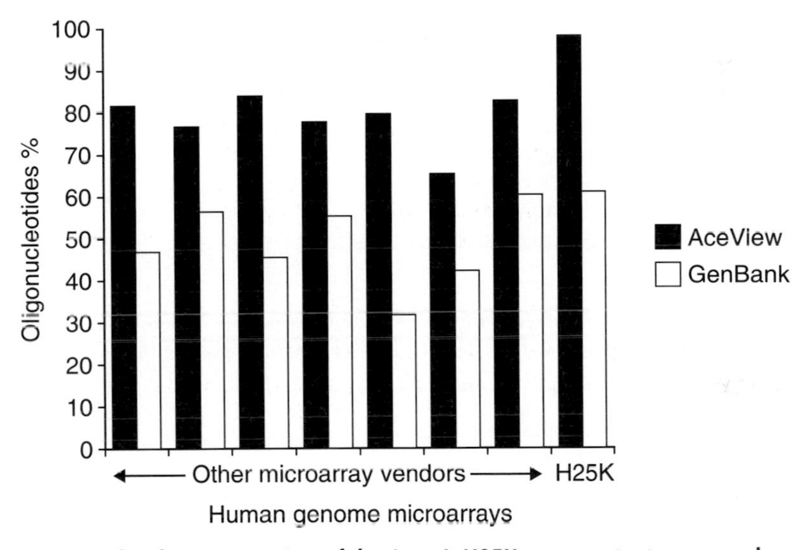

Figure 4. Graphical representation of the ArrayIt H25K gene content compared with the gene content of whole human genome microarrays from other leading vendors.
The *y*-axis shows the percentage of oligonucleotides in each microarray (*x*-axis) bearing an exact match to a transcript in the AceView (filled columns) or GenBank (open columns) database.

2.1 Recommended protocols

Protocol 1

First-strand cDNA synthesis for first-round amplification of small-quantity RNA

Equipment and Reagents

- RiboAmp HS RNA Amplification kit (MDS Analytical Technologies); Superscript™ III Enzyme (not included in the kit – Invitrogen, 200 U μl^{-1})
- NanoDrop ND-1000 (NanoDrop Technologies)
- Agilent 2100 bioanalyzer (Agilent Technologies)
- Polyinosinic acid (poly(I)) carrier RNA (Sigma)
- RNase-free dH_2O
- RNase-free 0.5 and 1.5 ml tubes
- Thermal cycler
- Microcentrifuge (Eppendorf 5415D or equivalent)
- Manual pipettor (2–20 µl)
- Manual pipettor (20–200 µl)
- RNase-free pipette tips

Method

1. To a 0.5 ml RNase-free microcentrifuge tube, add 100–500 pg of total cellular RNA and 100 ng of poly(I) carrier RNA to a total reaction volume of 10 µl[a].
2. Add 1 µl of Primer 1 and mix well. Spin the tube down briefly in a microcentrifuge.
3. Incubate the samples in a thermal cycler at 65°C for 5 min and place on ice for 1 min. Spin down briefly in a microcentrifuge.
4. Thaw all of the first-strand kit components except the enzymes. Place the thawed components on ice and prepare the first-strand master cocktail by mixing in a 1.5 ml tube 2 µl of enhancer, 5 µl of first-strand master mix, 1 µl of first-strand enzyme mix, and 1 µl of Superscript™ III Enzyme. Mix well.
5. To each 11 µl RNA sample, add 9 µl of first-strand master cocktail. Mix well by flicking the tube and spin down briefly.
6. Incubate the 20 µl samples in a thermal cycler at 42°C for 60 min to allow cDNA synthesis. Spin down briefly in a microcentrifuge.
7. To each 20 µl sample, add 2 µl of first-strand nuclease mix to degrade the RNA. Mix gently and spin down briefly.
8. Incubate the samples in a thermal cycler at 37°C for 30 min and 95°C for 5 min, and cool down to 4°C.
9. Proceed directly to *Protocol 2* or store at –20°C until ready to use. The total sample volume should be 22 µl at this step.

Note

[a]Volume adjustments to 10 μl should be made using RNase-free dH_2O. The quality of the total cellular RNA is essential for efficient amplification and should be checked for integrity using the Agilent 2100 bioanalyzer.

Protocol 2

Second-strand cDNA synthesis for first-round amplification of small-quantity RNA

Equipment and Reagents

- RiboAmp HS RNA Amplification kit (MDS Analytical Technologies)
- NanoDrop ND-1000 (NanoDrop Technologies)
- Agilent 2100 bioanalyzer (Agilent Technologies)
- RNase-free 0.5 and 1.5 ml tubes
- Thermal cycler
- Microcentrifuge (Eppendorf 5415D or equivalent)
- Manual pipettor (2–20 μl)
- Manual pipettor (20–200 μl)
- RNase-free pipette tips

Method

1. Obtain the 22 μl first-strand cDNA samples from *Protocol 1*, step 9, and place on ice.
2. To each 22 μl sample, add 1 μl of Primer 2 at 4°C and mix well. Spin down briefly.
3. Incubate the 23 μl samples in a thermal cycler at 95°C for 2 min to denature the cDNA. Chill on ice for at least 2 min.
4. Thaw all of the second-strand kit components on ice except for the enzyme. Prepare a second-strand master cocktail by mixing in a 1.5 ml tube 29 μl of second-strand master mix and 1 μl of second-strand enzyme mix.
5. To each 23 μl sample, add 30 μl of second-strand master cocktail and mix well to allow second-strand cDNA synthesis. Spin down briefly.
6. Incubate the samples in a thermal cycler at 25°C for 10 min, 37°C for 30 min, and 70°C for 5 min. Cool and hold the samples at 4°C and proceed to the next step[a]. The total sample volume at this step should be 53 μl.

Note

[a]Proceed to the next step (*Protocol 3*, step 1) as quickly as possible. The double-stranded cDNA samples should be held at 4°C for no more than 30 min.

Protocol 3

Purification of double-stranded cDNA products

Equipment and Reagents

- RiboAmp HS RNA Amplification kit (MDS Analytical Technologies)
- NanoDrop ND-1000 (NanoDrop Technologies)
- Agilent 2100 bioanalyzer (Agilent Technologies)
- RNase-free 0.5 and 1.5 ml tubes
- Microcentrifuge (Eppendorf 5415D or equivalent)
- Manual pipettor (2–20 μl)
- Manual pipettor (20–200 μl)
- Manual pipettor (10–1000 μl)
- RNase-free pipette tips

Method

1. Obtain the 53 μl double-stranded cDNA samples from *Protocol 2*, step 6.
2. Prepare a DNA/RNA purification column for each cDNA sample as follows: add 250 μl of DNA binding buffer to each column, incubate at room temperature for 5 min to equilibrate, and centrifuge the column tube assembly at full speed in a microcentrifuge (16 000 *g*) for 1 min.
3. To each 53 μl double-stranded cDNA sample, add 200 μl of DNA binding buffer and mix well. Apply the entire 253 μl cDNA sample to the equilibrated DNA/RNA purification column.
4. Centrifuge the column tube assembly at 100 ***g*** for 2 min, followed immediately by a second spin at 10 000 ***g*** for 1 min.
5. Add 250 μl of DNA wash buffer to each column and spin the column tube assembly at 16 000 ***g*** for 2 min. Recentrifuge at 16 000 ***g*** for 1 min (optional).
6. Discard the flow-through liquid and the collection tube, and place the column containing the purified double-stranded cDNA onto a new collection tube.
7. Add 12 μl of DNA elution buffer to the column and incubate for 1 min at room temperature.
8. Centrifuge the column tube assembly at 1000 ***g*** for 1 min, followed by 16 000 ***g*** for 1 min to elute the 12 μl purified double-stranded cDNA sample. Proceed to the *in vitro* transcription step or store the cDNA sample at –20°C.

Protocol 4

First-round *in vitro* transcription (IVT) and purification of amplified RNA (aRNA)

Equipment and Reagents

- RiboAmp HS RNA Amplification kit (MDS Analytical Technologies)
- RNase-free 0.5 and 1.5 ml tubes
- Thermal cycler
- Microcentrifuge (Eppendorf 5415D or equivalent)
- Manual pipettor (2–20 μl)
- Manual pipettor (20–200 μl)
- RNase-free pipette tips

Method

1. Obtain the 12 μl purified double-stranded cDNA samples from *Protocol 3*, step 8.
2. Thaw all of the IVT reaction components except the enzyme. Prepare an IVT master cocktail by mixing 2 μl of enhancer, 2 μl of IVT buffer, 6 μl of IVT master mix, and 2 μl of IVT enzyme mix.
3. To each 12 μl double-stranded cDNA sample, add 12 μl of the IVT master cocktail and mix well. Spin down briefly.
4. Incubate the samples in a thermal cycler at 42°C for 6 h to allow the *in vitro* transcription process to proceed, and then cool the samples to 4°C.
5. Add 1 μl of DNase mixture, mix gently, and spin down briefly. Incubate the samples at 37°C for 15 min to degrade the cDNA, leaving the aRNA intact. Cool the samples to 4°C. The total sample volume at this step should be 25 μl.
6. Prepare a DNA/RNA purification column for each aRNA sample as follows: add 250 μl of RNA binding buffer to each DNA/RNA purification column, incubate for 5 min at room temperature, and centrifuge at 16000 ***g*** for 1 min.
7. To each 25 μl aRNA sample, add 120 μl of RNA binding buffer, mix well, and apply the entire 145 μl sample to the equilibrated column.
8. Centrifuge the column tube assembly at 100 ***g*** for 2 min, followed immediately by a second spin at 10000 ***g*** for 30 s.
9. Add 200 μl of RNA wash buffer to the column and spin at 10000 ***g*** for 1 min.
10. Add an additional 200 μl of RNA wash buffer to the column and spin at 16000 ***g*** for 2 min. Recentrifuge the column tube assembly at 16000 ***g*** for 1 min (optional).
11. Discard the flow-through and the collection tube, and place the column containing the purified aRNA into a new collection tube.
12. Add 11 μl of RNA elution buffer to the column and incubate the column tube assembly at room temperature for 1 min.
13. Centrifuge the column tube assembly at 1000 ***g*** for 1 min and then at 16000 ***g*** for 1 min to elute the purified aRNA. The aRNA sample volume at this step should be 11 μl. Proceed to the next protocol or store at –20°C.

Protocol 5

First-strand cDNA synthesis for second-round amplification

Equipment and Reagents

- RiboAmp HS RNA Amplification kit (MDS Analytical Technologies); Superscript™ III Enzyme (not included in the kit – Invitrogen, 200 U μl^{-1})
- NanoDrop ND-1000 (NanoDrop Technologies)
- Agilent 2100 bioanalyzer (Agilent Technologies)
- RNase-free dH_2O
- RNase-free 0.5 and 1.5 ml tubes
- Thermal cycler
- Microcentrifuge (Eppendorf 5415D or equivalent)
- Manual pipettor (2–20 μl)
- Manual pipettor (20–200 μl)
- RNase-free pipette tips

Method

1. Obtain the 11 μl aRNA samples from *Protocol 4*, step 13. Thaw the samples and place 10 μl of each aRNA sample into a 0.5 ml RNase-free tube.
2. To each 10 μl aRNA sample, add 1 μl of Primer 2, mix well, and spin down briefly. Incubate the samples in a thermal cycler at 65°C for 5 min to denature the aRNA, then chill the samples on ice for 1 min and spin down briefly.
3. Thaw all of the first-strand kit components except the enzymes. Place the thawed components on ice and prepare a first-strand master cocktail by mixing in a 1.5 ml tube 2 μl of enhancer, 5 μl of first-strand master mix, 1 μl of first-strand enzyme mix, and 1 μl of Superscript™ III Enzyme. Mix well.
4. To each 11 μl sample of aRNA and primer, add 9 μl of first-strand master cocktail. Mix gently by tapping the tube, and spin down briefly. Incubate the 20 μl samples in a thermal cycler at 25°C for 10 min and then at 37°C for 60 min to allow first-strand cDNA synthesis to proceed.
5. Chill the samples on ice for 1 min and spin down briefly to consolidate the sample at the bottom of each tube. The samples can be use immediately for *Protocol 6* or stored overnight at –20°C. The total sample volume at this step should be 20 μl.

Protocol 6

Second-strand cDNA synthesis for second-round amplification

Equipment and Reagents

- RiboAmp HS RNA Amplification kit (MDS Analytical Technologies)
- RNase-free 0.5 and 1.5 ml tubes
- Thermal cycler

- Microcentrifuge (Eppendorf 5415D or equivalent)
- Manual pipettor (2–20 μl)
- Manual pipettor (20–200 μl)
- RNase-free pipette tips

Method

1. Obtain each 20 μl first-strand cDNA sample from *Protocol 5*, step 5.
2. To each 20 μl first-strand cDNA sample, add 1 μl of Primer 3, mix well, and spin down briefly.
3. Incubate the samples in a thermal cycler at 95°C for 5 min to denature the cDNA, place the samples on ice, and chill for 2 min.
4. Thaw all of the second-strand kit components except the enzymes and place them on ice. Prepare the second-strand master cocktail by mixing 29 μl of second-strand master mix and 1 μl of second-strand enzyme mix.
5. To each 21 μl sample of cDNA and primer, add 30 μl of second-strand master cocktail on ice, mix gently by tapping the tube, and spin down briefly.
6. Incubate each 51 μl sample at 37°C for 30 min to allow second-strand cDNA synthesis to proceed and then at 70°C for 5 min to inactive the enzyme. Hold each sample at 4°C until ready to proceed to the next step[a].
7. To equilibrate the DNA/RNA purification columns, add 250 μl of DNA binding buffer to each column, incubate the columns for 5 min at room temperature, and spin at full speed (16000 *g*) in a microcentrifuge for 1 min.
8. To each 51 μl double-stranded cDNA sample, add 200 μl of DNA binding buffer, mix well, and apply the entire 251 μl sample to an equilibrated column.
9. Centrifuge the column tube assembly in a microcentrifuge at 100 ***g*** for 2 min, followed immediately by a second spin at 10000 ***g*** for 30 s.
10. Add 250 μl of DNA wash buffer to the column and spin in a microcentrifuge at 16000 ***g*** for 2 min. Recentrifuge the column tube assembly at 16000 ***g*** for 1 min (optional).
11. Discard the flow-through and the collection tube, and place the column containing the double-stranded cDNA into a new collection tube.
12. Add 12 μl of DNA elution buffer to each column and incubate the assembly at room temperature for 1 min.
13. Centrifuge the column tube assembly at 1000 ***g*** for 1 min, followed by 16000 ***g*** for 1 min. Store the cDNA sample at –20°C until ready to use. The volume of purified double-stranded cDNA at this step should be 12 μl.

Note

[a]Proceed to the next step (*Protocol 7*, step 1) as quickly as possible. The double-stranded cDNA samples should be held at 4°C for no more than 30 min.

Protocol 7

Second-round IVT of double-stranded cDNA

Equipment and Reagents

- RiboAmp HS RNA Amplification kit (MDS Analytical Technologies)
- NanoDrop ND-1000 (NanoDrop Technologies)
- Agilent 2100 bioanalyzer (Agilent Technologies)
- RNase-free 0.5 and 1.5 ml tubes
- Thermal cycler
- Microcentrifuge (Eppendorf 5415D or equivalent)
- Manual pipettor (2–20 μl)
- Manual pipettor (20–200 μl)
- RNase-free pipette tips

Method

1. Obtain the 12 μl purified double-stranded cDNA samples from *Protocol 6*, step 13.
2. Thaw all of the IVT reaction components except the enzyme. Prepare an IVT master cocktail by mixing 2 μl of enhancer, 2 μl of IVT buffer, 6 μl of IVT master mix, and 2 μl of IVT enzyme mix.
3. To each 12 μl double-stranded cDNA sample, add 12 μl of the IVT master cocktail, mix well, and spin down briefly.
4. Incubate the 24 μl samples in a thermal cycler at 42°C for 6 h to allow IVT to proceed. Cool the samples to 4°C.
5. Add 1 μl of DNase mixture, mix gently, and spin down briefly. Incubate at 37°C for 15 min to digest the cDNA and then cool to 4°C.
6. To equilibrate the DNA/RNA purification columns, add 250 μl of RNA binding buffer to each column, incubate for 5 min at room temperature, and centrifuge each column assembly at 16000 ***g*** for 1 min.
7. To each 25 μl aRNA sample, add 120 μl of RNA binding buffer, mix well, and apply the entire 145 μl volume of each sample to an equilibrated column.
8. Centrifuge the column tube assembly at 100 ***g*** for 2 min, followed immediately by a second spin at 10000 ***g*** for 30 s.
9. Add 200 μl of RNA wash buffer to each column and spin at 10000 ***g*** for 1 min.
10. Add a second 200 μl volume of RNA wash buffer to each column and spin at 16000 ***g*** for 2 min. Recentrifuge the column tube assembly at 16000 ***g*** for 1 min (optional).
11. Discard the flow-through and the collection tube, and place the column containing the aRNA into a new collection tube.
12. Add 30 μl of RNA elution buffer and incubate the column tube assembly at room temperature for 1 min.
13. Centrifuge the column tube assembly at 1000 ***g*** for 1 min and then at 16000 ***g*** for 1 min to elute the aRNA.

14. Measure the aRNA concentration using a NanoDrop spectrophotometer and calculate the aRNA concentration and yield. The quality and length of the aRNA products can be determined using an Agilent bioanalyzer. The aRNA transcript length should be 200–1000 nt. The aRNA sample volume at this step should be 30 μl.

Protocol 8

Fluorescent labeling of aRNA using Cy3 and Cy5 dyes

Equipment and Reagents

- Turbo Labeling Cy3 kit (MDS Analytical Technologies)
- Turbo Labeling Cy5 kit (MDS Analytical Technologies)
- NanoDrop ND-1000 (NanoDrop Technologies)
- Agilent 2100 bioanalyzer (Agilent Technologies)
- RNase-free dH_2O
- RNase-free 0.5 and 1.5 ml tubes
- Thermal cycler
- Microcentrifuge (Eppendorf 5415D or equivalent)
- Manual pipettor (2–20 μl)
- Manual pipettor (20–200 μl)
- RNase-free pipette tips

Method

1. Obtain the 30 μl aRNA samples from *Protocol 7*, step 14.
2. Pipette 2–15 μg of each aRNA sample into an RNase-free 0.5 ml microcentrifuge tube and adjust the total volume of each sample to 40 μl with RNase-free dH_2O[a].
3. Briefly spin all of the kit components of the Turbo Labeling Cy3 and Cy5 kits to consolidate the contents at the bottom of each tube.
4. To each 40 μl aRNA sample, add 5 μl of Turbo Cy3 or Turbo Cy5 reagent and 5 μl of 10× labeling buffer. Mix well by gently flicking each tube.
5. Incubate the samples at 85°C for 15 min and at 4°C for 1–30 min to allow Cy3 or Cy5 labeling of the aRNA. The volume of each dye-labeled aRNA sample at this step should be 50 μl.

Note

[a]Ensure that the final concentration of aRNA in the labeling reaction is ≥50 ng/μl as suboptimal labeling is observed if the aRNA concentration is less than 50 ng/μl.

Protocol 9

Column purification of Cy3- and Cy5-labeled aRNA

Equipment and Reagents

- Turbo Labeling Cy3 kit (MDS Analytical Technologies)
- Turbo Labeling Cy5 kit (MDS Analytical Technologies)
- NanoDrop ND-1000 (NanoDrop Technologies)
- Agilent 2100 bioanalyzer (Agilent Technologies)
- RNase-free dH_2O
- RNase-free 0.5 and 1.5 ml tubes
- Microcentrifuge (Eppendorf 5415D or equivalent)
- Manual pipettor (20–200 µl)
- RNase-free pipette tips

Method

1. Obtain the 50 µl dye-labeled aRNA samples from *Protocol 8*, step 5.
2. Resuspend the purification column matrix by vortexing each column.
3. Loosen the column cap with a one-quarter turn and snap off the bottom closure.
4. Place the column in the 2 ml collection tube provided with the kit.
5. Spin each column for 1 min at 16000 *g*.
6. Discard the flow-through and retain the collection tube.
7. To each column, add 300 µl of RNase-free H_2O and centrifuge for 1 min at 16000 *g*. Discard the collection tube and flow-through.
8. Place the freshly prepared column in an RNase-free 1.5 ml microcentrifuge tube.
9. Transfer the entire 50 µl volume of the dye-labeled aRNA sample into the column, using a separate column for each sample.
10. Centrifuge the columns for 1 min at 16000 *g* to elute the purified Cy3- or Cy5-labeled aRNA samples. The volume of each purified dye-labeled aRNA sample at this step should be 50 µl.

Protocol 10

First-strand cDNA synthesis from total human RNA

Equipment and Reagents

- ArrayIt H25K Whole Human Genome Microarray (TeleChem International)
- SenseAMP RNA Amplification kit (Genisphere)
- NanoDrop ND-1000 (NanoDrop Technologies)
- Agilent 2100 bioanalyzer (Agilent Technologies)
- SuperScript II reverse transcriptase (Invitrogen)
- RNase-free dH_2O
- RNase-free 0.5 and 1.5 ml tubes
- Thermal cycler
- Microcentrifuge (Eppendorf 5415D or equivalent)
- Manual pipettor (2–20 μl)
- Manual pipettor (20–200 μl)
- RNase-free pipette tips
- 1× TE buffer (10 mM Tris/HCl, pH 8.0, 1 mM EDTA; Ambion)

Method

1. Thaw the components (vials 1–13) of the SenseAMP RNA Amplification kit on ice.
2. Prepare the RNA/primer mixture by adding the following components to a 0.5 ml RNase-free tube: 250 ng of total human RNA (1–7 μl), 2 μl of SenseAMP dT_{24} RT primer (50 ng/μl, vial 1), 2 μl of random primer (250 ng/μl, vial 2), and 0–6 μl of RNase-free H_2O (vial 10) to a total volume of 11 μl[a].
3. Heat the RNA/primer mixture in a thermal cycler to 80°C for 10 min to denature the RNA and place on ice immediately for 2 min. Spin briefly in a microcentrifuge and place on ice.
4. Prepare a master mix (9 μl per mRNA/primer mixture) by adding the following kit components to a 0.5 ml RNase-free tube on ice: 4 μl of 5× first-strand buffer, 2 μl of 0.1 M dithiothreitol (DTT)[b], 1 μl of Superase-In (vial 4), 1 μl of dNTP mix (vial 3), and 1 μl of SuperScript II reverse transcriptase[c] for a total master mix volume of 9 μl.
5. To each 11 μl of denatured RNA/primer mixture, add 9 μl of master mix for a total volume of 20 μl. Mix gently and spin briefly in a microcentrifuge.
6. Incubate each 20 μl sample at 42°C for 2 h to allow first-strand cDNA synthesis to proceed.
7. After the 2 h incubation, spin briefly in a microcentrifuge and add 80 μl of 1× TE buffer for a total sample volume of 100 μl.

Notes

[a]Make sure the integrity of the cellular RNA is checked using the Agilent 2100 bioanalyzer.
[b]Use 2 μl of 0.1 M DTT if this reagent is supplied with the reverse transcriptase enzyme; otherwise, use 2 μl of RNase-free H_2O.
[c]Use Superscript II reverse transcriptase or an equivalent reverse transcriptase.

Protocol 11

Purification of first-strand cDNA

Equipment and Reagents

- Microcentrifuge (Eppendorf 5415D or equivalent)
- Manual pipettor (1–20 μl)
- Manual pipettor (20–200 μl)
- Manual pipettor (10–1000 μl)
- RNase-free pipette tips
- MinElute PCR Purification kit (Qiagen)
- RNeasy MinElute kit (Qiagen)
- RNeasy Mini kit (Qiagen)
- RNase-free 0.5 and 1.5 ml tubes
- 100% Ethanol
- 80% Ethanol

Method

1. Obtain the 100 μl first-strand cDNA samples from *Protocol 10*, step 7, and purify using the MinElute PCR Purification kit as follows.
2. To each 100 μl sample, add 500 μl of buffer PB and mix well.
3. Apply the entire 600 μl cDNA mixture to a MinElute column and centrifuge for 1 min at 14 000 ***g*** in a microcentrifuge.
4. Discard the flow-through and place the MinElute column containing the cDNA products into the same collection tube.
5. Add 750 μl of buffer PE to the MinElute column and centrifuge for 1 min.
6. Discard the flow-through. Place the MinElute column back into the same collection tube.
7. Add 500 μl of 80% ethanol to the MinElute column and centrifuge for 2 min to wash the cDNA products.
8. Discard the flow-through. Place the MinElute column back into the same collection tube.
9. Open the column caps and place in a microcentrifuge with the cap opposite the direction of rotation of the rotor to avoid breaking the cap off. Centrifuge for 5 min.
10. Place the MinElute column into a new 1.5 ml microcentrifuge tube.
11. Elute the cDNA from the purification column as follows: add 10 μl of buffer EB to the center of the column membrane, incubate at room temperature for 2 min, centrifuge for 2 min, discard the column, and save the 10 μl of purified cDNA. If the eluted cDNA sample has a volume of less than 10 μl, add RNase-free H_2O (vial 10) to a final volume of 10 μl.

Protocol 12

First-strand cDNA tailing

Equipment and Reagents

- ArrayIt H25K Whole Human Genome Microarray (TeleChem International)
- SenseAMP RNA Amplification kit (Genisphere)
- Thermal cycler
- Microcentrifuge (Eppendorf 5415D or equivalent)
- Manual pipettor (2–20 µl)
- Manual pipettor (20–200 µl)
- RNase-free pipette tips
- RNase-free 0.5 and 1.5 ml tubes

Method

1. Obtain the 10 µl purified cDNA sample from *Protocol 11*, step 11.
2. Heat the purified cDNA samples in a thermal cycler to 80°C for 10 min to denature the cDNA. Place the samples immediately on ice for 2 min. Spin briefly and return to ice.
3. Prepare a master mix (10 µl per cDNA sample) by adding the following kit components to a 0.5 ml RNase-free tube on ice: 2 µl of 10× reaction buffer (vial 6), 2 µl of nuclease-free H_2O (vial 10), 4 µl of 10 mM dTTP (vial 5), and 2 µl of TdT enzyme (vial 7) for a final volume of 10 µl.
4. To each 10 µl denatured cDNA sample, add 10 µl of master mix for a volume of 20 µl. Mix gently and spin briefly in a microcentrifuge.
5. Incubate at 37°C for 2 min to allow the enzymatic tailing reaction to proceed[a].
6. Transfer the samples to 80°C immediately to stop the enzymatic tailing reaction. Spin briefly in a microcentrifuge and cool to room temperature for 1–2 min. The total sample volume at this step should be 20 µl.

Note

[a]Do not exceed a 2 min total duration for the enzymatic tailing step at 37°C!

Protocol 13

Conversion of tailed cDNA into T7 promoter-containing double-stranded cDNA

Equipment and Reagents

- ArrayIt H25K Whole Human Genome Microarray (TeleChem International)
- SenseAMP RNA Amplification kit (Genisphere)
- NanoDrop ND-1000 (NanoDrop Technologies)
- Agilent 2100 bioanalyzer (Agilent Technologies)
- Thermal cycler
- Microcentrifuge (Eppendorf 5415D or equivalent)
- Manual pipettor (2–20 µl)
- Manual pipettor (20–200 µl)
- RNase-free pipette tips
- RNase-free 0.5 and 1.5 ml tubes

Method

1. Obtain the 20 µl tailed cDNA samples from *Protocol 12*, step 6.
2. To each 20 µl tailed cDNA sample, add 2 µl of T7 template oligonucleotides (vial 8) for a total volume of 22 µl. Mix gently and spin briefly in a microcentrifuge.
3. Incubate at 37°C for 10 min to anneal the T7 oligonucleotides to the tailed cDNA templates.
4. To each 22 µl primed and tailed cDNA sample, add 1 µl of 10× reaction buffer (vial 6), 1 µl of dNTP mix (vial 3), and 1 µl of Klenow enzyme (vial 9). Mix gently and spin briefly in a microcentrifuge.
5. Incubate the 25 µl samples at room temperature for 30 min to allow double-stranded DNA synthesis to proceed.
6. Stop the Klenow enzyme reaction by heating the samples in a thermal cycler at 65°C for 10 min. Spin briefly in a microcentrifuge and place on ice for 2 min.
7. The volume of the double-stranded T7 promoter-containing cDNA sample at this step should be 25 µl. Proceed directly to the IVT protocol (*Protocol 14*) using 12.5 µl of sample and store the remaining 12.5 µl at –20°C for future use or for a parallel aRNA amplification reaction.

Protocol 14

IVT of T7 promoter-containing cDNA templates

Equipment and Reagents

- ArrayIt H25K Whole Human Genome Microarray (TeleChem International)
- SenseAMP RNA Amplification kit (Genisphere)
- NanoDrop ND-1000 (NanoDrop Technologies)
- Agilent 2100 bioanalyzer (Agilent Technologies)
- Thermal cycler
- Microcentrifuge (Eppendorf 5415D or equivalent)
- Manual pipettor (2–20 μl)
- Manual pipettor (20–200 μl)
- RNase-free pipette tips
- RNase-free 0.5 and 1.5 ml tubes

Method

1. Obtain the 12.5 μl T7 promoter-containing cDNA samples from *Protocol 13*, step 7.
2. Incubate the 12.5 μl cDNA samples at 37°C for 10 min to reanneal the strands.
3. Thaw the T7 nucleotide mix (vial 11) and 10× T7 reaction buffer (vial 12) at room temperature and keep at room temperature until ready to use. Mix the 10× T7 reaction buffer (vial 12) by vortexing to suspend any precipitated buffer components.
4. To each 12.5 μl cDNA sample, add the following room temperature components in the order listed: 8 μl of T7 nucleotide mix (vial 11), 2.5 μl of 10× T7 reaction buffer (vial 12), and 2 μl of T7 enzyme mix (vial 13). Mix gently and spin briefly in a microcentrifuge. The total reaction volume at this step should be 25 μl.
5. Incubate for 6–16 h at 37°C in a thermal cycler with a heated lid set to 37°C to allow the synthesis of sense-strand aRNA from the double-stranded T7 promoter-containing cDNA templates.
6. Stop the aRNA amplification reaction by placing the samples at -20°C until ready to proceed to the next step. The volume of aRNA at this step should be 25 μl.

Protocol 15

Purification of sense-strand amplified RNA

Equipment and Reagents

- ArrayIt H25K Whole Human Genome Microarray (TeleChem International)
- SenseAMP RNA Amplification kit (Genisphere)
- NanoDrop ND-1000 (NanoDrop Technologies)
- Agilent 2100 bioanalyzer (Agilent Technologies)
- RNase-free dH_2O
- Thermal cycler
- Microcentrifuge (Eppendorf 5415D or equivalent)
- Manual pipettor (2–20 µl)
- Manual pipettor (20–200 µl)
- Manual pipettor (20–1000 µl)
- RNase-free pipette tips
- RNeasy MinElute kit (Qiagen)
- RNase-free 0.5 and 1.5 ml tubes
- 100% Ethanol
- 80% Ethanol

Method

1. Obtain the RNeasy MinElute kit components.
2. Obtain the 25 µl aRNA samples from *Protocol 14*, step 6, and place each sample in a 1.5 ml RNase-free tube.
3. To each 25 µl aRNA sample, add 75 µl of RNase-free dH_2O for a final sample volume of 100 µl.
4. Add 350 µl of RLT buffer (no BME added) and mix well by pipetting up and down.
5. Add 250 µl of 100% ethanol and mix well by pipetting up and down. Transfer each 700 µl sample into an RNeasy spin column. Centrifuge at 13 000 ***g*** for 15 s in a microfuge.
6. Discard the flow-through and collection tube. Place each column containing the aRNA into a new collection tube supplied with the kit.
7. Add 500 µl of RPE buffer (prepared with ethanol as indicated in the kit) to each spin column. Centrifuge at 13 000 ***g*** for 15 s in a microfuge. Discard the flow-through and place each spin column containing the aRNA back in the same collection tube.
8. Add 500 µl of 80% ethanol to each spin column. Centrifuge at 13 000 ***g*** for 2 min in a microfuge. Discard the flow-through and the collection tube. Place each spin column containing the aRNA in a new collection tube supplied with the kit.
9. Centrifuge the spin columns with lids open for 5 min at 13 000 ***g*** in a microfuge to dry the column membrane and remove any residual ethanol. This will increase the yield of aRNA upon elution.
10. Place each spin column into a 1.5 ml collection tube supplied with the kit. Add 12 µl of 50°C RNase-free H_2O to each spin column, taking care to not touch the column matrix with the pipette tip. Incubate at room temperature for 2 min and centrifuge at 13 000 ***g*** for 1 min in a microfuge to elute the aRNA. The final volume of sense-strand aRNA at this step should be 12 µl.

11. Analyze the sense aRNA using a NanoDrop spectrophotometer to determine the concentration, yield and purity[a].

Note

[a]Mix 1 µl of aRNA with 9 µl of 1 mM Tris/HCl (pH 8.0) to dilute the aRNA tenfold. Determine the absorbance in 1 µl of tenfold diluted aRNA. An absorbance reading of 1.0 at 260 nm corresponds to an aRNA concentration of 40 ng/µl. Multiply the absorbance reading by 40 and then by 10 to determine the concentration of the original aRNA sample. Calculate the absorbance ratio at 260 and 280 nm ($A_{260/280}$ ratio) to determine the aRNA purity. Pure RNA has an $A_{260/280}$ ratio of 2.0. A ratio of 2.0–2.3 should be observed with the aRNA sample. Higher ratios may indicate that excessively long poly(A) tails were added during the terminal transferase step (*Protocol 12*). The minimum aRNA concentration required for labeling (*Protocol 16*) is 1.6 µg/µl. Samples of aRNA <1.6 µg/µl should be concentrated by vacuum centrifugation (e.g. Speedvac), lyophilization, evaporation in a heat block at 60°C, or with the use of concentrator columns (e.g. Microcon YM-3; Millipore).

Protocol 16

Synthesis of aminoallyl-modified cDNA from sense-strand aRNA

Equipment and Reagents

- ArrayIt H25K Whole Human Genome Microarray (TeleChem International)
- cDNA Synthesis kit (Genisphere)
- NanoDrop ND-1000 (NanoDrop Technologies)
- Agilent 2100 bioanalyzer (Agilent Technologies)
- SuperScript II reverse transcriptase (Invitrogen)
- Aminoallyl-dUTP (1 mM; Ambion)
- RNase-free dH_2O
- Thermal cycler
- Microcentrifuge (Eppendorf 5415D or equivalent)
- Manual pipettor (2–20 µl)
- Manual pipettor (20–200 µl)
- RNase-free pipette tips
- 1× TE buffer (10 mM Tris/HCl, pH 8.0, 1 mM EDTA; Ambion)
- RNase-free 0.5 and 1.5 ml tubes
- cDNA stop solution (0.5 M NaOH, 50 mM EDTA)
- 1 M Tris/HCl (pH 8.0)

Method

1. Obtain the 12 µl purified sense aRNA samples from *Protocol 15*, step 10. Transfer 10 µg of aRNA into a 0.5 ml RNase-free tube. Add RNase-free H_2O to a final volume of 6.2 µl for each sample.
2. To each 6.2 µl (10 µg) sample of aRNA, add 1.3 µl (19.5 µg) of random 12mer primer (green-capped tube, 15 µg/µl). The sample volume at this step should be 7.5 µl.

3. Incubate the 7.5 μl sense aRNA/primer mixtures in a thermal cycler at 80°C for 5 min to denature the aRNA.
4. During the 5 min denaturation step at 80°C (step 3), prepare a reverse transcription reaction cocktail (12.5 μl per sample) by mixing the following components in a 1.5 ml RNase-free tube: 4 μl of 5× first-strand buffer, 2 μl of 0.1 M DTT, 1 μl of Superase-In RNase inhibitor (white-capped tube), 1 μl of low-dTTP dNTP mix (purple-capped tube), 2.5 μl of 1 mM aminoallyl-dUTP, and 2 μl of SuperScript II reverse transcriptase. Mix gently and spin the tube briefly.
5. After the 5 min denaturation step, place the 7.5 μl sense aRNA/primer mixtures on ice for 2 min to allow priming of the aRNA. Spin briefly and incubate the primed aRNA samples at room temperature for 3 min.
6. To each 7.5 μl primed sense aRNA sample, add 12.5 μl of reverse transcription reaction cocktail for a final sample volume of 20 μl. Mix gently and spin briefly.
7. Incubate the 20 μl samples in a thermal cycler at 42°C for 2.5 h to allow aminoallyl-dUTP incorporation into the nascent cDNA.
8. After the 2.5 h cDNA synthesis step, add 3.5 μl of cDNA stop solution (0.5 M NaOH, 50 mM EDTA) to halt the cDNA synthesis process. Mix and spin each sample briefly.
9. Incubate the 23.5 μl samples at 65°C for 30 min to degrade the aRNA, leaving the cDNA intact.
10. Add 5 μl of 1 M Tris/HCl (pH 8.0) to adjust the pH of each sample to neutral.
11. Add 71.5 μl of 1× TE buffer to adjust the final volume of each aminoallyl-modified cDNA sample to 100 μl.

Protocol 17

Purification and concentration of aminoallyl-modified cDNA

Equipment and Reagents

- ArrayIt H25K Whole Human Genome Microarray (TeleChem International)
- cDNA Synthesis kit (Genisphere)
- NanoDrop ND-1000 (NanoDrop Technologies)
- Agilent 2100 bioanalyzer (Agilent Technologies)
- RNase-free dH_2O
- Microcentrifuge (Eppendorf 5415D or equivalent)
- Manual pipettor (2–20 μl)
- Manual pipettor (20–200 μl)
- RNase-free pipette tips
- 1× TE buffer (10 mM Tris/HCl, pH 8.0, 1 mM EDTA; Ambion)
- 10 mM Tris/HCl (pH 8.0)

Method

1. Obtain the 100 μl aminoallyl-modified cDNA samples from *Protocol 16*, step 11.
2. Place a Microcon YM-50 column into the collection tube provided[a].
3. Transfer the entire 100 μl volume of aminoallyl-modified cDNA sample into the column reservoir, making sure not to contact the column membrane with the pipette tip. Secure the tube cap and centrifuge for 6 min at 13 000 ***g***.
4. Add 200 μl of 1× TE buffer to the column reservoir and mix gently by pipetting up and down five times, making sure not to contact the column membrane with the pipette tip. Secure the tube cap and centrifuge for 6 min at 13 000 ***g***.
5. Carefully remove the YM-50 column containing the aminoallyl-modified cDNA from the collection tube. Discard the flow-through liquid and place the YM-50 column back into the same collection tube.
6. Add 200 μl of 1× TE buffer to the sample reservoir making sure not to contact the column membrane directly. Mix gently by pipetting up and down five times, making sure not to contact the column membrane with the pipette tip. Secure the tube cap and centrifuge for 6 min at 13 000 ***g***[b].
7. Carefully remove the YM-50 column containing the aminoallyl-modified cDNA from the collection tube and discard the collection tube.
8. Add 5 μl of 10 mM Tris/HCl (pH 8.0) to the sample reservoir making sure not to contact the column membrane with the pipette tip. Gently tap the side of the reservoir to mix.
9. Carefully place the sample reservoir *upside-down* in a *new collection tube* and centrifuge for 3 min at 13 000 ***g*** to elute the purified cDNA.
10. Measure the volume of aminoallyl-modified cDNA in the collection tube. The final volume should be 5–10 μl. If necessary, add RNase-free H_2O to a final aminoallyl-modified cDNA sample volume of 10 μl.

Notes

[a]Use Microcon YM-50 columns only for this step. Do not use alternative columns.

[b]If fluorescent deoxynucleotides instead of aminoallyl modified deoxynucleotides were used to generate the labeled cDNA, repeat steps 5 and 6 two additional times to remove unincorporated fluorescent deoxynucleotides from the sample.

Protocol 18

Cy3 and Cy5 dye labeling of aminoallyl-modified cDNA

Equipment and Reagents

- ArrayIt H25K Whole Human Genome Microarray (TeleChem International)
- Aminoallyl cDNA Labeling kit (Ambion)
- Cy3 mono-reactive dye pack (Amersham)
- Cy5 mono-reactive dye pack (Amersham)
- NanoDrop ND-1000 (NanoDrop Technologies)
- Agilent 2100 bioanalyzer (Agilent Technologies)
- Thermal cycler or heating block
- Microcentrifuge (Eppendorf 5415D or equivalent)
- Manual pipettor (2–20 μl)
- Manual pipettor (20–200 μl)
- RNase-free pipette tips
- 1× TE buffer (10 mM Tris/HCl, pH 8.0, 1 mM EDTA; Ambion)
- Dimethyl sulfoxide
- 4 M Hydroxylamine

Method

1. Obtain the 10 μl purified aminoallyl-modified cDNA samples from *Protocol 17*, step 10.
2. Reduce the volume of each cDNA sample from 10 μl to 2.5 μl by incubating in a thermal cycler or heating block for 10–15 min with the cap open at 65°C.
3. Suspend the Cy3 and Cy5 mono-reactive dyes each in 45 μl of dimethyl sulfoxide. Store the dyes in the dark at -20°C.
4. To each concentrated 2.5 μl aminoallyl-modified cDNA sample, add 4.5 μl of coupling buffer. The total sample volume should be 7 μl at this step.
5. To each 7 μl aminoallyl-modified cDNA sample, add 3 μl of Cy3 or 3 ml of Cy5 mono-reactive dye for a final reaction volume of 10 μl. Mix gently and spin briefly.
6. Incubate each 10 μl sample at room temperature (20–25°C) for 1 h to allow coupling of the Cy3 or Cy5 dyes to the aminoallyl-modified cDNA.
7. Quench the coupling reactions by adding 6 μl of 4 M hydroxylamine. Mix and spin briefly. The total sample volume at this step should be 16 μl.
8. Incubate each 16 μl sample at room temperature (20–25°C) for 15 min to quench the coupling reaction.
9. To each 16 μl quenched Cy3 or Cy5 dye reaction, add 84 μl of 1× TE buffer for a final sample volume of 100 μl.

Protocol 19

Purification of Cy3- and Cy5-labeled cDNA

Equipment and Reagents

- MinElute PCR Purification kit (Qiagen)
- Microcentrifuge (Eppendorf 5415D or equivalent)
- Manual pipettor (2–20 μl)
- Manual pipettor (20–200 μl)
- Manual pipettor (20–1000 μl)
- RNase-free pipette tips

Method

1. Obtain the 100 μl quenched Cy3 and Cy5 dye samples from *Protocol 18*, step 9.
2. To each 100 μl sample, add 500 μl of buffer PB from the MinElute PCR Purification kit and mix well.
3. Apply the entire 600 μl dye mixture to a MinElute column and centrifuge for 1 min at 14000 ***g*** in a microcentrifuge.
4. Discard the flow-through and place the MinElute column containing the dye-labeled cDNA into the same collection tube.
5. Add 750 μl of buffer PE to the MinElute column and centrifuge for 1 min at 14000 ***g*** in a microcentrifuge.
6. Discard the flow-through and place the MinElute column containing the dye-labeled cDNA back into the same collection tube.
7. Centrifuge for 2 min at 14000 ***g*** in a microcentrifuge to remove any residual ethanol.
8. Place the MinElute column containing the dye-labeled cDNA into a new 1.5 ml microcentrifuge tube.
9. Add 10 μl of buffer EB to the center of the column membrane and incubate at room temperature for 2 min. Spin in a microcentrifuge for 2 min and discard the column. The eluted sample contains the purified, dye-labeled cDNA, the final volume of which should be 10 μl.

Protocol 20

Hybridization of fluorescent cDNA mixtures to whole human genome microarrays

Equipment and Reagents

- ArrayIt H25K Whole Human Genome Microarray (TeleChem International)
- ArrayIt hybridization cassettes (TeleChem International)
- ArrayIt high-throughput wash station (TeleChem International)
- ArrayIt high-speed microarray centrifuge (TeleChem International)
- ArrayIt SpotLight microarray scanner (TeleChem International)
- ArrayIt 10× wash buffers A, B, and C (TeleChem International)
- 2× SDS-based hybridization buffer (Genisphere)
- 24 × 60 mm LifterSlips (Erie Scientific)
- DyeSaver 2 (Genisphere)
- Thermal cycler
- Microcentrifuge (Eppendorf 5415D or equivalent)
- Water bath (60°C)
- Manual pipettor (2–20 μl)
- Manual pipettor (20–200 μl)
- RNase-free pipette tips

Method

1. Obtain the 10 μl fluorescent cDNA samples from *Protocol 19*, step 9.
2. Thaw and suspend the 2× SDS-based hybridization buffer by heating in a thermal cycler at 80°C for 10 min. Mix vigorously with a vortex to ensure that the buffer is suspended completely. Spin for 1 min in a microcentrifuge to remove any insoluble material in the 2× hybridization buffer.
3. Prepare the 50 μl two-color fluorescent probe samples by mixing the following components: 10 μl of Cy3-labeled reference cDNA, 10 μl of Cy5-labeled test cDNA, 5.0 μl of nuclease-free H_2O, and 25 μl of 2× SDS-based hybridization buffer[a]. Mix the probe samples gently with a vortex and spin for 1 min in a microcentrifuge at full speed to remove any particulates.
4. Heat the 50 μl probe mixture at 75°C for 10 min, and then cool to 60°C.
5. Pre-heat the ArrayIt H25K Whole Human Genome Microarray to 60°C and place the 50 μl probe sample (60°C) onto the center of the printed microarray. Place a 24 × 60 mm glass LifterSlip onto the 50 μl probe sample and allow the probe mixture to sheet across the entire LifterSlip surface, making sure that the entire H25K microarray is covered with probe solution.
6. Place the H25K microarray into an ArrayIt hybridization cassette containing 10 μl of dH_2O for humidification, submerge the hybridization cassette into a 60°C water bath, and incubate in the dark for 16 h at 60°C.
7. After the 16 h hybridization, place the H25K microarray into the slide rack of the high-throughput wash station and wash immediately with moderate mixing as follows: (i) 2 min in 500 ml of 1× wash buffer A at 52°C until the LifterSlip falls off[b]; (ii) 10 min in 500 ml of 1× wash buffer A at 52°C, (iii) 10 min in 500 ml of 1× wash buffer B at room temperature, and (iv) 10 min in 500 ml of 1× wash buffer C at room temperature.

8. After the wash steps, transfer the H25K microarray to a high-speed microarray centrifuge and dry the microarray by spinning for 10 s[c].
9. Scan the H25K microarray in the green (Cy3) and red (Cy5) channels using the ArrayIt SpotLight microarray scanner and save the TIFF files for subsequent data quantification and analysis.

Notes

[a]This protocol is designed for two-color fluorescence analysis involving a reference cDNA sample and a test cDNA sample derived from RNA mixtures for which comparative gene expression information is sought. For single-color analysis, use a single cDNA only and compensate for the volume difference by adding an additional 10 µl (15 µl total) of nuclease-free H_2O.
[b]The LifterSlip should be removed from the wash station as quickly as possible to prevent it scratching the H25K surface.
[c]DyeSaver 2 coating can be applied at this step to preserve fluorescent signals.

3. TROUBLESHOOTING

The protocols presented herein have been optimized for use with the 44K Whole Human Genome Microarray from Agilent and the H25K Whole Human Genome Microarray from ArrayIt. For best results, carefully follow the instructions provided by the manufacturers. Poor results are most often due to degraded RNA or to kit components that have lost activity due to extended storage or use. Wear nitrile gloves at all times when handling RNA and make sure always to use RNase-free materials and solutions. Samples should be examined frequently during the amplification and labeling procedures to make sure that each step is proceeding as expected. The NanoDrop and BioAnalyzer instruments provide an excellent means of verifying aRNA and cDNA quality during the amplification and labeling processes. The same instruments can also be used to identify faulty steps in these processes.

Acknowledgements

The authors wish to thank their colleagues for critical comments and helpful suggestions on the manuscript. The authors also wish to thank Dr Robert C. Getts and his colleagues at Genisphere, Inc. (Hatfield, PA, USA) for developing and optimizing the Genisphere SenseAmp and cDNA Synthesis kit protocols. This work was supported by a generous grant from the ArrayIt Life Sciences Division at TeleChem International, Inc. and by MDS Analytical Technologies, Sunnyvale, California. M.S. is a Visiting Scholar at TeleChem.

4. REFERENCES

★★★ 1. **Schena M, Shalon D, Davis RW & Brown PO** (1995) *Science*, **270**, 467–470. – *A landmark paper by virtue of being the first microarray publication.*

★★ 2. **Schena M** (1999) *DNA Microarrays: a Practical Approach.* Edited by M. Schena. Oxford University Press, UK. – *The first book on DNA microarrays.*

★ 3. **Schena M** (2000) *Microarray Biochip Technology.* Edited by M. Schena. BioTechniques Book Division, Eaton Publishing, MA, USA. – *The first concepts book on DNA microarrays.*

★★ 4. **Schena M** (2003) *Microarray Analysis.* J. Wiley & Sons, NJ, USA. – *The first textbook on DNA microarrays.*

★★★ 5. **Shi L, Reid LH, Jones WD, *et al.*** (2006) *Nat. Biotech.* **24**, 1151–1161. – *A landmark project establishing FDA quality control guidelines for DNA microarrays.*

★ 6. **Stears RL, Martinsky T & Schena M** (2003) *Nat. Med.* **9**, 140–145. – *Explains the uses of single- and multi-patient DNA microarrays for genotyping.*

★ 7. **Schena M, Shalon D, Heller R, Chai A, Brown PO & Davis RW** (1996) *Proc. Natl. Acad. Sci. U.S.A.* **93**, 10614–10619. – *First publication showing the use of DNA microarrays for human gene expression studies.*

★ 8. **DeRisi JL, Iyer VR, Brown PO** (1997) *Science*, **278**, 680–686. – *One of two initial publications of DNA microarray analysis of the entire yeast genome.*

★ 9. **Wodicka L, Dong H, Mittmann M, Ho MH & Lockhart DJ** (1997) *Nat. Biotech.* **15**, 1359–1367. – *One of two initial publications of DNA microarray analysis of the entire yeast genome.*

★ 10. **Cho RJ, Campbell MJ, Winzeler EA, *et al.* (1998)** *Mol. Cell*, **2**, 65–73. – *First publication of DNA microarray analysis of the yeast cell cycle on a genomic level.*

★★★ 11. **Venter JC, Adams MD, Myers EW, *et al.*** (2001) *Science*, **291**, 1304–1351. – *Landmark publication reporting a draft sequence of the human genome.*

★ 12. **Goff LA, Bowers J, Schwalm J, Howerton K, Getts RC & Hart RP** (2004) *BMC Genomics*, **5**, 76–84. – *Important publication reporting robust protocols that can be used for DNA microarray expression analysis of the entire human genome.*

CHAPTER 2

Toxicogenomics of dioxin

Craig R. Tomlinson, Saikumar Karyala, Danielle Halbleib, Mario Medvedovic, and Alvaro Puga

1. INTRODUCTION

Genomics is defined as the determination of the sequence, location, and annotation of the genes in a genome. Functional genomics is the study of how a particular genotype gives rise to a particular phenotype, and the primary method of carrying out functional genomics studies is to determine global gene RNA expression patterns by microarray analysis. Toxicogenomics is a specialized field of functional genomics and is the study of how environmental toxicants affect gene RNA expression patterns. The purpose of this chapter is to discuss some of the methods used in toxicogenomics research. The methods that are discussed utilize a two-color microarray approach in which two differently labeled mRNA probe samples are applied to the same microarray slide and then scanned to obtain signals in each color channel. Microarrays containing long oligonucleotides (rather than cDNAs or short oligonucleotides) are used for the approaches described herein. The effect on gene expression resulting from treatment with the classic environmental toxicant 2,3,7,8-tetrachlorodibenzo-*p*-dioxin (TCDD or dioxin) is used as an instructive example of how toxicogenomics studies are performed.

TCDD is a halogenated polycyclic aromatic hydrocarbon and is ubiquitous in the atmosphere, soil, water, and food (1). The average daily adult human intake of TCDD is estimated to be 47 pg/day (2). In certain human populations, such as those that regularly consume fish, the daily intake of TCDD is estimated to be 390–8400 pg/day. Combustion of fossil fuels and byproducts of commercial chemical syntheses are the major sources of TCDD contamination (3). A major concern regarding TCDD is not only its environmental longevity but its biological persistence. TCDD accumulates in the adipose tissues (4), and the elimination half-life of TCDD in humans is approximately 7–11 years (5).

DNA Microarrays: *Methods Express* (M. Schena, ed.)

TCDD is the prototypical dioxin and causes a large number of harmful biological effects and diseases. In humans, TCDD causes chloracne, a scarring skin disease (6, 7), endometriosis (8), immunotoxicity and liver damage (9), cardiovascular disease and diabetes (10, 11), and cancer (12–15). Furthermore, TCDD is one of the most potent tumor promoters known (16, 17). Exposure of TCDD to mice and other rodents during embryogenesis causes teratogenic abnormalities (18) including hydronephrosis and cleft palate (19, 20). The goal of all toxicogenomics studies is to learn how pre-natal, neonatal, childhood, and adult gene expression is affected by omnipresent and dangerous environmental toxicants such as TCDD so that preventative and curative measures can be implemented.

1.1 The aromatic hydrocarbon receptor signaling pathway

Most, if not all, of the biological effects of TCDD exposure are mediated by the aryl hydrocarbon receptor (AHR) (21). The AHR signaling pathway is schematized in *Fig. 1* (also available in the color section). The AHR is a major player in the molecular defense of polycyclic aromatic hydrocarbon environmental toxicants and is a member of the PER-ARNT-SIM or PAS family of basic-region helix–loop–helix or bHLH transcription factors (22).

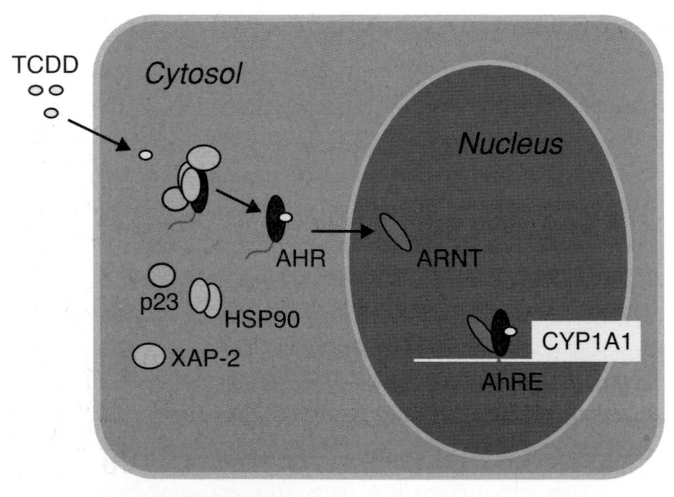

Figure 1. AHR signaling pathway (see page xx for color version).
The chemical toxin TCDD moves across the plasma membrane into the cytosol where it binds to a high-affinity AHR that exists in the unbound state as a complex with other cellular proteins. TCDD-bound AHR moves into the nucleus, binds specifically to high-affinity chromosomal sites, and regulates the expression of cellular genes.

The vertebrate AHR is found in the cytosol in association with HSP90 chaperones, immunophilin-like proteins (XAP2/ARA9/AIP), and p23 (23). Upon ligand binding, the AHR translocates to the nucleus where it complexes with the aryl hydrocarbon receptor nuclear translocator (ARNT) (24). The AHR/ARNT heterodimer activates the transcription of a battery of genes encoding proteins involved in xenobiotic metabolism, which in turn catabolize the AHR ligand (toxicant) to complete an elegant biological feedback loop. Thus, gene transactivation mediated by the AHR represents an adaptive response required for the detoxification of foreign compounds. Some of the major genes transcriptionally activated by the toxicant/AHR/ARNT complex include the cytochrome P450 *Cyp1* family (phase I genes), as well as several phase II detoxification genes (25, 26), via transactivation through enhancer domains known as AHR-, dioxin-, or xenobiotic-response elements (AhREs, DREs, or XREs) (27). However, the halogen side groups on TCDD render it highly resistant to degradation by the phase I and II gene products, leading to chronic activation of the AHR and the harmful biological effects discussed above (28).

1.2 The AHR in homeostasis and development

Mice with homozygous loss of *Ahr* gene function have been generated in four laboratories (29–32) and are resistant to most of the consequences of TCDD toxicity (33). However, mice with the *Ahr* gene deleted have numerous developmental defects including cardiomegaly, immune system deficiencies, hepatic fibrosis (29), liver retinoid accumulation and decreased retinoic acid metabolism (34), heart hypertrophy, fibrosis of the liver, vascular hypertrophy, mineralization in the uterus, and gastric hyperplasia (34, 35). In mouse embryonic fibroblasts derived from $Ahr^{-/-}$ mice, there is a lower cell proliferation rate and a higher rate of apoptosis (36). One of the most obvious effects of AHR deletion on the mouse embryo is a drastically reduced liver blood supply due to the failure of the shunt from the portal vein to the inferior vena cava to close properly, which normally occurs perinatally. AHR deletion also results in the loss of liver sinusoids and vascular anomalies in the eye and kidney (37). Thus, the AHR, in addition to its role in xenobiotic metabolism, appears to play vital developmental roles in vascular patterning, organ modeling, extracellular matrix deposition, cell proliferation, apoptosis, and the development of the heart and immune system.

The gene coding for AHR has been cloned and examined functionally in several vertebrates and invertebrates (reviewed in 38). For example, it was shown recently that the orthologous *Caenorhabditis elegans* AHR protein directs the fate and gene expression of two of the 26 GABAergic neurons (39). Similar to all invertebrate AHRs tested to date, the *C. elegans* AHR does not bind TCDD or any of the classic AHR ligands, nor does the *C. elegans* AHR activate xenobiotic metabolism (40). This and similar observations have led to the idea that perhaps the ancestral function of the AHR was unrelated to xenobiotic toxicity and instead primarily played a developmental role. The AHR, in addition to its role as a regulator of xenobiotic metabolizing genes in vertebrates, plays a significant role in development in both vertebrates and invertebrates.

1.3 Toxicogenomics questions

A genomic profiling study that we completed recently illustrates how a typical high-throughput toxicogenomics inquiry is performed (41). Armed with the knowledge that (i) TCDD causes a plethora of diseases and defects, (ii) the AHR is the major mediator of the ill effects of TCDD, and (iii) the AHR has developmental roles, we asked two questions: first, whether the AHR carries out additional functions in the cell independent of its role as a xenobiotic receptor, and secondly, whether, in addition to the AHR pathway, another cellular mechanism or signaling pathway responds to TCDD. The microarray studies were carried out using aortic smooth muscle cells derived from wild-type and $Ahr^{-/-}$ mice. To summarize the results briefly, analysis of the gene expression profiles allowed us to propose the hypothesis that, in addition to xenobiotic metabolism, the AHR is involved in the regulation of the transforming growth factor-β (TGF-β) signaling pathway, and that TGF-β signaling may act as an alternative pathway to AHR signaling in response to TCDD (41, 42).

2. METHODS AND APPROACHES

2.1 Overall approach to a toxicogenomics study: three phases

High-throughput approaches to biological problems such as a microarray-based functional genomics study are often expensive and inherently produce large volumes of data. It is therefore imperative that the studies be carried out as efficaciously and economically as possible. In order to meet these goals, three important phases of a toxicogenomics study must be completed successfully (see *Fig. 2*). Phase I is the experimental design of the microarray project. The successful completion of phase I is realized when the microarray results are relevant to the scientific questions asked. We discuss some of the important considerations of proper experimental design below. Phase II is the microarray experimental process in which

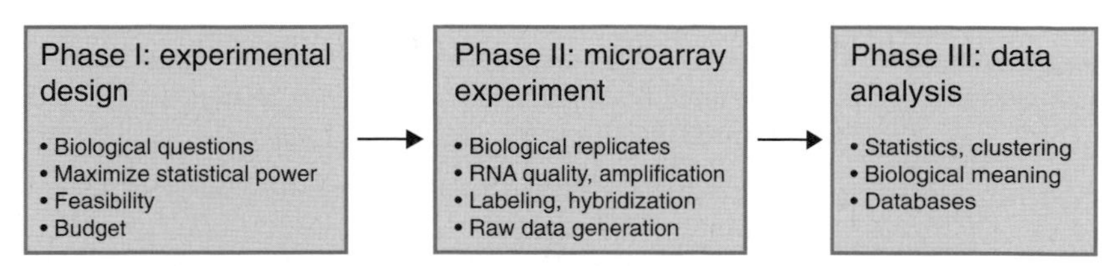

Figure 2. Successful completion of a functional genomics experiment using microarrays involving three distinct phases (I–III).

the raw data are generated. We provide preferred protocols for RNA isolation, RNA quality control, RNA amplification, cDNA synthesis, cDNA and amplified RNA labeling, microarray slide production, microarray hybridization and wash conditions, microarray slide scanning, and data generation. Phase III covers the analyses of the voluminous amounts of data generated in a typical set of microarray experiments. We briefly cover some of the statistical and clustering analyses used, how these analyses can be applied to extract biological meaning, and how these data are stored and accessed in databases.

2.1.1 Phase I: experimental design

There are four primary intertwined considerations in devising an appropriate experimental plan. First and foremost, it is imperative to design the microarray experiments so that the forthcoming results properly address the scientific questions being asked. The well-prepared design eliminates unnecessary comparisons, and as a result irrelevant data are minimized. Secondly, the best possible design is one that offers the greatest statistical power. For example, a major means of increasing statistical power is to use at least four biological replicates representing at least four microarray comparisons for each experimental condition. The third consideration pertains to the feasibility of carrying out the experimental plan. Typical feasibility questions include some of the following. Can sufficient amounts of RNA be isolated from a selected cell type or tissue? Can a particular experimental treatment or time frame be applied in the desired way to affect gene transcription? Can a sufficient number of biological replicates be generated? The experimental design may need to be modified depending on how the feasibility questions are answered. The last consideration involves the best way to design an experiment within a monetary budget, i.e. the intent is to design the most efficacious and statistically powerful experiment within a funding limit. Again, sometimes in order to meet this intent, the experimental design must be modified or trimmed.

Fig. 3 illustrates a simple experimental plan of appropriate design in response to two toxicological questions (41). What other role(s) does the AHR carry out in the wild type, in which gene expression profiles from untreated wild-type and *Ahr*$^{-/-}$ cells are compared (see *Fig. 3a*)? What genes respond similarly to TCDD regardless of the AHR status of the cell (see *Fig. 3b*)? For the second question, the gene expression profiles of TCDD-treated wild-type and *Ahr*$^{-/-}$ cells were compared with each other. The experimental plan in *Fig. 3* shows how multiple questions can be investigated simultaneously from a single experimental design, which in turn can generate very different gene expression datasets whilst maintaining an economy of microarray slides.

2.1.2 Phase II: microarrays

The microarray methods that are discussed here (see section 2.2) utilize dual-channel hybridization approaches on the long oligonucleotide (70mer) platform. The use of oligonucleotide microarrays is widespread, and generally such approaches employ either short oligonucleotides (e.g. 25mers) (43, 44) or long oligonucleotides (50- to 80mers) (45, 46). Although recent studies have shown that the quality of microarray results can be platform-independent (47, 48), the long oligonucleotide platform

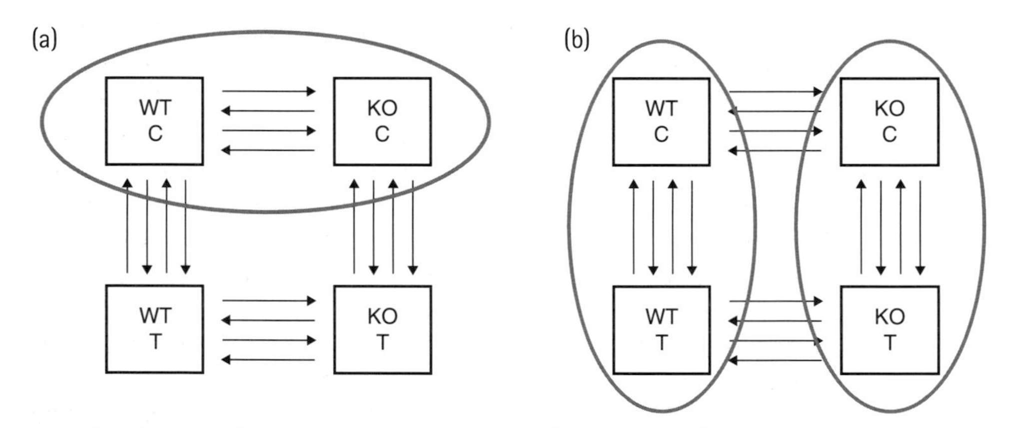

Figure 3. Schematic of toxicogenomics experimental design.
High-throughput gene expression analysis of vehicle-treated control (C) and 10 nM TCDD-treated (T) cultured mouse wild-type (WT) and *Ahr*$^{-/-}$ knockout (KO) null aorta smooth muscle cells. Boxes represent the different experimental conditions. TCDD was added at 80–90% confluence and the cells were exposed for 24 h. Arrows between the boxes denote a quadruplicate experiment with two 'dye flips'. The experimental design was set up to investigate two questions: (i) What other role(s) does the AHR carry out in the wild type? (ii) What genes respond similarly to TCDD, regardless of the AHR status of the cell? (Note that TCDD is a highly toxic compound. All personnel should be instructed as to safe handling procedures. Laboratory coats, gloves, and masks are required when using stock solutions of this compound.)

may have the advantage of increased sensitivity and specificity (49, 50), and the protocols in this chapter describe how to maximize sensitivity and specificity (41, 42, 51).

2.1.3 Phase III: data analysis

A major concern in the analysis of large data sets is proper statistical analysis of the microarray results and data interpretation. The probability of detecting differential expression between two experimental samples depends on the magnitude of the differential expression, experimental variability, and the number of measurements made for each treatment combination. If comparisons among some treatment combinations are made indirectly, such as when elements of the 'loop' design are utilized (52), the probability also depends on the distance between two treatment combinations in the experimental design.

The statistical analyses are performed in three steps. First, the data are normalized with the goal of removing systematic biases and thus reducing experimental variability (53). In the case of spotted dual-channel microarrays, two major sources of the systematic variability are the spot-specific local background fluorescence and the difference in the overall intensities of the two fluorescent dyes used in the study. The process of normalization generally proceeds by subtracting the local background and centering the log ratios of two channel intensities near zero. The centering is performed by subtracting the fitted local regression curve value from the corresponding log ratios. Secondly, analysis of variance (ANOVA) for each gene is carried out to identify those that were affected by one or more experimental treatments and/or their interactions. Statistical significance of changes in the expression level is performed using a Gaussian distribution model, and extensive model checking is performed to assess the appropriateness of the model. If it turns out that the data deviate significantly from this model, nonparametric resampling approaches (54) are applied. Significance levels for individual genes are adjusted using the false discovery rate approach to multiple hypotheses testing (55).

Following ANOVA analysis, cluster analysis is used to organize groups of differentially expressed genes further, based on the similarity of their expression profiles. In our experience, the Bayesian infinite mixture-based approach (56) usually produces the most meaningful groupings. The method allows the identification of groups of genes with similar expression profiles across different experimental conditions in a statistically significant fashion (57, 58). Cluster analysis complements ANOVA by identifying groups of genes that show similar expression patterns across different treatments and experimental conditions. Although the statistical power of ANOVA can be relatively high, if there are a large number of differentially expressed genes detected, a significant portion of them may be missed using ANOVA alone. One aim of cluster analysis is to reduce further the set of undetected genes in which expression is affected by the specific experimental treatments examined.

Once the statistical analyses have identified the significantly differentially expressed genes, the next challenge is to determine what the data mean in the context of biology. Such large datasets can be daunting, and open 'shareware' software packages downloadable via the Internet are helpful in interpreting the data. We recommend the EASE program, which allows the merging of data sets into various software packages (59). EASE is used to merge statistically analyzed data sets from our studies into the Gene Ontology (GO) program, which describes gene products in the context of biological processes, cellular components, and molecular functions in a species-independent manner (60). The GO program provides a mechanistic approach for determining the biological relevance and roles of the gene products identified in the differentially expressed gene data sets.

After the datasets have been analyzed statistically and biologically, the datasets must be made available to the scientific community. Microarray technology has introduced new challenges regarding the manipulation, storage, and accessibility of data. A large experiment can easily generate several million measurements. Data analysis on this scale requires the researcher to employ computer-based tools on data archived in an appropriate database. A database is generally not only an electronic depository but a representation of the data that models and implements the structure and relationships among the data items.

Designing a database that efficiently organizes all relevant microarray data and related experimental information is not a trivial task. A system known as the Minimum Information About a Microarray Experiment (MIAME) has been defined (61) and is generally accepted by the microarray community. The related Microarray Gene Expression Object Model (MAGE-OM) and the XML-based format (MAGE-ML) have been developed to facilitate database description, and the exchange of fully annotated datasets has also been achieved. Based on these standards and data models, relational databases have been designed, one of which is the well-known public repository of microarray data called ArrayExpress (62). Typical of what is required to manipulate the database, we have adopted a MySQL version of ArrayExpress called MaxD (http://www.bioinf.man.ac.uk/microarray/maxd/index.html) as a local solution for managing microarray data. All microarray data generated locally are stored in this database. In addition to the MIAME-compliant schema, the MaxD system contains tools for the automatic creation of MAGE-ML-compliant representation of the data for easy dissemination.

2.2 Recommended protocols

The quality of microarrays is heavily dependent on the quality of the RNA used to make the labeled probe. Therefore, RNA isolation methods and RNA quality control assays are critical. We recommend the protocol below to ensure valid and reliable microarray data and polymerase chain reaction (PCR) results from RNA samples. The basic procedure is as follows:

1. Isolate total RNA using TRI Reagent (MRC, Inc.). One can visit the MRC website for the most recent protocols (http://www.mrcgene.com; select RNA Isolation). Note that RNAzol and RNAzol B reagents are based on an older patent and their use may result in higher DNA contamination levels.
2. Treat the isolated RNA with a DNase/proteinase K step, which digests any contaminating DNA and protein. Contaminating DNA causes variable microarray and PCR results, and contaminating protein can inhibit labeling and cause RNA degradation.
3. Extract the RNA with TRI Reagent LS to remove the digested DNA and proteins. We also recommend isolating total RNA rather than poly(A)$^+$ RNA for use in microarrays. The protocols described here can also be viewed at http://microarray.uc.edu.

Protocol 1

Isolation of total RNA from tissue and cells[a]

Equipment and Reagents

- 15 ml Round-bottomed tubes
- 1.5 ml Microfuge tubes
- Liquid nitrogen
- TRI Reagent
- TRI Reagent LS
- Kontes homogenizer and pestle
- 75% Ethanol
- Water bath
- Microcentrifuge
- Bromochloropropane (BCP; MRC)
- Isopropanol
- RNase-free/low-endonuclease water
- DNase buffer (Ambion)

- TURBO DNase (2 units/μl; Ambion)
- Proteinase K (20 mg/ml; New England BioLabs)
- NanoDrop ND-1000
- Agilent 2100 Bioanalyzer

Method

1. For tissues, use pre-chilled long tweezers to place the tissue directly into a 15 ml round-bottomed tube containing 1 ml of liquid nitrogen (N_2). Store the tube without the cap for 24 h at –80°C. Frozen tissue can be stored indefinitely at –80°C.
2. Transfer the tissue into a fresh 15 ml round-bottomed tube containing TRI Reagent. We recommend 1 ml of TRI Reagent per 50 mg of tissue. Increasing the amount of tissue increases the chance of DNA and protein contamination.
3. Immediately homogenize the tissue with a Kontes homogenizer and pestle. Be sure that all of the tissue is homogenized. Rinse three times with water between samples.
4. Transfer 750 μl of the homogenate to a 1.5 ml microfuge tube. To thaw homogenates, place tubes briefly in a water bath set at room temperature and swirl.
5. Spin the homogenates at 12000 ***g*** for 10 min. Transfer the supernatants into new microfuge tubes. For cultured cells only, remove the culture medium by aspiration and add at least 1 ml of TRI Reagent per 10 cm^2 of culture flask area. Pass the cell lysate through the pipette several times to complete cell lysis[b].
6. For tissue and cell lysates, incubate for 5 min at room temperature.
7. It is beneficial to perform phase separation with BCP instead of chloroform. Use 100 μl of BCP per ml of TRI Reagent. The use of BCP instead of chloroform improves the quality of the isolated RNA. In addition, BCP is less toxic and less volatile than chloroform. If chloroform is used, the volume of chloroform should always be 20% of the initial volume of TRI Reagent. Vortex for a full 15 s and incubate for 3 min at room temperature. Using a larger volume of chloroform will increase the chance of DNA contamination.
8. Centrifuge at 12000 ***g*** for 15 min at 4°C.
9. Transfer the aqueous phase to a new centrifuge tube. Do not disturb the interface between the two layers.
10. Add 0.5 vols of isopropanol and incubate the samples for 10 min at room temperature. Incubations longer than 10 min or isopropanol volumes in excess of 0.5 vols may cause unwanted protein precipitation.
11. Pellet the RNA by centrifugation at 12000 ***g*** for 10 min at 4°C.
12. Remove the supernatant carefully as the RNA pellet may be loose and fragile. Centrifuge briefly and remove residual supernatant using a pipette fitted with a 10 μl tip.
13. To the RNA pellet, add 1 ml of 75% ethanol, vortex, and centrifuge at 7500 ***g*** for 5 min at 4°C.
14. Remove the ethanol carefully as the RNA pellet may be loose and fragile. Centrifuge briefly and remove residual supernatant using a pipette fitted with a 10 μl tip.
15. Air dry on ice for 5 min. Do not dry in a hood as the turbulence may contaminate the RNA or dislodge the pellet.
16. Redissolve the RNA pellet completely in 87 μl of RNase-free/low-endonuclease water.

17. For the DNase/proteinase K treatment, add 10 μl of DNase buffer and 3 μl (6 units) of TURBO DNase (Ambion). Mix and incubate for 15 min at 37°C. Add 4 μl of proteinase K (20 mg/ml). Mix and incubate for 30 min at 37°C.
18. Use TRI Reagent LS to extract the RNA. Add RNase-free/low-endonuclease water to adjust the volume to 250 μl.
19. Add 750 μl of TRI Reagent LS and shake vigorously for 15 s. Incubate the lysates for 5 min at room temperature.
20. Add 100 μl of BCP, cover the sample tightly, and shake vigorously. Store the mixture at room temperature for 8 min. Centrifuge the mixture at 12 000 ***g*** for 15 min at 4°C.
21. Transfer the top (aqueous) phase into a fresh tube. Precipitate the RNA by adding 0.5 ml of isopropanol. Store the samples at room temperature for 8 min and centrifuge at 12 000 ***g*** for 8 min at 4°C.
22. Remove the supernatant and wash with 1 ml of 75% ethanol. Centrifuge at 12 000 ***g*** for 8 min at 4°C. Carefully discard the supernatant and air dry the pellet for 5 min.
23. Dissolve the pellet in a small volume of RNase-free/low-endonuclease water.
24. For microarray labeling reactions, the final concentration of RNA should be 2 μg/μl.
25. To determine accurate RNA concentration, the absorbance ratio at 260 and 280 nm is assayed using a NanoDrop ND-1000. RNA quality (i.e. the degree of degradation) is assayed on an Agilent 2100 Bioanalyzer.
26. For both the NanoDrop and Agilent 2100 Bioanalyzer analyses, 1 μg of RNA sample is diluted to 0.1 μg/μl in Tris/EDTA buffer. For RNA of high purity, the absorbance ratio (260/280) should be greater than 1.8.

Notes

[a]As a precautionary note, the importance of careful technique cannot be overemphasized when working with RNA. RNaseAWAY (Molecular BioProducts) is used to treat all tubes, pestles, benches, and any other implement or surface that may come into contact with RNA-containing samples. Nuclease-free water is used for all suspensions and in the preparation of all buffers. Pre-label all tubes prior to use to prevent unnecessary manipulations of RNA-containing tubes. Change gloves often during the RNA isolation procedure.

[b]For smaller amounts of tissue or a smaller number of cells, the protocol can be scaled down proportionately.

Although for optimal labeling we typically recommend 20–30 μg of total RNA, 10 μg or even as little as 5 μg of superior-quality RNA can be converted efficiently into cDNA and labeled using the indirect labeling method described below. This protocol is a slightly modified version of a protocol developed in Patrick Brown's laboratory at Stanford University (http://cmgm.stanford.edu/pbrown/protocols/amino-allyl.htm). The Brown protocol is based on a protocol developed by DeRisi (45) available at http://www.microarray.org.

Protocol 2

cDNA synthesis and indirect aminoallyl labeling

Equipment and Reagents

- Oligo(dT) (2 μg/μl)
- SuperScript III reverse transcriptase (RT) kit (Invitrogen)
- SuperScript III RT (200 units/μl; Invitrogen)
- 50× dNTP stock (25 mM each of dATP, dCTP, and dGTP, 7.5 mM dTTP, and 17.5 mM aminoallyl-dUTP, using a 7:3 ratio of aminoallyl-dUTP to dTTP. The 50× stock is made by mixing 10 μl each of 100 mM dATP, dCTP, and dGTP (Phizer-Pharmacia), 3 μl of 100 mM dTTP, and 7 μl of 100 mM aminoallyl-dUTP (Sigma). The aminoallyl-dUTP (1 mg) is dissolved in 19.1 μl of 0.1 M KPO_4)
- RNasin (40 units/μl; Fisher)
- 0.1 M NaOH
- 0.1 M HCl
- 3 M Sodium acetate
- 80 and 100% Ethanol
- CyDye Post-Labeling Reactive Dye Packs (Amersham)
- Coupling buffer (0.1 M sodium carbonate adjusted to pH 9.0. A 1 M stock solution of coupling buffer is made as follows: dissolve 8.4 g of $NaHCO_3$ in 70 ml of water, adjust the pH to 9.0, and bring the total volume to 100 ml. The coupling buffer should be pH tested before every use, discarded when the pH exceeds 9.15, and made fresh every 4 weeks)
- QIAquick PCR Purification kit (Qiagen)

Method

Reverse transcription reaction[a]

For optimal labeling, 20–30 μg of total RNA (2–3 mg/ml) is required per reaction. As little as 5 μg of total RNA can be used if the RNA is of very high quality.

1. Set up a 20 μl annealing reaction for each RNA sample as follows: 2.5 μl of 2 μg/μl oligo(dT) (5 μg total), 5.5 μl RNase-free dH_2O, and 10.0 μl of 2 μg/μl total RNA (20 μg total, from *Protocol 1*). For multiple annealing reactions, prepare a master mix of oligo(dT) and RNase-free dH_2O, using 8.0 μl of master mix per reaction.
2. Heat to 70°C for 10 min to denature the RNA. Cool on ice for 5 min to allow annealing of the oligo(dT) to the RNA template.
3. To each annealing reaction, add 11.6 μl of RT master mix to allow reverse transcription of the annealed templates. The RT master mix is prepared as follows: mix 6.0 μl of 5× First Strand Buffer (in the SuperScript III RT kit), 0.6 μl of 50× dNTP stock solution, 3.0 μl of 0.1 M dithiothreitol (DTT) (in the SuperScript III RT kit), 0.5 μl of 40 units/μl RNasin and mix well. Then add 1.5 μl of 200 units/μl SuperScript III RT and mix well. For multiple reactions, calculate the master mix volume to accommodate an extra one or two reactions to allow for cumulative transfer errors if they exist. Separate reactions should be prepared for samples labeled with Cy3 and Cy5 (see below).
4. Incubate the RT reactions at 50°C for 1 h.

5. Following the RT step, degrade the RNA templates away from the newly synthesized cDNA strands by adding 15 µl of 0.1 M NaOH and incubating at 70°C for 10 min.
6. Cool the samples by incubating them at room temperature for 1 min and neutralize the NaOH by adding 15 µl of 0.1 M HCl. At this stage, if the reactions have proceeded properly, the samples will contain nascent cDNA templates with highly degraded polyribonucleotides.
7. Add 6 µl of 3 M sodium acetate (pH 5.2) to each sample.
8. Add 150 µl cold absolute ethanol to each sample. Proceed to the next step or store at -20°C overnight if a convenient stopping point is needed in the procedure.
9. Pellet the cDNA products by centrifugation (12 000 *g*) at room temperature (20–25°C) for 15 min.
10. Pour off the supernatant and add 750 µl of cold 80% ethanol. Flick each tube gently to allow some mixing.
11. Centrifuge the tubes at 12 000 *g* for 10 min at 4°C to repellet the cDNA.
12. Repeat steps 10 and 11 once.
13. Dry the cDNA pellets by vacuum centrifugation at room temperature (20–25°C) for 10 min.

Dye coupling and cDNA purification

1. Resuspend the dry aminoallyl-labeled cDNA pellets by adding 5 µl of dH_2O and 5 µl of coupling buffer. Mix well by vortexing.
2. Incubate at 42°C for 10 min to dissolve the cDNA.
3. Mix well by vortexing and centrifuge each tube briefly to collect the samples at the bottom of each tube.
4. Transfer the 10 µl aminoallyl-labeled cDNA mixture into a dry Cy3 or Cy5 dye pack sample and mix well by pipetting to dissolve.
5. Incubate for 1 h at room temperature in the dark. Mix by gentle vortexing every 15–30 min.
6. Purify the fluorescent cDNA products using Qiagen QIAquick columns. Add 250 µl of Buffer PB and 35 µl of 0.1 M sodium acetate (pH 5.2) to each reaction tube. Vortex and centrifuge briefly.
7. Apply the fluorescent mixture to a QIAquick column and centrifuge at 12 000 *g* for 1 min. The cDNA products are bound to the column membrane at this point. Decant and discard the flow-through.
8. Add 750 µl of Buffer PE to the column and centrifuge at 12 000 *g* for 1 min. Decant and discard the flow-through. Centrifuge at 12 000 *g* for 1 min to remove residual Buffer PE.
9. Transfer the QIAquick column containing the bound cDNA products to a fresh 1.5 ml microfuge tube.
10. Add 35 µl of Buffer EB to the center of the filter and incubate for 5 min at room temperature. Centrifuge at 12 000 *g* for 1 min to elute the bound cDNA products.
11. Dry the purified cDNA products using vacuum centrifugation. The samples are ready for hybridization at this point.

Note

[a]Total RNA samples that are degraded or present in amounts <5 μg require one to two rounds of amplification to obtain sufficient material for labeling. The probe amplification protocol is provided with the Amino Allyl MessageAmp kit (Ambion) and is used in accordance with the instructions of Ambion. The kit produces 50–120 μg of amplified RNA (aRNA) with 2 μg of total RNA as the starting material. A total of 10 μg of aRNA is required per microarray slide. As little as 100 ng of total RNA is sufficient if two rounds of amplification are used. The kit is based on a modified Eberwine procedure (63), in which dsDNA is generated from the RNA template using RT and a specialized oligonucleotide containing an oligo(dT) sequence fused to a T7 promoter. The presence of the T7 promoter in each template allows linear amplification by DNA polymerase. Aminoallyl-UTP is incorporated into the aRNA during the linear amplification process by T7 polymerase. The cyanine dye coupling is carried out as described above for cDNA dye coupling, except that the 5 μl volume of water is not added to the reaction because the aRNA is in the aqueous phase. A total of 5 μl of aRNA (2 μg/μl) is added to the coupling buffer and the 10 μl mixture is used to suspend the cyanine dye. Studies have shown that the aRNA obtained using this amplification procedure accurately reflects the original total cellular RNA (64).

Protocol 3

Microarray slide production

Equipment and Reagents

- 3× saline sodium citrate (SSC)
- Aminosilane-coated slides (CEL Associates)
- Stratalinker (Stratagene)
- Mouse 70mer Oligonucleotide library (version 3.0) (Operon Biotechnologies)[a]

Method

1. The 70mer oligonucleotides are suspended in 3× SSC at an oligonucleotide concentration of 30 μM and printed on aminosilane-coated slides.
2. The oligonucleotides are coupled to the slide surface using cross-linking with UV radiation. The slides are placed microarray side facing upward in a Stratalinker and irradiated at 60 mJ.
3. Following cross-linking, the slides are stored at room temperature in vacuum-sealed packaging.

Note

[a]The mouse 70mer Oligonucleotide Library (version 3.0) from Operon Biotechnologies representing 24 878 annotated genes and 32 829 transcripts was used in the studies described here. The 70mers are printed at 22°C and 65–85% relative humidity at 100–125 μM center-to-center spacing using a high-speed robotic microarrayer (OmniGrid-100 model; GeneMachines). The microarrayer is fitted with a 48-pin holder and 48 Stealth SMP3 pins from TeleChem International, Inc. ArrayIt Division. Printed spot volumes are 0.5 nl and spot diameters are 75–85 μM.

Protocol 4

Microarray hybridization and wash conditions (65)

Equipment and Reagents

- Bovine serum albumin (BSA) (Sigma)
- 20× SSC
- Sodium dodecyl sulfate (SDS)
- Isopropanol
- Formamide
- DTT
- 1 µg/µl Cot-1 DNA (Roche Diagnostics)
- 10 µg/µl Poly(A)$^+$ DNA (Sigma)
- 4 µg/µl Yeast tRNA (Sigma)
- Hybridization chamber (Corning)
- Water bath
- Microarray high-speed centrifuge (TeleChem International)
- Axon GenePix 4000A or 4000B and GENEPIX PRO v5.0 software (Axon Instruments)

Method

Pre-hybridization

1. Prepare the pre-hybridization buffer by making a solution that contains 5× SSC, 0.1% sodium dodecyl sulfate (SDS), and 1% BSA. Heat to 50°C while stirring to dissolve the BSA.
2. Pre-hybridize the slides by placing them in a staining jar containing pre-hybridization buffer, and incubate at 48°C for 45–60 min. The pre-hybridization step is used to reduce nonspecific probe binding and fluorescent background.
3. Following the pre-hybridization step, wash away the pre-hybridization buffer by dipping the slides up and down ten times in two consecutive staining jars containing dH_2O. Remove excess dH_2O by shaking the slide rack up and down several times.
4. Dehydrate the microarray surface by dipping the slides up and down ten times at room temperature (20–25°C) in isopropanol. Make sure the slides are used within 60 min of pre-hybridization, as hybridization efficiency decreases if the slides are allowed to dry for >60 min.

Hybridization

1. Prepare 2× hybridization buffer by making a solution that contains 50% formamide, 10× SSC, and 0.2% SDS. Heat the solution to 48°C to make sure all of the components suspend completely.
2. Suspend the Cy3- and Cy5-labeled probes in 9 µl of water and heat to 95°C for 3 min to denature the nucleic acids completely.
3. Add 12 µl of blocking solution master mix to each probe sample and vortex. Blocking solution master mix contains 8 µl of 1 µg/µl Cot-1 DNA, 2 µl of 10 µg/µl poly(A)$^+$ DNA, and 2 µl of 4 µg/µl yeast tRNA.

4. Add 21 μl of 2× hybridization buffer (pre-heated to 48°C) to each probe sample, mix by vortexing gently, and spin at 12 000 ***g*** for 1 min in a microcentrifuge to pellet any debris in the sample.
5. Carefully remove and transfer the 42 μl of probe supernatant to a fresh tube and incubate for 2 min at 48°C to pre-heat the probe mixture.
6. Pipette the 42 μl of probe mixture onto a pre-hybridized microarray slide and place a 22 × 60 mm cover slip directly onto the droplet to create a thin layer of probe mixture across the microarray surface.
7. Place the slide in a hybridization chamber with the microarray and cover slip facing upward and seal the chamber. A total of 12 μl of H_2O is added to the small reservoirs at each end of the chamber to prevent dehydration of the probe mixture during hybridization.
8. Place the hybridization chamber into a water bath set to 48°C and incubate for 40–60 h (51).

Post-hybridization washes

1. Remove the microarray from the hybridization chamber taking care not to disturb the cover slip.
2. Place the slide in a slide rack and transfer into a staining dish containing 1× SSC, 0.1% SDS, and 0.1 mM DTT at 48°C.
3. Gently remove the cover slip while the slide is submerged and agitate for 15 min to remove unbound probe molecules from the microarray surface. This step represents a low-stringency wash.
4. Transfer the slides into a staining dish containing 0.1× SSC, 0.1% SDS, and 0.1 mM DTT at 48°C and agitate for 5 min to remove unbound probe molecules from the microarray surface. This step represents a high-stringency wash.
5. Repeat step 4 two more times.
6. Transfer the slides into a staining dish containing 0.1× SSC and 0.1 mM DTT at room temperature (20–25°C) and agitate for 5 min. This step removes SDS from the slide surface to prevent interference in the scanning step.
7. Repeat step 6 once.
8. Spin the slides for 10 s in a microarray high-speed centrifuge (or equivalent) to dry the surface completely[a].

Note

[a]Microarray slides are scanned using an Axon Scanner and image files are 'gridded' using GENEPIX PRO v5.0 software. As described earlier, raw data from the microarray study are statistically analyzed in three steps: (i) the data are normalized, (ii) ANOVA for each gene is determined, and (iii) cluster analysis is performed. The statistically analyzed datasets are merged into EASE computer programs to determine biological meaning. The datasets are then placed into databases to allow the data to be accessed by the scientific community.

3. TROUBLESHOOTING

We have found that 90% of all failed microarray experiments are due to poor-quality RNA. Thus, the importance of careful RNA isolation and quality control analysis cannot be overemphasized. Other far less common sources of failure include missing or merged spots on the microarray slide, insufficient pre-hybridization or washing, and hybridization chamber leaks.

Acknowledgements

We sincerely thank Larry Polintan, Jennifer A. Schwanekamp, Maureen A. Sartor, Subramaniam Venkatesan, and Srinivasan Raghuraman for their technical assistance and critical reading of the manuscript.

4. REFERENCES

1. **Zedeck MS** (1980) *J. Environ. Pathol. Toxicol.* **3**, 537–567.
2. **US Department of Health and Human Services, and Public Health Service, National Toxicology Program** (2001).
3. **Kimbrough RD** (1987) *Annu. Rev. Pharmacol. Toxicol.* **27**, 87–111.
4. **Geyer HJ, Scheunert I, Rapp K, Gebefugi I, Steinberg C & Kettrup A** (1993) *Ecotoxicol. Environ. Saf.* **26**, 45–60.
5. **van den Berg M, de Jongh J, Poiger H & Olson JR** (1994) *Crit. Rev. Toxicol.* **24**, 1–74.
6. **Suskind RR** (1985) *Scand. J. Work Environ. Health.* **11**, 165–171.
7. **Zugerman C** (1990) *Dermatol. Clin.* **8**, 209–213.
8. **Yoshida K, Ikeda S & Nakanishi J** (2000) *Chemosphere,* **40**, 177–185.
9. **Pohjanvirta R & Tuomisto, J** (1994) *Pharmacol. Rev.* **46**, 483–549.
10. ★ **Flesch-Janys D, Berger J, Gurn P, *et al.*** (1995) *Am. J. Epidemiol.* **142**, 1165–1175. *– A convincing study linking TCDD exposure to cardiovascular disease in a human population.*
11. **Steenland K, Piacitelli L, Deddens J, Fingerhut M & Chang LI** (1999) *J. Natl. Cancer Inst.* **91**, 779–786.
12. **Manz A, Berger J, Dwyer JH, Flesch-Janys D, Nagel S & Waltsgott H** (1991) *Lancet,* **338**, 959–964.
13. ★★ **Fingerhut MA, Halperin WE, Marlow DA, *et al.*** (1991) *N. Engl. J. Med.* **324**, 212–218. *– An epidemiology study showing that TCDD causes cancer in humans.*
14. **Bertazzi PA** (1991) *Sci. Total Environ.* **106**, 5–20.
15. **Bertazzi PA, Pesatori AC, Consonni D, Tironi A, Landi MT & Zocchetti C** (1993) *Epidemiology,* **4**, 398–406.
16. **Graham MJ, Lucier GW, Linko P, Maronpot RR & Goldstein JA** (1988) *Carcinogenesis,* **9**, 1935–1941.
17. **Pitot HC, Goldsworthy T, Campbell HA & Poland A** (1980) *Cancer Res.* **40**, 3616–3620.
18. **Birnbaum LS** (1995) *Environ. Health Perspect.* **103** (Suppl. 7), 89–94.
19. **Abbott BD** (1995) *Toxicology,* **105**, 365–373.

★ 20. **Couture LA, Abbott BD & Birnbaum LS** (1990) *Teratology*, **42**, 619–627. – *A concise and still one of the best reviews of the teratogenic effects of TCDD.*

21. **Lucier GW, Portier CJ & Gallo MA** (1993) *Environ. Health Perspect.* **101**, 36–44.

22. **Swanson HI & Bradfield CA** (1993) *Pharmacogenetics*, **3**, 213–230.

23. **Carlson DB & Perdew GH** (2002) *J. Biochem. Mol. Toxicol.* **16**, 317–325.

24. **Hoffman EC, Reyes H, Chu FF, *et al.*** (1991) *Science*, **252**, 954–958.

25. **Hankinson O** (1995) *Annu. Rev. Pharmacol. Toxicol.* **35**, 307–340.

26. **Sutter TR & Greenlee WF** (1992) *Chemosphere*, **25**, 223–226.

★ 27. **Whitlock JP Jr** (1993) *Chem. Res. Toxicol.* **6**, 754–763. – *Presents an excellent early review of the molecular actions of TCDD-activated AHR.*

28. **Uno S, Dalton TP, Sinclair PR, *et al.*** (2004) *Toxicol. Appl. Pharmacol.* **196**, 410–421.

★★★ 29. **Fernandez-Salguero P, Pineau T, Hilbert DM, *et al.*** (1995) *Science*, **268**, 722–726. – *One of the first papers describing the developmental roles of the AHR. The authors used a mouse model in which expression of the* Ahr *gene was deleted.*

30. **Brenner DA** (1996) *Hepatology*, **23**, 379–380.

31. **Schmidt JV, Su GH, Reddy JK, Simon MC & Bradfield CA** (1996) *Proc. Natl. Acad. Sci. U.S.A.* **93**, 6731–6736.

32. **Mimura J, Yamashita K, Nakamura K, *et al.*** (1997) *Genes Cells*, **2**, 645–654.

33. **Fernandez-Salguero PM, Hilbert DM, Rudikoff S, Ward JM & Gonzalez FJ** (1996) *Toxicol. Appl. Pharmacol.* **140**, 173–179.

34. **Andreola F, Fernandez-Salguero PM, Chiantore MV, Petkovich MP, Gonzalez FJ & de Luca LM** (1997) *Cancer Res.* **57**, 2835–2838.

35. **Fernandez-Salguero PM, Ward JM, Sundberg JP & Gonzalez FJ** (1997) *Vet. Pathol.* **34**, 605–614.

36. **Elizondo G, Fernandez-Salguero P, Sheikh MS, *et al.*** (2000) *Mol. Pharmacol.* **57**, 1056–1063.

★★ 37. **Lahvis GP, Lindell SL, Thomas RS, *et al.*** (2000) *Proc. Natl. Acad. Sci. U.S.A.* **97**, 10442–10447. – *An excellent paper detailing the developmental effects resulting from the loss of the* Ahr *gene in the mouse.*

38. **Hahn ME** (2002) *Chem. Biol. Interact.* **141**, 131–160.

39. **Huang X, Powell-Coffman JA & Jin Y** (2004) *Development*, **131**, 819–828.

40. **Powell-Coffman JA, Bradfield CA & Wood WB** (1998) *Proc. Natl. Acad. Sci. U.S.A.* **95**, 2844–2849.

41. **Guo J, Sartor M, Karyala S, *et al.*** (2004) *Toxicol. Appl. Pharmacol.* **194**, 79–89.

42. **Karyala S, Guo J, Sartor M, *et al.*** (2004) *Cardiovasc. Toxicol.* **4**, 47–73.

43. **Lipshutz RJ, Fodor SP, Gingeras TR & Lockhart DJ** (1999) *Nat. Genet.* **21**, 20–24.

44. **Singh-Gasson S, Green RD, Yue Y, *et al.*** (1999) *Nat. Biotechnol.* **17**, 974–978.

★★ 45. **DeRisi J, Penland L, Brown PO, *et al.*** (1996) *Nat. Genet.* **14**, 457–460. – *An early high-impact paper demonstrating the power of microarrays.*

46. **Schena M, Heller RA, Theriault TP, Konrad K, Lachenmeier E & Davis RW** (1998) *Trends Biotechnol.* **16**, 301–306.

★ 47. **Larkin JE, Frank BC, Gavras H, Sultana R & Quackenbush J** (2005) *Nat. Methods*, **2**, 337–344. – *An excellent and complete study comparing and contrasting different microarray platforms.*

★ 48. **Irizarry RA, Warren D, Spencer F, *et al.*** (2005) *Nat. Methods*, **2**, 345–350. – *Provides a companion report to reference 47 comparing different microarray laboratory procedures.*

49. **Kane MD, Jatkoe TA, Stumpf CR, Lu J, Thomas JD & Madore SJ** (2000) *Nucleic Acids Res.* **28**, 4552–4557.

50. **Bozdech Z, Zhu J, Joachimiak MP, Cohen FE, Pulliam B & DeRisi JL** (2003) *Genome Biol.* **4**, R9.

51. **Sartor M, Schwanekamp J, Halbleib D, *et al.*** (2004) *Biotechniques*, **36**, 790–796.

★★★ 52. **Churchill GA** (2002) *Nat. Genet.* **32** (Suppl.), 490–495. – *Provides a seminal report on microarray experimental designs.*

53. Nadon R & Shoemaker J (2002) *Trends Genet.* **18**, 265–271.
54. Kerr MK & Churchill GA (2001) *Biostatistics,* **2**, 183–201.
55. Benjamini Y & Hochberg Y (1995) *J. Royal Statist. Soc. B,* **57**, 289–399.
56. Medvedovic M & Sivaganesan S (2002) *Bioinformatics,* **18**, 1194–1206.
57. Yeung KY, Medvedovic M & Bumgarner RE (2003) *Genome Biol.* **4**, R34.
★★ **58. Medvedovic M, Yeung KY & Bumgarner RE** (2004) *Bioinformatics,* **20**, 1222–1232. *– Presents a description of a superb clustering method.*
59. Hosack DA, Dennis G Jr, Sherman BT, Lane HC & Lempicki RA (2003) *Genome Biol.* **4**, R70.
60. Ashburner M, Ball CA, Blake JA, *et al.* (2000) *Nat. Genet.* **25**, 25–29.
61. Brazma A, Hingamp P, Quackenbush J, *et al.* (2001) *Nat. Genet.* **29**, 365–371.
62. Brazma A, Parkinson H, Sarkans U, *et al.* (2003) *Nucleic Acids Res.* **31**, 68–71.
★ **63. van Gelder RN, von Zastrow ME, Yool A, Dement WC, Barchas JD & Eberwine JH** (1990) *Proc. Natl. Acad. Sci. U.S.A.* **87**, 1663–1667. *– Describes the basic method for mRNA amplification used by many microarray laboratories.*
64. Iscove NN, Barbara M, Gu M, Gibson M, Modi C & Winegarden N (2002) *Nat. Biotechnol.* **20**, 940–943.
★ **65. Hegde P, Qi R, Abernathy K, *et al.*** (2000) *Biotechniques,* **29**, 548–550, 552–544, 556. *– A techniques paper that is the basis of the microarray protocols used in microarray laboratories.*

CHAPTER 3

Amplified differential gene expression microarrays

Zhijian J. Chen and Kenneth D. Tew

1. INTRODUCTION

Numerous methods have been developed to detect differential expression of genes, including differential display (1), suppression subtractive hybridization (2), serial analysis of gene expression or SAGE (3), and the DNA microarray (4). These methods were designed to compare inherent differences in gene expression between two closely related samples. The DNA microarray also allows comparisons to be made between large numbers of similar or different samples, but this application of microarrays is beyond the scope of this chapter. In comparisons of two samples, the threshold of detection ('sensitivity') varies among these methods and this serves to limit the detection of small changes (e.g. less than twofold) in expression. When the lower boundary of sensitivity cannot be superseded, an alternative approach is to include a step for quantitative enhancement of differences in gene expression before detecting them. Amplified differential gene expression (ADGE) was designed to magnify in a quadratic manner the ratios of gene expression values between two samples, and then to display the magnified differences (5). In one application, the ratios of HL60/ADR to HL60 cells for DNA-dependent protein kinase and MRP1 were detected as 3.8 and 47.6, respectively, using ADGE, compared with 2.2 and 9.8 using a reverse transcriptase polymerase chain reaction (PCR) (5).

As ADGE is a method of pair-wise comparison, it can be coupled seamlessly with the two-dye system of the DNA microarray. The ADGE microarray combines the ratio magnification of ADGE and the high-throughput profiling capabilities of the DNA microarray. When compared with a standard microarray approach, the ADGE microarray magnified ratios of differential expression between HL60 and HL60/TLK286 cells, leading to an increase in both detection sensitivity and fidelity (6). The

DNA Microarrays: *Methods Express* (M. Schena, ed.)

ADGE microarray was used to profile differential expression of ABCA2-transfected cell lines, revealing two clusters of transport-related and oxidative stress response genes (7). The principles underlying the ADGE procedure, integration of ADGE with DNA microarrays, and detailed protocol information are addressed in this chapter.

2. METHODS AND APPROACHES

Many methods exist for comparing gene expression levels in two related biological samples, but to date ADGE is the only procedure that allows a magnification of the ratios of differentially expressed genes. This novel procedure is described in detail below.

The schematic procedure of the ADGE microarray approach is shown in *Fig. 1*. Two comparative nucleic acid samples are selected, a control sample and a tester sample. Total RNA or mRNA is used as the starting material from which to make double-stranded cDNA from the two samples. The control and tester cDNA mixtures are then digested with the *Taq*I enzyme to produce a mixture of fragments containing the *Taq*I-generated 2 nt overhanging sequences. *Taq*I-digested control and tester DNA fragments are ligated to their respective CT (control) and TT (tester) adapters. The adapter-linked control and tester DNAs are reassociated by mixing equivalent amounts by weight or molar ratios, denaturing the templates, and then allowing annealing to occur to form the reassociated DNA. As the DNAs are both digested with *Taq*I and have the same molecular weight, equivalent amounts of control and tester DNA are obtained by using equivalent mass amounts (i.e. the same number of micrograms) in the annealing reaction. The control DNA probe is amplified by using PCR with the CT primer complementary to the CT adapter, whilst the tester DNA probe is amplified with the TT primer complementary to the TT adapter. Both control and tester probes can be labeled during or after PCR amplification. The labeled probes are mixed together and hybridized to the microarray, which is then washed and scanned in two colors to generate the data.

The quadratic magnification of differential expression ratios is achieved by combining DNA reassociation and PCR amplification (see *Fig. 1*). DNA reassociation occurs when the adapter-linked control and tester cDNA fragments are mixed in equivalent amounts, denatured, and annealed. Five different duplexes are formed after DNA reassociation: the control DNA with the CT adapters on both ends, tester DNA with the TT adapters on both ends, hybrid DNA with the CT adapter on one end and the TT adapter on the other end, the end fragment of control DNA with the CT adapter on a single end, and the end fragment of tester DNA with the TT adapter

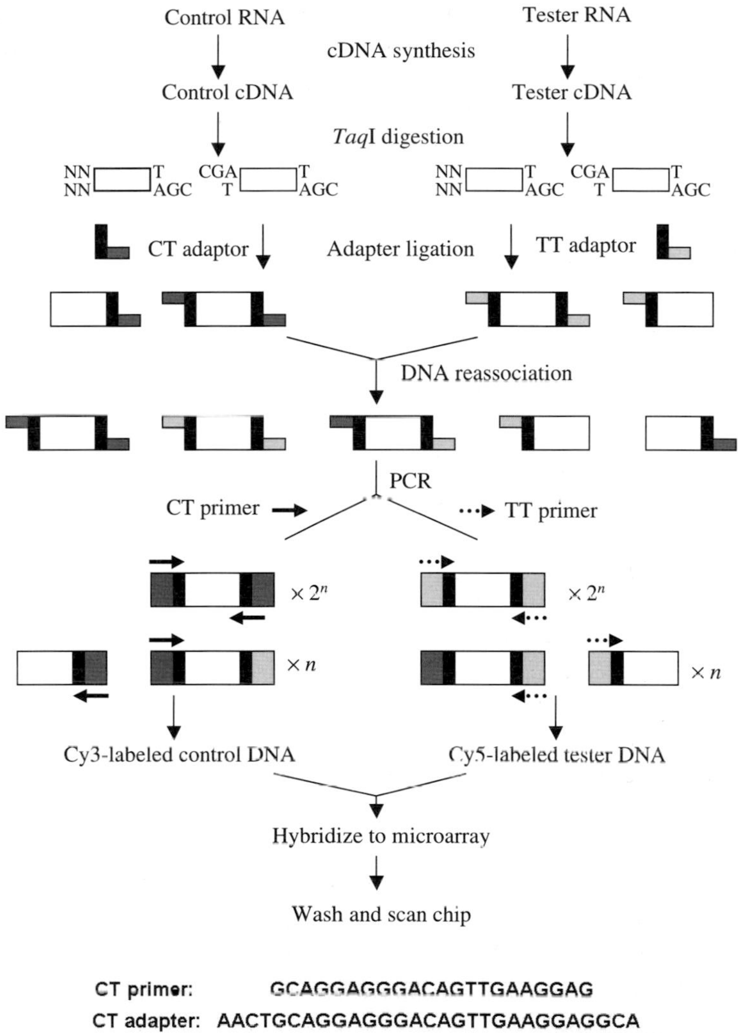

CT primer: GCAGGAGGGACAGTTGAAGGAG

CT adapter: AACTGCAGGAGGGACAGTTGAAGGAGGCA
CCTCCGTGC

TT primer: CAGAGGTGAGACAGGAGTGGAG

TT adapter: AACTCAGAGGTGAGACAGGAGTGGAGGCA
CCTCCGTGC

Figure 1. Schematic representation of the ADGE microarray process.
The control and tester cDNAs are synthesized from RNA and digested with *Taq*I. The control and tester *Taq*I fragments are ligated to the CT and TT adapters, respectively. The adapter-linked control and tester DNAs are reassociated by mixing equivalent amounts, denaturing the double-stranded molecules, and allowing annealing to occur. The reassociated DNA is used as template to generate Cy3-labeled control DNA via the CT primer and Cy5-labeled tester DNA via the TT primer. The labeled probes are mixed together and hybridized on a DNA microarray, which is then washed and scanned in two colors to provide the data. The CT and TT primers and adapters shown at the bottom of the figure are examples of oligonucleotides that work well for the *Taq*I restriction site.

on a single end (see *Fig. 1*). The relative amounts of the first three types of duplexes for each gene are theoretically determined by the algebraic formula:

$$(a + b)(a' + b') = aa' + bb' + a'b + ab'$$

where a and a′ are the number of sense and antisense strands of the control DNA; b and b′ are the number of sense and antisense strands of the tester DNA; aa′is the number of double strands of the control DNA; bb′ is the number of double strands of the tester DNA; and a′b and ab′ are the number of double strands of the hybrid DNA.

For example, for a gene overexpressed fivefold in the tester over the control, $bb'/aa' = 5$. Thus, the formula is:

$$(a + 5b)(a' + 5b') = aa' + 25bb' + 5a'b + 5ab'$$

After DNA reassociation, the ratio of bb'/aa' is magnified from 5 to 25. If another gene is downregulated twofold in the tester, $aa'/bb' = 2$. The formula is:

$$(2a + b)(2a' + b') = 4aa' + bb' + 2a'b + 2ab'$$

Thus, the ratio of aa'/bb' increases from 2 to 4 after DNA reassociation. For a gene with the same transcription level between tester and control, the ratio remains the same after DNA reassociation. For the last two types of duplexes, the hybrid duplex of end fragments cannot be distinguished from the control or tester duplex. Thus, the relative amounts of these two types of duplexes remain the same.

The ratio of control and tester DNA with adapters on the two ends is magnified quadratically for each gene after DNA reassociation. However, they are not separated from each other or from the hybrid DNA and end fragments. Subsequently, selective PCR is used to separate control DNA from tester DNA. The CT primer complementary to the CT adapter amplifies the control DNA exponentially, whilst the hybrid and end fragment DNA are amplified only linearly, as the control DNA has the CT adapter on both ends and the hybrid and end fragment DNA have the CT adapter on only one end. The TT primer complementary to the TT adapter amplifies the tester DNA exponentially and the hybrid and end fragment DNA linearly (see *Fig. 1*). After 20 or more cycles of PCR, the exponentially amplified control or tester DNA is more than a million times more abundant than the linearly amplified DNA. Therefore, the ratio of each gene between PCR products with the CT and TT primers is quadratically magnified from the ratio between the control and tester samples. A test experiment shows that an input ratio of 2 was detected as 3.6 with the ADGE microarray compared with 1.5 with a standard microarray (see *Fig. 2*, available in the color section).

The correlation between the detected ratio (y) and the input ratio (x) for an ADGE microarray is $y = 1.05x^{1.55}$ with $R^2 = 0.97$, whilst the correlation is $y = 0.56x + 0.39$ with $R^2 = 0.96$ for standard microarray (see *Fig. 2*).

The design of the adapter and primer is critical for achieving the quadratic magnification of ratios. The basic structure of the adapter should be the same, even though the sequences may differ depending on the type of restriction enzyme used and other factors. In the example shown, the CT and TT adapters and primers are designed for the *Taq*I restriction enzyme (see *Fig. 1*). The adapters are composed of long and short oligonucleotides. The short oligonucleotides have the same sequence between the CT and TT adapters in order to permit the formation of hybrid DNA molecules. The complementary region between the long and short oligonucleotides is usually 7 nt. If the complementary region is too short, the adapters may not form stable hybrids. If the complementary region is too long, cross-priming becomes a possible complication. The adapters have cohesive ends complementary to *Taq*I. As the nucleotide T is changed to an A in the adapters, the *Taq*I site is not recovered after ligation of the adapters to the *Taq*I-digested DNA. The CT and TT primers are complementary to partial or entire regions of the CT and TT adapters, respectively. Besides the general rules of primer design, the length of the unique 5′ region between the CT and TT primers (at least 10 nt) should be sufficient to prevent cross-priming. No selective nucleotide is added to the 3′ end of the CT and TT primers in ADGE microarray, as all genes are expected to amplify in a single PCR. The CT primer is used to generate the control DNA probe, whilst the TT primer generates the tester DNA probe.

Fluorescent labeling of the control and tester DNA probes can be achieved using either direct incorporation of dye-linked primers or dye-linked nucleotides during the PCR amplification, or by indirectly coupling of dyes to aminoallyl groups after the PCR amplification. The advantage of the primer-labeling approach is that the microarray signal intensity is independent of the length of the PCR products. The disadvantage is that only two fluorescent molecules are attached per PCR product, which reduces the 'specific activity' of the fluorescent probe molecules. Nucleotide labeling can be implemented during the PCR amplification by directly incorporating Cy3–dCTP into the control DNA and Cy5–dCTP into the tester DNA. With indirect labeling, aminoallyl–dUTP is incorporated into both the control and tester PCR products, which is in turn coupled to monoreactive Cy3 and Cy5 dyes. As both strands of the probe DNA are labeled, nucleotide-based labeling typically generates strong signals because of the high specific activity of the probe molecules.

The ADGE microarray combines the strengths of ADGE and DNA microarray. It magnifies the ratios of differential gene expression as a power function, improves detection sensitivity, enhances signal intensity, and reduces the amount of starting material required, whilst maintaining the high-throughput aspects of DNA microarray analysis. The magnification of the power function ($y=1.05x^{1.55}$) is valid up to a sixfold input ratio or 30-fold detected ratio for up- and downregulated genes of variable abundance (6). The ADGE magnification raises the magnitude of expression ratios above the lower limit of detection, thus improving the sensitivity of a DNA microarray assay. The signal intensity is improved by labeling both DNA strands, and as little as 125 ng of total RNA is enough for hybridization to a high-density microarray.

2.1 Recommended protocols

Protocol 1

First-strand cDNA synthesis

Equipment and Reagents

- 2 μg/μl oligo(dT) (12- to 18mers)
- SuperScript Double-Stranded cDNA Synthesis kit, containing:
  - 5× First-strand buffer
  - 0.1 M DTT
  - 10 mM dNTPs
  - Superscript II RT
- Water bath at 70°C
- Microcentrifuge

Method

1. Obtain 11 μl of control RNA and 11 μl of tester RNA containing equivalent mass amounts of total RNA (0.2–10 μg). For the control RNA[a], set up the following reaction: 11 μl of control total RNA and 1 μl of 2 μg/μl oligo(dT) (12- to 18mers). For the tester RNA[a], set up the following reaction: 11 μl of tester total RNA and 1 μl of 2 μg/μl oligo(dT) (12- to 18mers).
2. Incubate the mixture at 70°C for 10 min to denature the RNA. Chill on ice for 1 min and centrifuge briefly to consolidate the sample in the bottom of the tube.
3. Preparing the master mix as follows: 8 μl of 5× First-strand Buffer, 4 μl of 0.1 M DTT, and 2 μl of 10 mM dNTPs. Add 7 μl of master mix to each reaction and mix well by gentle vortexing.
4. Add 1 μl of Superscript II reverse transcriptase (RT) to each reaction and mix gently by flicking the tube several times with your fingers. Do not vortex as this can mechanically damage the Superscript II RT enzyme.
5. Incubate the reaction for 1 h at 42°C to allow first-strand cDNA synthesis to proceed[b]. Store the reactions on ice before proceeding to *Protocol 2*.

Notes

[a]In terms of the starting material required for first-strand cDNA synthesis, it is not necessary to use mRNA, as total RNA works well. The preferred amount of total RNA is 5–10 µg, although as little as 0.2 µg of total RNA is sufficient for one microarray hybridization reaction. If a small amount of total RNA is used, the eluate after DNA reassociation should be concentrated by drying down the volume.

[b]A PCR thermal cycler with a heated lid can be used for the 1 h reaction to prevent condensation droplets forming in the reaction tube.

Protocol 2

Second-strand cDNA synthesis

Equipment and Reagents

- SuperScript Double-Stranded cDNA Synthesis kit (Invitrogen) containing:
  - DEPC-treated dH_2O
  - 10 mM dNTPs
  - 5× Second-strand buffer
  - *E. coli* DNA polymerase
  - *E. coli* RNase H
  - *E. coli* DNA ligase
- 0.5 M EDTA (pH 8.0)
- Phenol : chloroform : isoamyl alcohol (25 : 24 : 1, v/v), stored at 4°C
- 20 µg/µl Glycogen (Sigma), stored at −20°C
- 3 M Sodium acetate (pH 5.2)
- 70 and 100% Ethanol
- Water bath at 16°C
- Centrifuge and vacuum centrifuge

Method

1. To each first-strand reaction, add 130 µl of second-strand cDNA synthesis mixture prepared as follows: 182 µl of DEPC-treated dH_2O, 6 µl of 10 mM dNTPs, 60 µl of 5× Second-strand buffer, 8 µl of *E. coli* DNA polymerase, 2 µl of *E. coli* RNase H, and 2 µl of *E. coli* DNA ligase. Mix gently by flicking the tube several times with your fingers. Do not vortex as this can mechanically damage the enzymes.
2. Incubate at 16°C for 2 h to allow second-strand synthesis to proceed. Place on ice.
3. Add 10 µl of 0.5 M EDTA (pH 8.0) to each tube to stop the enzymatic reaction.
4. Add 160 µl of phenol : chloroform : isoamyl alcohol (25 : 24 : 1, v/v) to each reaction and mix thoroughly by vortexing. This step inactivates and removes proteins from the reaction.
5. Centrifuge at room temperature (20–25°C) for 5 min at 14 000 ***g*** and carefully transfer the upper (aqueous) layer containing the cDNA products to a fresh 1.5 ml microfuge tube.
6. To the aqueous cDNA samples, add 1 µl of 20 µg/µl glycogen, 20 µl of 3 M sodium acetate (pH 5.2), and 300 µl of 100% ethanol. Mix the samples thoroughly by vortexing. The cDNA products are insoluble at this point and can be isolated by centrifugation.

7. Centrifuge at 4°C for 20 min at 14000 *g* to pellet the cDNA products. Carefully remove and discard the supernatant.
8. Add 500 μl of 70% ethanol to the pellet, centrifuge for 5 min at 14000 *g*, and carefully remove and discard the supernatant. This step removes contaminating salts and other impurities from the cDNA pellet.
9. Dry the cDNA pellet at 37°C for 10 min using vacuum centrifugation. This step evaporates and removes residual ethanol that can interfere with downstream steps.
10. Resuspend each cDNA pellet in 25 μl of ddH_2O.

Protocol 3

*Taq*I digestion of double-stranded cDNA templates

Reagents

- *Taq*I enzyme (10 units/μl) and 10× REact 2 buffer (Invitrogen)
- Water bath at 65°C

Method

1. Obtain equivalent mass amounts of control and tester cDNA and set up the following *Taq*I digestion reactions: 25 μl of control or tester cDNA, 3 μl of 10× REact 2 Buffer, and 2 μl of *Taq*I enzyme. Ensure that the cDNA and restriction buffer are mixed thoroughly before adding *Taq*I enzyme. After adding *Taq*I, mix the tube gently by flicking it with your fingers. Do not vortex as this can damage the *Taq*I enzyme.
2. Digest the cDNAs with *Taq*I by incubating the reactions at 65°C for 2 h.

Protocol 4

Ligation of CT and TT adapters to *Taq*I-digested cDNA templates

Equipment and Reagents

- 1–3 units/μl T4 DNA ligase and 10× T4 DNA ligase buffer (Promega)
- 60 μM CT adapter: 5′-AACTGCAGGAGGGACAGTTGAAGGAGGCA-3′
  3′-CCTCCGTGC-5′
- 60 μM TT adapter: 5′-AACTCAGAGGTGAGACAGGAGTGGAGGCA-3′
  3′-CCTCCGTGC-5′
- Water bath at 14°C

Method

1. Set up the following ligation reactions for the CT and TT adapters. The control template contains the following components: 30 μl of *Taq*I digested control DNA fragments, 4 μl

of 10× T4 DNA ligase buffer, 4 µl of 60 µM CT adapter, and 2 µl of 1–3 units/µl T4 DNA ligase. The tester template contains the following components: 30 µl of *Taq*I digested tester DNA fragments, 4 µl of 10× T4 ligase buffer, 4 µl of 60 µM TT adapter, and 2 µl of T4 DNA ligase.

2. Incubate the ligation mixtures for 12–16 h at 14°C.

Protocol 5

Reassociation of control and tester DNA

Equipment and Reagents

- 6× EE buffer (60 mM EPPS, 3 mM EDTA, adjusted to pH 8.0)
- 3 M NaCl
- QIAquick PCR purification kit (Qiagen)
- Water baths at 95 and 68°C
- Vacuum centrifuge

Method

1. Set up the following DNA reassociation reaction using mass-equivalent amounts[a] of control and tester DNA: 10 µl of CT adapter-control cDNA[b], 10 µl of TT adapter-tester cDNA[b], and 20 µl of 6× EE buffer[c]. Mix well by vortexing.
2. Denature the mixture by heating at 95°C for 5 min. Chill on ice immediately.
3. Add 20 µl of 3 M NaCl and mix well by vortexing.
4. Incubate at 68°C for 12–16 h to allow reassociation to occur.
5. Purify the reassociation reactions using a QIAquick PCR purification kit.
6. Elute using 50 µl of ddH_2O and reduce the volume by vacuum centrifugation if a small amount of starting material was used for the reassociation reaction.

Notes

[a]DNA reassociation requires equivalent amounts of control and tester DNA. As the templates possess the same molecular weight, mass equivalence (e.g. the same number of micrograms) ensures molar equivalence between the templates. One way of checking molar (mass) equivalence between the control and tester cDNAs is to quantify actin sequence levels using actin-ADGE primers. Actin-ADGE primers are CT and TT primers with selective nucleotides complementary to the actin gene added to the primer sequence. Following the adapter ligation step, remove 1 µl aliquots of the control and tester DNA and make a 10-fold and 100-fold dilution of each. The diluted DNA is used as template for PCR using the actin-ADGE primers for amplification. The actin sequence levels are compared between the control and tester samples and, where necessary, concentrations are adjusted to equivalence by diluting one of the samples with ddH_2O.

[b]The volume of cDNA can be 5–15 µl. Make up any volume differences with ddH_2O.

[c]The volume of 6× EE buffer should always exceed the volume of cDNA by a factor of two.

Protocol 6

Microarray probe preparation by direct labeling[a]

Equipment and Reagents

- Advantage cDNA polymerase mix and 10× cDNA PCR buffer (Clontech)
- dNTP mix (10 mM each of dGTP, dATP, and dTTP, and 6 mM dCTP)
- 60 μM CT primer (5′-GCAGGAGGGACAGTTGAAGGAG-3′)
- 60 μM TT primer (5′-CAGAGGTGAGACAGGAGTGGAG-3′)
- cDNA polymerase (Clontech)
- Cy3-dCTP (GE Healthcare)
- Cy5-dCTP (GE Healthcare)
- QIAquick PCR purification kit (Qiagen)
- Thermocycler
- Water bath at 42°C
- Vacuum centrifuge

Method

1. Set up PCRs to allow direct labeling of the control and tester reassociated DNAs[b]. For the control DNA, the reaction contains the following: 37 μl of reassociated DNA, 5 μl of 10× PCR buffer, 1 μl of dNTP mix, 4 μl of Cy3-dCTP, 2 μl of 60 μM CT primer, and 1 μl of cDNA polymerase. For the tester DNA, the reaction contains the following: 37 μl of reassociated DNA, 5 μl of 10× PCR buffer, 1 μl of dNTP mix, 4 μl of Cy5-dCTP, 2 μl of 60 μM TT primer, and 1 μl of cDNA polymerase. Mix the reactions gently by flicking the tube with your fingers. Do not vortex as this can mechanically damage the enzyme.
2. Cycle the PCRs using the following regime: 72°C for 5 min; 94°C for 1 min; 30 cycles of 94°C for 30 s, 62°C for 30 s, and 72°C for 90 s; 72°C for 5 min. Store in the dark at 4°C.
3. Following the PCR step, purify the reaction products using a QIAquick PCR purification kit.
4. Elute the labeled products twice using 40 μl of ddH_2O (pre-warmed to 42°C) per elution.
5. Reduce the 80 μl sample to 5 μl using vacuum centrifugation.

Notes

[a]PCR labeling using direct (*Protocol 6*) and indirect (*Protocol 7*) procedures both suffice for microarray applications. Direct labeling follows a simpler protocol and usually produces stronger hybridization signals, but requires expensive Cy3- and Cy5-dCTP reagents.

[b]A single PCR per template is usually sufficient for most experiments, although two to three reactions can be set up and used to increase the microarray signal intensities.

Protocol 7

Microarray probe preparation by indirect labeling

Equipment and Reagents

- Advantage cDNA polymerase mix and 10× cDNA PCR buffer (Clontech)
- 60 μM CT primer (5′-GCAGGAGGGACAGTTGAAGGAG-3′)
- 60 μM TT primer (5′-CAGAGGTGAGACAGGAGTGGAG-3′)
- dNTP mix (10 mM each dGTP, dATP, and dTTP, 6 mM dCTP)
- cDNA polymerase (Clontech)
- QIAquick PCR purification kit (Qiagen)
- dNTPs/aminoallyl-dUTP mix containing 10 mM each dGTP, dATP, and dCTP, 2 mM dTTP, and 8 mM aminoallyl-dUTP (Sigma). Store at −20°C
- 3 M Sodium acetate (pH 5.2)
- Glycogen (20 μg/μl)
- Absolute ethanol (100% or 200 proof)
- Cy3 monoreactive dye (Amersham). Minimize light exposure at all times. Resuspend each pack in 45 μl of DMSO and store at −20°C
- Cy5 monoreactive dye (Amersham). Minimize light exposure at all times. Resuspend each pack in 45 μl of DMSO and store at −20°C
- 2× Coupling buffer: 0.2 M sodium bicarbonate ($NaHCO_3$) at pH 9.0
- Thermocycler
- Centrifuge
- Vacuum centrifuge

Method

1. Set up the following PCRs to amplify the control and tester DNA templates. The control template PCR contains: 34 μl of reassociated DNA[a], 5 μl of 10× PCR buffer, 1 μl of 10 mM dNTP mix, 1 μl of 60 μM CT primer, and 1 μl of cDNA polymerase. The tester template PCR contains: 42 μl of reassociated DNA, 5 μl of 10× PCR buffer, 1 μl of 10 mM dNTPs, 1 μl of 60 μM TT primer, and 1 μl of cDNA polymerase.
2. Carry out the PCR using the following regime: 72°C for 5 min to fill in the ends; 94°C for 1 min; 25 cycles of 94°C for 30 s, 66°C for 30 s, and 72°C for 90 s; 72°C for 5 min. Store at 4°C.
3. Following the PCR process, purify the PCR products using a QIAquick PCR purification kit and elute the purified products with 42 μl of ddH_2O.
4. Set up the following PCRs to incorporate aminoallyl-dUTP into the control and tester templates. The control template PCR contains: 42 μl of DNA template, 5 μl of 10× PCR buffer, 1 μl of dNTP/aminoallyl-dUTP mix, 1 μl of 60 μM CT primer, and 1 μl of cDNA polymerase. The tester template PCR contains: 42 μl of DNA template, 5 μl of 10× PCR buffer, 1 μl of dNTP/aminoallyl-dUTP mix, 1 μl of 60 μM TT primer, and 1 μl of cDNA polymerase.
5. Carry out the PCR process using the following regime: 94°C for 1 min, 6 cycles of 94°C for 30 s, 64°C for 30 s, 72°C for 90 s; 72°C for 1 min. Store at 4°C.
6. Following the PCR process, add the following to each tube: 1 μl of 20 μg/μl glycogen, 10 μl of 3 M sodium acetate (pH 5.2), and 100 μl of absolute ethanol. Mix thoroughly by vortexing for 30 s.

7. Pellet the PCR products by centrifugation at 4°C for 20 min at 14000 *g*. Carefully remove and discard the supernatant, taking care not to disrupt the DNA pellet at the bottom of the tube.
8. Wash the DNA pellet by adding 500 μl of 70% ethanol. Centrifuge for 5 min at 14000 *g* to re-pellet the DNA.
9. Carefully remove and discard the supernatant and dry the DNA pellet completely using vacuum centrifugation at 37°C for 10 min. The DNA pellets should be completely dry at this point. If any residual ethanol is observed, repeat the vacuum centrifugation step for another 5–10 min.
10. Resuspend the DNA pellet in 5 μl of 2× coupling buffer. Add 5 μl of Cy3 monoreactive dye to the control DNA and 5 μl of Cy5 monoreactive dye to the tester DNA. Incubate at room temperature (20–25°C) in the dark for 1 h to allow coupling of the monoreactive dye molecules to the aminoallyl-containing DNA.
11. Add 50 μl of 100 mM sodium acetate (pH 5.2) to each reaction.
12. Purify the dye-coupled DNA products using a QIAquick PCR purification kit. Elute each sample twice with 40 μl of ddH_2O (pre-warmed to 42°C). Dry each 80 μl product down to 7 μl using vacuum centrifugation. The volume of certain samples may be less than 7 μl. In this case, adjust the volume of each sample to 7 μl using ddH_2O.

Note

[a]The volume of reassociated DNA should be 1–34 μl depending on the amount of starting material. Sample DNA volumes less than 34 μl must be compensated for by the addition of dH_2O to a total volume of 34 μl. A total of two to three reactions can be used if a stronger microarray signal is needed.

Protocol 8

Microarray hybridization, washing, and scanning

Equipment and Reagents

- 2× Hybridization buffer (10× SSC, 0.2% SDS, 2 μg/μl salmon DNA, 2 μg/μl poly(A), 25× Denhardt's solution)
- Water baths at 42, 58 and 95°C
- Wash buffer I (2× SSC, 0.1% SDS)
- Wash buffer II (1× SSC)
- Wash buffer III (0.2× SSC)
- ArrayIt Microarray Hybridization Cassette (TeleChem International, Inc., ArrayIt Division) or equivalent
- Microarray High-speed Centrifuge (TeleChem International, Inc., ArrayIt Division)
- Microarray scanner

Method

1. Add 7 μl of 2× hybridization buffer (pre-warmed to 42°C) to the 7 μl control and tester probes. Mix well by vortexing. This provides 14 μl of control and tester probe at a hybridization buffer concentration of 1×.

2. Denature the probes by incubation at 95°C for 5 min. Chill on ice for 5 min and incubate at 42°C for 10 min to pre-warm the probe mixtures.
3. Mix the denatured control and tester probes together, pre-heat to 58°C, and pipette directly onto a microarray. Place a cover slip over the droplet to form a thin film of hybridization solution across the entire printed microarray surface.
4. Quickly place the microarray inside a hybridization cassette containing a suitable volume of dH_2O for hydration, and incubate in a water bath set at 58°C for 6–16 h.
5. Following the 6–16 h hybridization step, remove the hybridization cassette from the water bath, remove excess water by blotting with laboratory towels, disassemble the cassette, and remove the microarray slide.
6. Transfer the slide quickly to a staining jar containing wash buffer I and allow the cover slip to float free from the microarray surface.
7. Transfer the slide to a fresh staining jar containing wash buffer I and incubate with agitation for 5 min.
8. Transfer the slide to a staining jar containing wash buffer II and incubate with agitation for 5 min.
9. Transfer the slide to a staining jar containing wash buffer III and incubate with agitation for 5 min.
10. Dry the slide by centrifugation at 650 r.p.m. for 5 min using a Microarray High-speed Centrifuge or equivalent device.
11. Scan with a microarray scanner in both the Cy3 and Cy5 channels.

3. TROUBLESHOOTING

Weak microarray signals globally at all spot locations is one pitfall observed from time to time. The cause of weak signals is typically either too little probe or inefficient incorporation of Cy dyes into the probe molecules. The following steps may be taken to circumvent this problem: (i) reamplify or make additional probe samples; (ii) improve elution efficiency from purification columns by using pre-warmed (65°C) ddH_2O as the elution buffer; and (iii) make sure that the DNA is dissolved completely and mixed thoroughly in the dye-coupling reaction. These troubleshooting hints suffice to eliminate weak signals in >95% of all cases where weak signals are observed.

High nonspecific background is also observed on occasion. Several steps may contribute to this problem including insufficient blocking of the microarray surface, failure to use carrier DNA in the hybridization mixture, insufficient hybridization temperature, and insufficient washing. Background can be reduced by: (i) increasing the salmon sperm and poly(A) DNA concentration in the hybridization buffer; (ii) increasing the hybridization temperature; and (iii) extending the washing times, particularly that of wash buffer III (see *Protocol 8*).

Skewed ratios between control and tester samples after normalization can also occur. A major reason for this problem is the use of nonequivalent amounts of control and tester DNA for the DNA reassociation step. The equivalence of control and tester DNA can be assessed and adjusted using the method outlined in section 2.1.

Acknowledgements

This work was supported by an NIH grant CA083778 to K.D.T.

4. REFERENCES

1. **Liang P & Pardee A** (1992) *Science*, **257**, 967–970.
2. **Diatchenko L, Lau Y, Campbell A, *et al.*** (1996) *Proc. Natl. Acad. Sci. U.S.A.* **93**, 6025–6030.
3. **Zhang L, Zhou W, Velculescu EV, *et al.*** (1997) *Science*, **276**, 1268–1272.
4. ★★★ **Schena M, Shalon D, Davis RW & Brown PO** (1995) *Science*, **270**, 467–470. – *Landmark paper in biology presenting the original paper on microarrays.*
5. ★★★ **Chen ZJ, Shen H & Tew KD** (2001) *Nucleic Acids Res.* **29**, e46. – *Landmark paper in the microarray field providing the original description of ADGE.*
6. **Chen ZJ, Gaté L, Davis W, Ile KE & Tew KD** (2003) *BMC Genomics*, **4**, 28.
7. **Chen ZJ, Vulevic B, Ile KE, *et al.*** (2004) *FASEB J.* **18**, 1129–1131.

CHAPTER 4

DNA microarray detection and genotyping of human papillomavirus

TaeJeong Oh, SookKyung Woo, MyungSoon Kim, and Sungwhan An

1. INTRODUCTION

Accurate detection and identification of pathogens causing infectious disease is of critical importance in environmental surveillance, clinical medicine, and bio-defense settings (1, 2). Existing methods including PCR-based testing to screen a broad range of viruses have inheritable limitations in overall throughput, and are thereby restricted to detecting a limited number of candidate viruses. PCR-amplified nucleic acid hybridization to DNA microarrays has demonstrated great promise in obviating this problem. Several DNA microarray-based tests for detecting and genotyping viruses including severe acute respiratory syndrome coronavirus (SARS) (3), human immunodeficiency virus (4), group A rotavirus (5), and human papillomavirus (HPV) (6–8) have been developed recently. In the case of virus diagnosis, multiple DNA targets printed on a glass substrate can be used for specific recognition of different viruses and can discriminate among different virus subtypes.

Although the first DNA microarray publication described cDNA microarrays fabricated by spotting PCR products onto the chemically coated surface of glass slides, there is a growing trend towards replacing cDNA microarray assays with the long-oligonucleotide microarray format. Various detection sequences used to date include long-PCR products, short oligonucleotides (20–30 nt), and long oligonucleotides (50–80 nt). Although microarrays containing each type of target element are capable of detecting fundamental differences among different viruses, each type of binding sequence offers advantages and limitations.

This chapter presents experimental details of DNA microarrays containing both PCR products and oligonucleotides as sequences for

DNA Microarrays: *Methods Express* (M. Schena, ed.)

pathogen detection using the HPV detection and genotyping assay as an example. We describe DNA microarray-based HPV genotyping mainly using PCR-amplified DNAs as targets. PCR target-based microarrays have an advantage over oligonucleotide targets in that different double-stranded targets can be maintained in a common cloning vector, allowing the use of a single pair of vector-specific primers to amplify every target by PCR. In addition, it is possible to immobilize PCR-amplified sequences onto modified glass surfaces without the use of expensive N-terminal oligonucleotide modifications. In our studies, PCR-amplified sequences also appear to show a longer shelf life post-printing compared with oligonucleotides.

On the other hand, if conventional cloning is deemed laborious and inconvenient, one can use oligonucleotide microarrays for genotyping with very good results. Owing to its robustness, simplicity, and the general utility of microarray assays, diagnostic microarrays will undoubtedly be highly useful for the clinical diagnosis of viral infection.

2. METHODS AND APPROACHES

There are two main categories of method for viral genotyping: serial and parallel. In the serial approach, individual samples are examined one sequence at a time for the presence of a single virus or variant. Popular serial approaches utilize DNA blotting, the polymerase chain reaction (PCR), reverse transcriptase PCR (RT-PCR), and other approaches. In the parallel approach, samples are assayed for the presence of many different viral sequences or variants in a single test. The most powerful and popular parallel method utilizes the DNA microarray, a chip containing a collection of different target sequences located at unique addresses on a glass surface. Each target element of the microarray is capable of binding to and assessing the amount of a given sequence in a complex viral probe mixture. A microarray containing ten binding elements, for example, would allow the examination of ten different viral sequences. This section provides a detailed discussion of the DNA microarray methods and approaches we use for detection and genotyping of HPV.

2.1 Target construction and microarray fabrication

HPV detection and classification have been carried out successfully using PCR-based viral DNA amplification, and PCR approaches using either type-specific or general primers for the amplification have been the most widely used methods for HPV detection in cervical cancer (9). Amplified PCR products can be used for viral classification based on sequence differences, but such approaches tend to be time-consuming and laborious if performed in a serial manner.

In the case of DNA microarrays, PCR products are printed on glass substrates or slides such that a single PCR can be used to manufacture thousands of diagnostic microarrays. Microarrays provide quick and convenient tools for sequence variation detection in so-called genotyping assays. The target sequences for viral genotyping are obtained primarily from annotated databases of fully sequenced viral genomes in GenBank or the published literature. In the case of HPVs, more than 80 HPV types have been identified (10). For target construction for HPV genotyping, we selected type-specific sequences (30 nt) from the hypervariable region within the L1 gene of HPV. Jacob *et al.* (11) and the public HPV sequence database (http://hpv-web.lanl.gov/stdgen/virus/hpv/) are useful resources for sequence selection. Every target sequence we selected contained more than 6 nt mismatches across the 30 nt region when compared with the corresponding regions of the other HPVs.

To manufacture DNA microarrays for HPV detection and genotyping using the HPV sequence as targets, there are several options including: short oligonucleotides (<25–30 nt), which are usually modified to incorporate an amine or thiol linker at the N-terminus to permit immobilization onto a treated glass surface (12–14); long oligonucleotides (50–70 nt), which can be immobilized by printing either with or without an N-terminal modification; and PCR products that are amplified from cloned vectors and can be printed with or without an N-terminal modification. Due to the current cost of N-terminal modifications, the most cost-effective DNA microarray strategies utilize target sequences of sufficient length (50–5000 nt) to allow the use of unmodified long oligonucleotides and PCR products. PCR targets for microarray manufacturing (see *Protocol 1*) are the most affordable and hence are utilized extensively herein, as well as the more expensive long-oligonucleotide microarray format.

To produce dsDNA targets for HPV genotyping, complementary-sense and antisense 30mer oligonucleotides specific for 27 HPV genotypes (15) were synthesized, annealed, and subjected to a 3′ end Adenine tailing reaction using a thermocycler (see *Protocol 1*). The target sequences are listed in *Table 1*. The 30 bp dsDNA oligonucleotides were then cloned into the pGEM-T Easy vector for ease of propagation and manipulation. As each target sequence is flanked by vector-derived sequences including the T7 and SP6 promoter regions on the 5′ and 3′ ends, respectively, PCR products (210 bp) containing each 30 nt type-specific target sequence can be amplified using a common pT7 and pSP6 primer set (see *Table 2*). These PCR products are then easily immobilized on glass slides using the same procedures that are used for cDNA targets.

Table 1. HPV type-specific targets (30 bp sequences)

HPV type	Sequence (5′→3′)	Nucleotide position
6	ATCCGTAACTACATCTTCCACATACACCAA	6815–6844
11	ATCTGTGTCTAAATCTGCTACATACACTAA	6799–6828
16	GTCATTATGTGCTGCCATATCTACTTCAGA	6662–6691
18	TGCTTCTACACAGTCTCCTGTACCTGGGCA	6647–6676
31	TGTTTGTGCTGCAATTGCAAACAGTGATAC	6583–6612
33	TTTATGCACACAAGTAACTAGTGACAGTAC	6622–6651
34	TACACAATCCACAAGTACAACTGCACCATA	6539–6568
35	GTCTGTGTGTTCTGCTGTGTCTTCTAGTGA	6502–6531
39	TCTACCTCTATAGAGTCTTCCATACCTTCT	6673–6702
40	GCTGCCACACAGTCCCCCACACCAACCCCA	6808–6837
42	CTGCAACATCTGGTGATACATATACAGCTG	6876–6905
43	TCTACTGACCCTACTGTGCCCAGTACATAT	94–123 (MY)[a]
44	GCCACTACACAGTCCCCTCCGTCTACATAT	6720–6749
45	ACACAAAATCCTGTGCCAAGTACATATGAC	6658–6687
51	AGCACTGCCACTGCTGCGGTTTCCCCAACA	6554–6583
52	TGCTGAGGTTAAAAAGGAAAGCACATATAA	6692–6721
54	TACAGCATCCACGCAGGATAGCTTTAATAA	6633–6662
56	GTACTGCTACAGAACAGTTAAGTAAATATG	6627–6656
58	ACTGAAGTAACTAAGGAAGGTACATATAAA	6678–6707
59	CTACTACTCTCTATTCCTAATGTATACACA	6647–6676
62	CTGCTGCAGCAGAATACACGGCTACCAACT	101–130 (MY)[a]
66	TTAACTAAATATGATGCCCGTGAAATCAAT	6694–6723
67	AATCAGAGGCTACATACAAAAATGAAAACT	6664–6693
68	AATCAGCTGTACCAAATATTTATGATCCTA	101–130 (MY)[a]
69	TCTGCATCTGCCACTTTTAAACCATCAGAT	6594–6623
70	GCCATACCTGCTGTATATAGCCCTACAAAG	6634–6663
72	TCTGTATCAGAATATACAGCTTCTAATTTT	6843–6872

[a]Nucleotide sequences indicate the positions in the 0.45 kb DNA fragment derived from MYH PCR (PCR using primers MY09, MY11, and HMB01).

Alternatively, oligonucleotide targets can also be used for HPV genotyping (see *Protocol 2*). In one approach, we synthesized 50–60 nt oligonucleotides without any end modification (see *Fig. 1*). The oligonucleotides contain 30 nt of HPV type-specific sequences bordered by 10–15 nt of flanking vector sequences. These long oligonucleotide targets can be directly immobilized onto silane-treated glass slides using the same method used to attach double-stranded PCR products (see *Protocol 3*). In addition to type-specific sequences, it is also useful to have generic positive-control HPV sequences that provide positive microarray signals for most HPV virus

Table 2. PCR primers used in this study

Primer name	Sequence (5′→3′)	Nucleotide position
MY09[a]	CGTCCMARRGGAWACTGATC	7033–7014 in HPV16
MY11[a]	GCMCAGGGWCATAAYAATGG	6582–6601 in HPV16
GP6+	GAAAAATAAACTGTAAATCA	6765–6746 in HPV16
HMB01[b]	GCGACCCAATGCAAATTGGT	
GAPDH-LEFT	TCAACGGATTTGGTCGTATT	77–96
GAPDH-RIGHT	TAGAGGCAGGGATGATGTTC	664–645
pSP6	ATTTAGGTGACACTATAGAA	139–158 in pGEM-T Easy
pT7	TAATACGACTCACTATAGGG	2999–3018 in pGEM-T Easy
TOPO-LEFT	AGCTTGGTACCGAGCTCGGAT	255–235 in pCR2.1-TOPO
TOPO-RIGHT	CGAATTGGGCCCTCTAGATGC	363–343 in pCR2.1-TOPO
MY09GAP[a]	CGTCCMARRGGAWACTGATCTCAACGGATTTGGTCGTATT	
MY11GAP[a]	GCMCAGGGWCATAAYAATGGTAGAGGCAGGGATGATGTTC	

[a]Nucleotide abbreviations are according to the IUPAC system of nomenclature: M = A or C; R = A or G; W = A or T; Y= C or T.
[b]HMB01 is directed against the minus strand of HPV type 51.

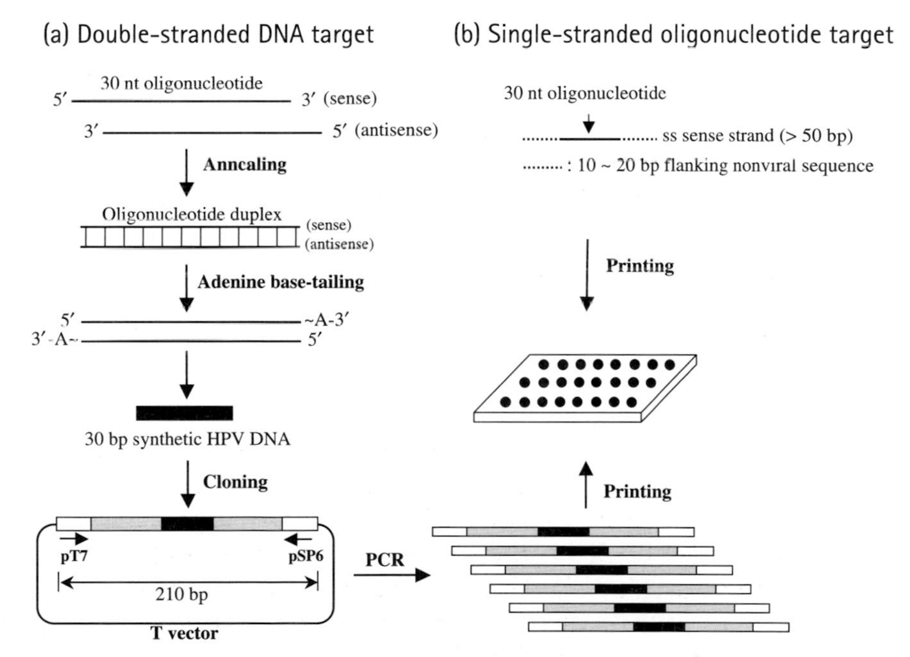

Figure 1. Outline of the protocol for HPV-specific target synthesis by oligonucleotide shuffling. Oligonucleotides were synthesized using the standard phosphoramidite method. Additional details are provided in *Protocol 1*.

types. The highly conserved 180 bp sequence within the HPV L1 gene is present among most known types of genital HPV and serves as a good positive-control sequence. To isolate the positive-control sequence, the 180 bp fragment was amplified using the MY11 and GP6+ primer set (see *Table 2*) from the HPV16-positive cervical cancer cell Caski.

General PCR control sequences are also valuable as these provide an indication of successful PCR results. The PCR control target, a ~550 bp region of the human glyceraldehyde 3-phosphate dehydrogenase (GAPDH) cDNA, was amplified by PCR using GAPDH-specific primers (GAPDH-LEFT and -RIGHT; see *Table 2*) from genomic DNA of the Caski cell line. The 550 bp GAPDH sequence was then cloned into a pGEM-T Easy vector. A negative-control target carrying a partial *lacZ* gene from *Escherichia coli* was amplified from pGEM-T using the pT7 and pSP6 primers.

To carry out genotyping assays of multiple samples on a single glass slide, we developed an eight-well hybridization reaction chamber that accommodates eight samples and allows eight hybridization reactions at once (see *Fig. 2*). For DNA microarray manufacturing, we printed the DNA targets on silane-treated glass slides using a robotic microarrayer operating in a highly controlled clean-room environment. The offsets of the printing pins matched the hybridization cassette, allowing eight microarrays per slide to be readily manufactured (see *Fig. 2*).

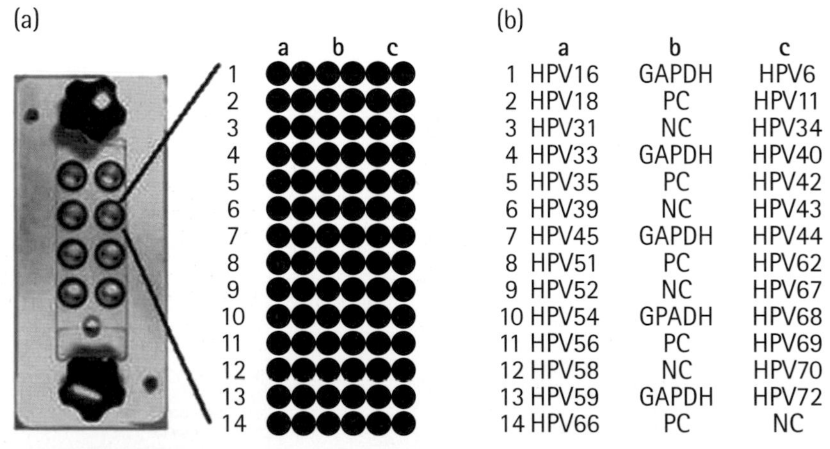

Figure 2. Microarray hybridization cassette and content map for HPV genotyping. (*a*) Eight-well platform hybridization reaction chamber. (*b*) Schematic diagram of the HPV DNA microarray target positions. The PCR control (GAPDH), HPV-positive control (PC) and negative-control *lacZ* gene (NC) were printed on the center of the slide. All targets were printed in duplicate.

2.2 Sensitivity and specificity of the HPV genotyping microarray

PCR is generally regarded as the most widely used nucleic acid-based diagnostic method, in large part due to its sensitivity. In our HPV genotyping microarray assay, however, the sensitivity of the assay is approximately 100-fold higher than that of the conventional PCR method for HPV detection (see *Fig. 3*, also available in the color section). To assess the sensitivity of the HPV DNA microarray, a serially diluted HPV16 plasmid template and a fixed amount of PCR control template were used in an end-point dilution test. Following amplification, a 20% volume of each PCR was analyzed either by gel electrophoresis or by microarray. We did not observe any amplified products below 1000 copies of starting targets on the agarose gel, whereas the corresponding targets labeled with Cy5 showed significant hybridization signals on the HPV genotyping microarray (see *Fig. 3*). These results demonstrate that PCR-based HPV genotyping by microarray is

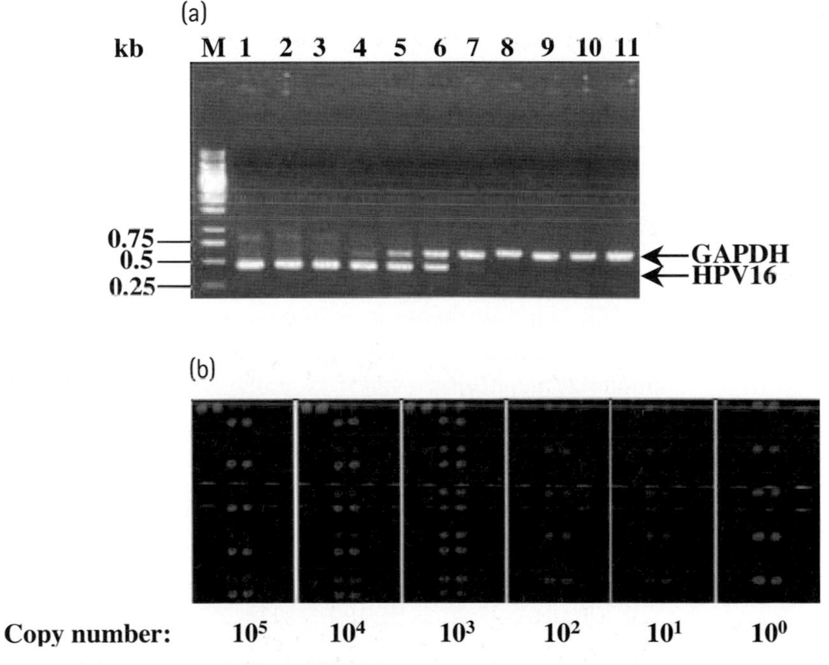

Figure 3. Comparison of the sensitivity of PCR versus an HPV DNA microarray (see page xxii for color version).
(*a*) Agarose gel electrophoresis of MYH PCR products with a serial dilution of the HPV16 plasmid. Lanes: M, 1 kb ladder; 1, 10^{10} copies; 2, 10^9 copies; 3, 10^8 copies; 4, 10^7 copies; 5, 10^6 copies; 6, 10^5 copies; 7, 10^4 copies; 8, 10^3 copies; 9, 10^2 copies; 10, 10^1 copies; 11, 1 copy. (*b*) HPV DNA microarray hybridization results using amplicons generated by MYH PCR starting with the designated number of copies of the HPV16 plasmid.

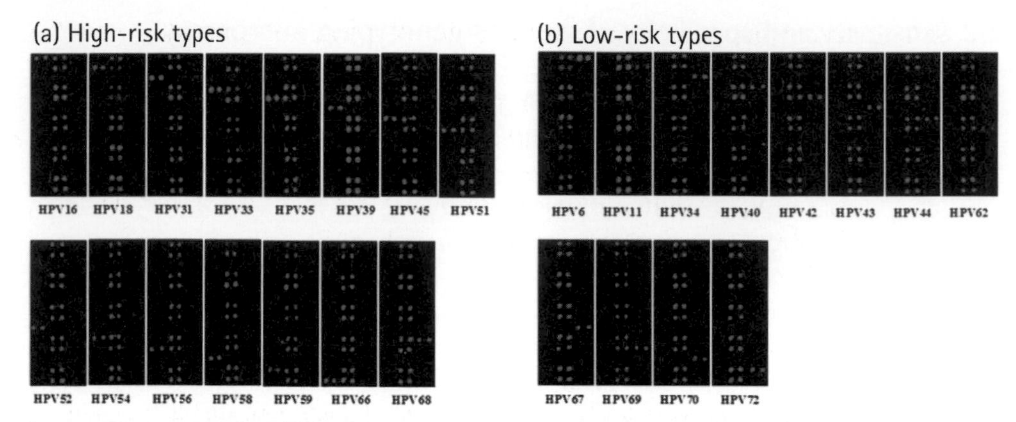

Figure 4. Microarray hybridization specificity of 27 HPV types amplified from plasmids (see page xxii for color version).
(*a*) Hybridization results with high-risk HPV types. (*b*) Hybridization results with low-risk HPV types. The HPV type is indicated below each microarray scan.

much more sensitive than the traditional PCR test. In this aspect, it should be noted that the sensitivity of the cDNA microarray assay is comparable to that of the TaqMan-based quantitative PCR assay (16).

To evaluate the specificity of the HPV type-specific microarray targets, we performed HPV DNA microarray hybridization with Cy5-labeled amplified probes from each plasmid containing sequences for the 27 HPV types. Each HPV type-specific probe mixture hybridized specifically to its corresponding target on the microarray (see *Fig. 4,* also available in the color section). Although excessive amounts of target DNA used for PCR could in theory lead to cross-hybridization, the HPV genotyping microarrays described here did not show any cross-hybridization signals among 27 genotypes, even if as many as 10^8 HPV-specific plasmid copies were used as the PCR template (data not shown). These findings indicate that the microarray elements used in this study are capable of specifically discriminating a broad spectrum of HPVs, even if the concentration of the probe mixture spans a considerable range.

2.3 Probe labeling, hybridization, and detection

In microarray hybridization experiments for viral genotyping applications, probe mixtures are usually prepared by PCR amplification. Reporter molecules (e.g. Cy3 and Cy5) can be incorporated directly into the amplified products. In addition to Cy3 and Cy5, there are many different fluorescent dyes available for direct incorporation into PCR probes. In *Protocol 2,* simple labeling conditions have been developed for Cy5-dUTP using a consensus primer set MY09, MY11, and HMB01 (see *Table 2*) in the L1 open reading frame to generate HPV probes. As a positive control to assess PCR efficiency, the human GAPDH cDNA was amplified using a primer pair with sequences specific for MY11 and MY09 at each end, respectively. The GAPDH PCR product carrying the MY11 and MY09 primer sequences (see *Table 2*) was cloned into pGEM-T Easy, which then was used as a template for the PCR control. The concentration of the PCR control template was determined empirically using a serial dilution test.

Once the HPV probes are labeled with Cy5, they are hybridized to the genotyping microarray in an aqueous salt buffer (see *Protocol 5*). The hybridization temperature and salt concentration can be varied to achieve the desired degree of specificity for a given set of targets. We found that the strongest hybridization signals for HPV genotyping are observed using *Protocol 5*. In this protocol, microarray hybridization and detection can be completed in 2–3 h after probe labeling (see *Protocol 4*). Microarray imaging is accomplished using a commercial microarray laser scanner equipped for Cy5 detection.

2.4 HPV genotyping microarrays for clinical analysis

We validated the HPV genotyping microarray assay using virus-infected tissue culture cells and clinical specimens isolated from female cervical scrapes. Initial validation was accomplished using DNA samples extracted from virus-infected and control (virus-negative) tissue culture cells (see *Fig. 5a*, also available in the color section). As expected, the HPV-negative cell lines A549 (lung cancer) and C33A (cervical cancer) failed to show positive signals for any of the 27 HPV genotypes, whilst the HPV-positive cell lines Caski (HPV16), SiHa (HPV16), and HeLa (HPV18) showed strong hybridization signals on their respective HPV-specific targets and to the positive-control sequences.

To assess the performance of the HPV microarray test in a clinical setting, we tested cervical scrapes collected from women diagnosed as HPV positive. The HPV DNA microarray successfully typed both single and multiple HPV infections (see *Fig. 5b*). In the HPV DNA microarray hybridization experiments with cervical scrapes, no false-positive or false-negative signals were observed when compared with the results of PCR-based detection. Oligonucleotide HPV genotyping microarrays were also tested for signal strength, specificity, and genotyping accuracy (see *Fig. 5c*). Compared with the PCR products used for the cDNA microarrays, the oligonucleotides used in these studies were slightly different in terms of target position for both the type-specific and positive controls; none the less, the oligonucleotide HPV genotyping microarrays also successfully detected single and multiple infections in clinical samples (see *Fig. 5c*).

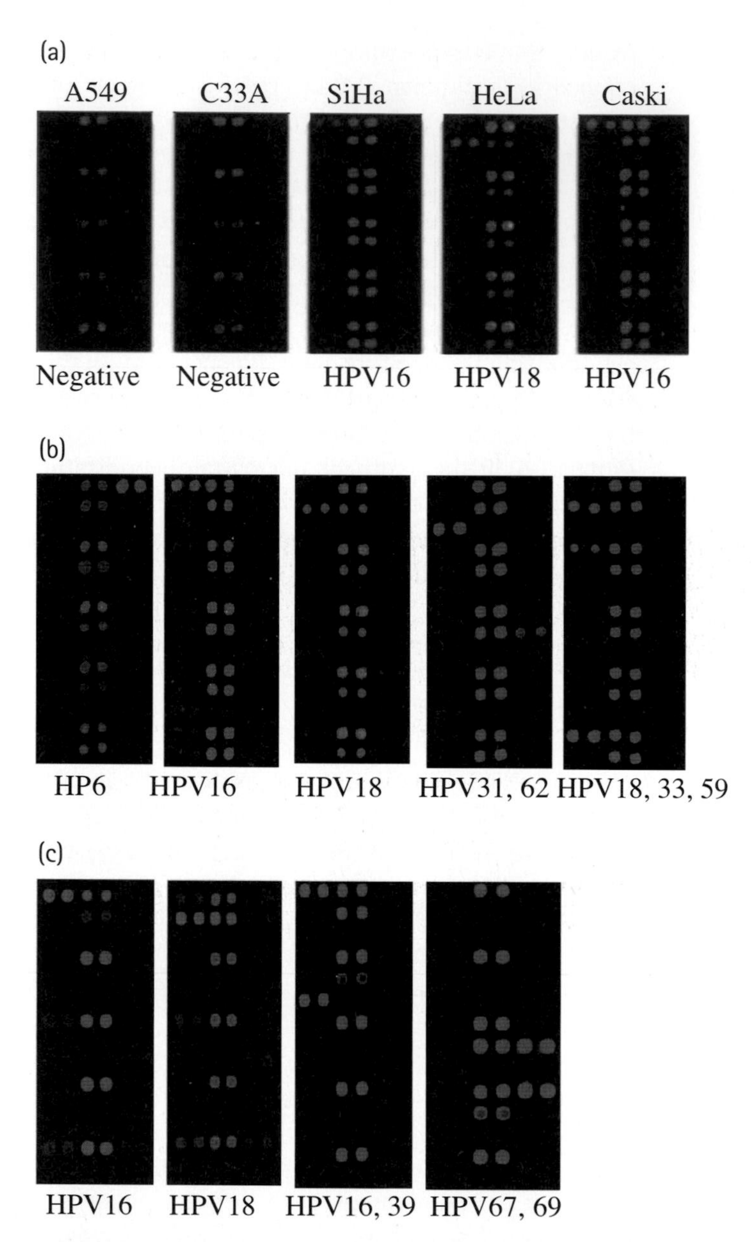

Figure 5. HPV genotyping assay using cell lines and cervical scrapes as the starting samples (see page xxiii for color version).
(*a*) The HPV-negative cell lines A549 and C33A showed hybridization signals with PCR control targets only, whereas the HPV-positive cell lines SiHa, HeLa, and Caski showed positive signals on the PCR controls, HPV-positive controls and each cognate type-specific target. (*b*) HPV genotyping results using HPV DNA microarrays and cervical scrapes as the starting samples. (*c*) HPV genotyping results using 30 nt type-specific oligonucleotide targets with cervical scrapes as the starting samples. Each HPV type is designated below each microarray scan.

Taken together, the HPV genotyping data suggest the usefulness of microarray assays for HPV detection in clinical diagnostics and large-scale epidemiology studies.

2.5 Recommended protocols

Protocol 1

Construction of dsDNA targets

Equipment and Reagents

- pGEM-T Easy vector (Promega)
- dNTP mix (2.5 mM each; Invitrogen)
- 10 units/µl T4 DNA ligase (Promega)
- 10× Ligation buffer (Promega)
- 10× PCR buffer (Solgent)
- 2.5 mM dATP (Invitrogen)
- *Taq* DNA polymerase (Solgent)
- PCR primers: custom-synthesized pT7 (20 pmol/µl) and pSP6 (20 pmol/µl)
- PCR thermal cycler (model PC802; Astec)
- *E. coli* competent cells DH5α
- Luria–Bertani (LB) agar plates containing 40 µg/ml 5-bromo-4-chloro-3-indolyl-β-D-galactopyranoside (X-gal) and 100 µg/ml ampicillin
- ABI 3700 DNA sequencer (Applied Biosystems)
- QIAquick PCR purification kit (Qiagen)
- Dimethylsulfoxide (DMSO) (molecular biology grade)
- 384 Well conical microplates (Genetix)

Method

1. Mix 5 µg of each complementary oligonucleotide target in a total of 10 µl of dH_2O. Denature completely by heating at 100°C for 3 min.
2. Reanneal the denatured oligonucleotides by incubating at room temperature for 60 min to form duplex DNA.
3. Extend 2 µg of each annealed mixture by PCR for 2 h in the presence of dATP only to add an adenine base (A residue) to both ends of the duplex products.
4. Ligate the adenine-tailed products into the pGEM-T Easy vector using 10 µl reactions containing the following: 1 µl of 10× ligation buffer, 1 µl of T4 ligase, 5 ng of pGEM-T Easy vector, and 1 µg of adenine-tailed DNA target. Incubate the ligation mixtures at 16°C for 8 h.
5. Transform 5 µl of each ligation mixture into DH5α cells and select clones using LB agar plates supplemented with 100 µg/ml ampicillin and 40 µg/ml X-gal.
6. Transfer single bacterial colonies into separate tubes and set up 25 µl PCRs as follows: 3 ng of plasmid DNA, 2.5 µl of 10× PCR buffer, 1 µl of dNTP mix, 1 µl of *Taq* DNA polymerase, 1 µl of 20 pmol/µl pT7 primer, 1 µl of 20 pmol/µl pSP6 primer, and 18.5 µl of dH_2O.

Amplify by PCR using the following thermal cycling program: 1 cycle at 94°C for 5 min; 30 cycles of 94°C for 45 s, 55°C for 45 s, and 72°C for 45 s; and 1 cycle at 72°C for 5 min.

7. Following the PCR step, confirm the presence of a 210 bp amplicon by agarose gel electrophoresis.
8. Following the agarose gel step, confirm the correct primary sequence of all of the HPV type-specific clones using an ABI 3700 DNA sequencer.
9. To increase the amount of target material, amplify each of the mixtures according to step 6 and purify the amplified products using a QIAquick PCR purification kit.
10. Suspend the purified DNA targets in 10 μl of 50% DMSO at a final concentration of 200–250 ng/μl.
11. Transfer the DNA target solutions to a 384-well microplate using an eight-channel pipetting device.

Protocol 2

Preparation of HPV oligonucleotide samples containing nonviral flanking sequences

Equipment and Reagents

- 2× Micro Spotting Solution Plus (TeleChem International, Inc.)
- 384-Well conical microplates (Genetix)

Method

1. Suspend 50–60 nt synthesized oligonucleotide targets carrying 30 nt of type-specific viral sequences in 5 μl of dH_2O at 100 pmol/μl.
2. Mix each oligonucleotide sample with 5 μl of 2× Micro Spotting Solution Plus printing buffer for a final oligonucleotide concentration of 50 pmol/μl. For the HPV-positive control targets in which five to six different sequences are mixed together, the final concentration of oligonucleotide is 50 pmol/μl and the final concentration of each individual oligonucleotide is 10 pmol/μl.
3. Transfer each target solution to a 384-well microplate using an eight-channel pipetting device.

Protocol 3

Microarray fabrication

Equipment

- 1″ x 3″ silanized glass slides (Corning)
- OmniGridII Robotic Microarrayer (GeneMachines)

Method

1. Using a robotic microarrayer, print the target samples from 384-well plates onto 50–100 silanized glass slides with eight microarrays per slide in a pattern compatible with the eight-well hybridization chamber. Printing should be performed in a dust-free environment at 40% relative humidity.
2. Following printing, irradiate the slides with UV light (350 mJ) to cross-link the DNA molecules to the silanized glass surface.
3. Once printed, place the microarrays in a sealed container and store in a desiccator at room temperature. Printed slides are stable for several months in these storage conditions.

Protocol 4

Direct probe labeling by PCR for HPV genotyping

Equipment and Reagents

- MYH primer mix containing 20 pmol/μl MY09 primer, 20 pmol/μl MY11 primer, 5 pmol/μl HB101 (Genotech) (see *Table 2*)
- 10× PCR buffer (Solgent)
- *Taq* DNA polymerase (5 units/μl; Solgent)
- dNTP mix containing 2.5 mM dATP, 2.5 mM dCTP, 2.5 mM dGTP, and 1.0 mM dTTP (Invitrogen)
- 1 mM Cy5-dUTP (Amersham Biosciences)
- PCR QIAquick purification kit (Qiagen)
- PCR machine (PC802; Astec)

Method

1. Prepare 25 μl PCRs as follows: 2.5 μl (10–50 ng) of template DNA, 1 μl (100 pg) of control DNA, 2.5 μl of 10× PCR buffer, 1 μl of MYH primer mix, 1 μl of dNTP mix, 1 μl of Cy5-dUTP, 1 μl of *Taq* DNA polymerase, and 15 μl of dH_2O. Amplify the reactions by PCR using the following thermal cycling conditions: 1 cycle at 94°C for 10 min; 30–35 cycles at 94°C for 45 s, 55°C for 45 s, and 72°C for 45 s; and 1 cycle at 72°C for 10 min.
2. Purify the Cy5-labeled target using a Qiagen QIAquick PCR purification kit or equivalent.
3. Elute the purified labeled products with 100 μl of dH_2O.

Protocol 5

Hybridization and detection of HPV genotyping by DNA microarray

Equipment and Reagents

- Blocking solution (3.5× SSC, 0.1% SDS, and 10 mg/ml bovine serum albumin (BSA); filter with 0.25 μM cellulose acetate filter prior to use)
- 20× SSC (molecular biology grade)
- 10% SDS (molecular biology grade)
- BSA (molecular biology grade)
- Isopropanol (molecular biology grade)
- 5 μg/μl Salmon sperm DNA (Invitrogen)
- Array Centrifuge (GenomicTree)
- Eight-well platform hybridization reaction chamber (GenomicTree)
- Axon 4000B microarray scanner (Axon Instruments) or equivalent
- Water baths set at 42 and 55°C

Method

1. Pre-heat the blocking solution to 42°C for 30 min.
2. Insert the microarray into 50 ml of pre-heated blocking solution and pre-hybridize for 45 min at 42°C. Plastic slide boxes (~50 ml volume) work well as hardware for the pre-hybridization step.
3. Wash the microarray three times for 2 min each with dH_2O to remove the blocking solution.
4. Plunge the microarray into isopropanol several times for 1 min to desiccate the microarray surface.
5. Dry the microarray by centrifugation in a microarray centrifuge. Following this step, HPV DNA microarrays should be used within 60 min for the best hybridization results.
6. Prepare a 100 μl hybridization probe mixtures as follows: 20 μl of purified labeled target DNA, 17.5 μl of 20× SSC, 3 μl of 10% SDS, 1 μl of salmon sperm DNA, and 58.5 μl of dH_2O.
7. Insert the printed microarray into the eight-well hybridization chamber and make sure assembly is complete before adding the hybridization probe mixtures to each well.
8. Heat the hybridization mixtures at 100°C for 2 min and then spin at 12 000 ***g*** for 1 min in a microcentrifuge to pellet any insoluble material. Apply the hybridization probe mixture supernatant to each well of the HPV microarray. Incubate at 55°C for 1.5–2 h to allow duplexes to form between the targets on the glass and the labeled probe molecules in solution.
9. After the hybridization reaction, remove the microarray and wash in 500 ml buffer volumes as follows: 2 min in 2× SSC at room temperature with stirring; transfer to a new beaker and wash for 5 min in 0.1× SSC + 0.1% SDS at 55°C (pre-warmed); transfer to a new beaker and wash twice for 1 min in 0.1× SSC.
10. Following the washing steps, dry the microarrays quickly by spinning in a microarray centrifuge.
11. Scan the HPV genotyping DNA microarrays using an Axon 4000B microarray scanner.

3. TROUBLESHOOTING

The protocols presented above have been highly optimized to ensure ease of use in the laboratory and clinic. Although we rarely have experimental difficulties implementing these procedures, there are a few complicated steps in the procedure that deserved special attention.

The steps that comprise *Protocol 1* are traditional cloning procedures that may not be familiar to some users, particularly young scientists who may not use cloning methods on a regular basis. Most cloning failures in *Protocol 1* are due to inactive enzymes, a problem that can be detected by using an independent assay to confirm enzymatic activity. Monitoring the correct fragment size by agarose electrophoresis and confirming the primary sequence by DNA sequencing are essential steps that must not be omitted for the sake of speed or convenience.

The preparation of microarray samples in *Protocol 2* is a straightforward procedure that is easily implemented as long as care is taken to make sure that the suspended oligonucleotides are mixed thoroughly both at the resuspension step and after adding Micro Spotting Solution Plus. Vortexing the microplates for at least 30 s is a particularly efficient mixing procedure. Failure to mix the microarray samples thoroughly will result in heterogeneous signals from one microarray to another, introducing unwanted noise into the hybridization data.

There are two aspects of *Protocol 3* that can lead to variable results: printing and cross-linking. For optimal printing results, make sure that the microarray pins are clean and free of debris to ensure efficient sample loading. Sonication according to the instructions of the manufacturer works well to clean pins that have accumulated debris. Printed microarrays can be inspected using a dissecting microscope to confirm that each microarray contains the expected number of printed elements. The process of attaching the printed DNA to functional groups on the microarray slide (i.e. cross-linking) occurs much more efficiently if the printed elements have been allowed to dry on the slide. Drying occurs spontaneously at low humidity and elevated temperatures. Weak microarray signals or printed elements of poor geometry are common signs of poor cross-linking efficiency.

One common cause of weak hybridization signals is poor dye quality, due either to photobleaching or to degraded dye-labeled nucleotides. Dyes should be stored frozen and in the dark at –80°C at all times. Incorporation efficiency in labeled samples can be measured spectrophotometrically by examining absorption at the two wavelengths that reflect DNA concentration (260 nm) and dye presence (e.g. 550 or 650 nm). Samples that show strong DNA absorption and weak dye absorption reflect either poor labeling efficiency or poor dye quality. Commercial fluorescent labeling reagents should be replaced at regular intervals (2–3 months) for the best results.

The most common pitfall in *Protocol 5* is high background, due either to the use of impure reagents or to poor blocking and hybridization conditions. Background that is punctate in nature (i.e. bright 'speckles' against an otherwise dark background) is due to the presence of contaminating particulates in either the reagents or the fluorescent samples. All reagents should be filtered (0.25 μM pore size) prior to use. Fluorescent samples should be spun hard by centrifugation to pellet any particulates before using the fluorescent supernatant.

Acknowledgements

This work was supported by a grant (Project No. 10012714) from the Ministry of Commerce, Industry and Energy, South Korea. The authors wish to thank Chiwang Yoon and Daekyung Yoon (GenomicTree) for their contributions to the HPV DNA microarray manufacturing.

4. REFERENCES

★★ 1. **Wang D, Coscoy L, Zylberberg M, *et al.*** (1999) *Proc. Natl. Acad. Sci. U.S.A.* **99**, 15687–15692. – *Presents a general viral genotyping method using a DNA microarray.*

2. **Call D** (2001) *Vet. Sci. Tomorrow*, **Issue 3**, 1–9.

3. **Long WH, Xiao HS, Gu XM, *et al.*** (2004) *J. Virol. Methods*, **121**, 57–63.

4. **Shieh B & Li C** (2004) *Retrovirology*, **1**, 11–14.

5. **Lovmar L, Fock C, Espinoza F, Bucardo F, Syvanen AC & Bondeson K** (2003) *J. Clin. Microbiol.* **41**, 5153–5158.

★ 6. **An HJ, Cho NH, Lee SY, *et al.*** (2003) *Cancer*, **97**, 1672–1680. – *A nice example of an HPV genotyping DNA microarray for clinical cervical samples.*

7. **de Roda Husman AM, Walboomers JM, van den Brule AJ, Meijer CJ & Snijders PJ** (1995) *J. Gen. Virol.* **76**, 1057–1062.

★★★ 8. **Oh TJ, Kim CJ, Woo SK, *et al.*** (2004) *J. Clin. Microbiol.* **42**, 3272–3280. – *A good collection of technical details and protocols for HPV genotyping DNA microarrays.*

9. **Klaassen CH, Prinsen CF, Valk HA, Horrevorts AM, Jeunink MA & Thunnissen FB** (2004) *J. Clin. Microbiol.* **42**, 2152–2160.

10. **Kleter B, Doorn LJ, Schrauwen L, *et al.*** (1999) *J. Clin. Microbiol.* **37**, 2508–2517.

★★ 11. **Jacobs MV, Husman AM de R, van den Brule AJC, Snijders PJF, Miejer CJL & Walboomers JMM** (1995) *J. Clin. Microbiol.* **33**, 901–905. – *Presents details of HPV type-specific probes for genotyping.*

12. **Beattie WG, Meng L, Turner SL, Varma RS, Dao DD & Beattie KL** (1995) *Mol. Biotechnol.* **4**, 213–225.

13. **Doktycz MJ & Beattie KL** (1997) In *Automation Technologies for Genome Characterization*, pp. 205–225. John Wiley & Sons, New York.

14. **Beier M & Hoheisel JD** (1999) *Nucleic Acids Res.* **27**, 1970–1977.

★★★ 15. **Gravitt PE, Peyton CL, Alessi TQ, *et al.*** (2000) *J. Clin. Microbiol.* **38**, 357–361. – *A nice paper for MYH PCR-based HPV genotyping assays and PCR protocols.*

16. **Iyer VR, Eisen MB, Ross DT, *et al.*** (1999) *Science*, **283**, 83–87.

CHAPTER 5

Expression profiling of transcriptional start sites

Mutsumi Kanamori-Katayama, Shintaro Katayama, Harukazu Suzuki, Yoshihide Hayashizaki

1. INTRODUCTION

The findings from mammalian genome sequencing projects and cDNA encyclopedia projects have revealed that the genomes of higher organisms, including humans and mice, encode much more information than is suggested simply by the number of genes (1–3). Almost all genes have a transcriptional regulatory region that allows control via complex transcriptional regulatory complexes. Proper cellular function requires the correct expression of all genes at the appropriate times during development and homeostasis. Understanding the complex transcriptional mechanisms that underlie gene networks requires investigating transcriptional regulatory sites on a whole-genome basis. With the genome sequences of humans and many model organisms released, a high-priority issue is to identify the regulatory regions that mediate gene expression regulation. Various transcription factors bind to transcription factor binding sites within the regulatory region (promoter region plus additional regulatory sequences). The transcription factors and regulatory regions work together to determine when, where, and how much of the gene product (RNA) is transcribed and eventually translated into protein.

Eukaryotic gene regulatory proteins can control transcription when bound to DNA sites (*cis* sequences) far away from the promoter by acting *in trans*. We use the term 'transcriptional regulatory region' to refer to the whole expanse of DNA involved in regulating transcription of a gene, including the promoter (where the general regulating transcription factors and the polymerase assemble), and all of the regulatory sequences to which gene regulatory proteins bind to control the rate of the assembly processes at the promoter (see *Fig. 1*, also available in the color section). As

DNA Microarrays: *Methods Express* (M. Schena, ed.)

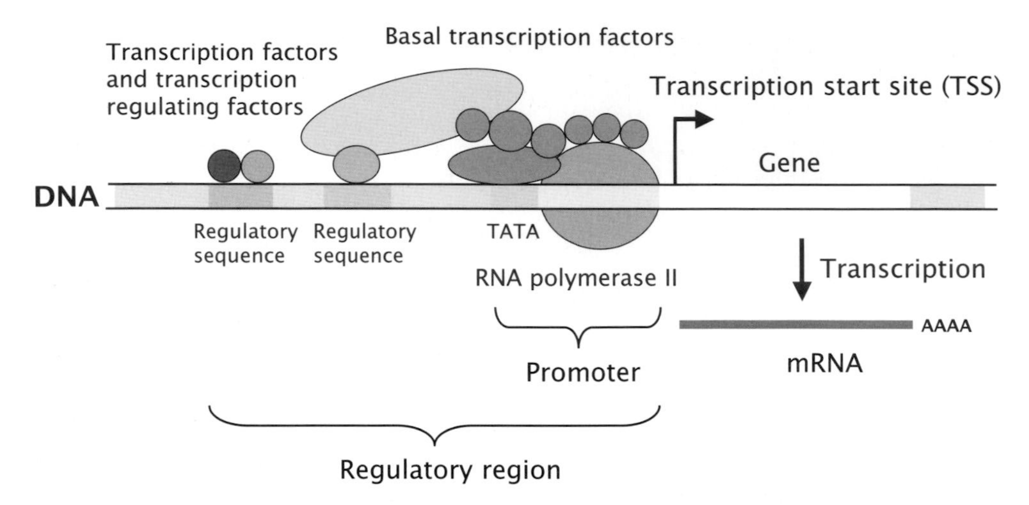

Figure 1. Schematic illustration of the regulatory region of a typical eukaryotic gene (see page xxiv for color version).
The promoter is the *cis*-acting DNA sequence where the general transcription factors and the polymerase assemble. Promoters typically contain one or more canonical core promoter sequences such as the TATA box. Gene regulatory proteins including transcription factors and transcription regulating factors bind to distal regulatory or enhancer sites and modulate the rate of transcription initiation. The transcription start site (TSS) defines the nucleotide position wherein mRNA synthesis initiates. Cellular mRNAs are often polyadenylated (AAAA) at the 3′ end of the nascent transcript.

the regulatory region generally maps in the vicinity (e.g. –1000 to +200 nt) of the transcription start site (TSS) (4), the position of the TSS must be identified accurately in order to study the regulatory region. Transcription factors allow individual genes to be turned on or off specifically by the cell. Regulatory regions containing promoters allow differences in protein expression levels to be achieved in each cell as required for particular physiological states. As the result, this feature of expression regulation is thought to give each cell its unique character and phenotype. It is therefore very important to study how transcriptional regulatory proteins bind to gene control regions on DNA and how such binding events influence the rate of transcription initiation.

Although computer-based promoter prediction tools have proven unsatisfactory for identifying the TSSs of genes (5), a range of other technologies have been utilized to identify TSS positions including rapid amplification of cDNA ends (RACE), 5′ serial analysis of gene expression (5′-SAGE) and full-length cDNA cloning. Recently, a systematic 5′-end analysis of the mouse and human transcriptome using a cap analysis of gene expression (CAGE) approach was carried out (6). This analysis identified a large number of TSS and core promoter elements (236 498 in mouse and 190 513 in human) (P. Carninci *et al.*, unpublished data). The data show that the majority of coding and noncoding genes have at least two alternative promoters. In these genes, different mRNA isoforms can be produced directly through different transcription start sites or indirectly by promoter-directed exon splicing (7, 8). For example, the gelsolin gene has two alternative promoters, where the gene produces the same protein in macrophages and liver, and an additional alternative promoter generates a protein with a different function in heart and cerebellum (P. Carninci *et al.*, unpublished data). In a second example, mouse *otx2* transcripts have been shown to derive from three different promoters, wherein one promoter was found to be used preferentially in early embryogenesis and the promoter usage was switched to the other promoters in the adult brain (9). Furthermore, differences in the 5′ untranslated regions may affect translation efficiency and mRNA stability. Of course, having alternative promoters suggests a possibility for these genes of being regulated differently at the level of transcription initiation (10). Moreover, it was suggested that alternative promoters of the *SRC* gene could play a role in *SRC* transcriptional activation in liver cancer cells, providing one example of how alternative promoter usage might be associated with diseases and cancer cells.

In view of the discussion above, the importance of developing new cost-effective technologies to investigate differences in multiple promoter usage in normal and disease tissues should be obvious. Approaches based on DNA sequence determination of short sequence tags require an enormous amount of sequencing, time, and money to analyze absolute expression levels. As a much more efficient and cost-effective alternative, novel experimental approaches for identifying promoter usage based on microarray technology have been developed.

2. METHODS AND APPROACHES

2.1 Bioinformatic identification of promoters, target sequence design, and microarray manufacturing

TSS microarrays differ from standard gene expression oligonucleotide microarrays in that TSS microarrays contain specific oligonucleotide targets designed to identify promoters, which in turn supplies information about promoter usage and transcripts produced by alternative promoters. This is in contrast to standard microarrays that use targets designed to the 3′ ends of transcripts and as such fail to assay promoter function at the 5′ gene ends.

2.2 Transcripts and tags

The source for the mouse TSS microarrays is a publication (2) describing a comprehensive polling of transcription start sites from the mouse genome using 102 597 FANTOM3 cDNA sequences, 56 006 GenBank (11) mouse mRNA sequences (Release 139.0 + daily 27 January, 2004), 25 803 RefSeq (12) mouse sequences (NM and NR at cumulative 25 January, 2004, XM and XR at 1 September, 2004), 35 247 EMSEMBL (13) mouse sequences (Release 25.33a), 11 567 973 CAGE (6) mouse tags, 385 797 mouse gene identification signature (GIS) ditags, 2 079 652 mouse genome signature cloning (GSC) ditags (14) and 558 686 RIKEN 5′ expressed sequence tags (ESTs). The transcripts and ESTs were mapped to the UCSC mm5 assembly (15) using BLAT (16) and SIM4 (17). The CAGE and ditags were mapped to the same assembly using BLAST (18).

2.3 Tag cluster definition

For the 5′-end tag cluster definition, so-called CAGE tags, the 5′ ends of GIS and GSC ditags were used together with 20 nt in the 5′-end of the RIKEN 5′-EST and FANTOM3 transcripts. Tags whose sequence overlapped on the same strand were clustered into tag clusters. The representative position of a tag cluster is the most frequent start site of the clustered tags in the following order of priority: CAGE, GIS, GSC, RIKEN 5′-ESTs, and FANTOM3 transcripts. The most upstream site is chosen as the representative site for the tag cluster if two or more start sites occur on the same tag number.

2.4 Target sequence preparation and target design

DNA targets for transcription start sites are obtained by excising the fragment from the genomic sequence, starting from the representative position of the tag cluster and extending 120 bp downstream (see *Fig. 2*, also available in the color section). A 60 bp region is chosen within the 120 bp fragment using hybridization efficiency and specificity as considerations. If the fragment overlaps the 5′ end of an intron, an additional fragment is excised according to the splice site. If the fragment does not overlap any known transcript, intergenic/intragenic probes designed from such fragments might not capture transcripts initiating from novel promoters. These fragments are turned reverse-complementary and formatted using the instructions provided in the 'Sequence Submission Guidelines' of Agilent (http://www.chem.agilent.com/). Candidate target sequences in these fragments were designed as proposed by the Agilent Probe Design Service.

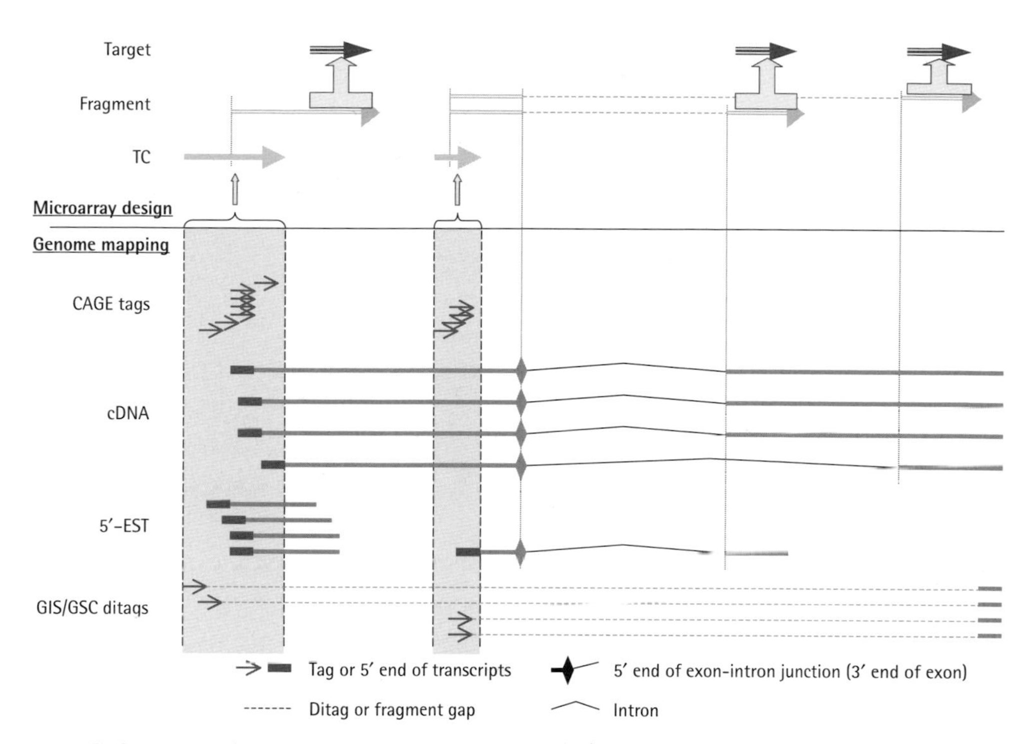

Figure 2. Concepts of target design for TSS microarrays (see page xxiv for color version). Lines in the bottom half of the figure depict full-length transcripts or cDNAs (orange) and the 5′ ends (CAGE, 5′-ESTs or GIS/GSC) aligned to the genomic sequence. The upper portion of the figure shows the regions used for microarray design including the actual microarray target sequences (dark blue arrows). The tag cluster represents overlapping regions of transcripts and 5′-end tags. The fragment for target design (turquoise arrow) begins from the major transcription start site and extends 120 bp from that site. If the fragment overlaps the 5′ end of an exon–intron junction, the fragment 'jumps' the intron to the closest 3′ exon. In the target design phase, double-stranded molecules are chosen using primer specificity and amplification efficiency as criteria. According to the Agilent Probe Design Service, appropriate 60 bp regions in each fragment would be suggested for use as microarray targets.

2.5 Synthesis and amplification of first-strand cDNA

As 5′ mRNA ends are needed to investigate promoter usage, standard microarray amplification and labeling methods based on 3′ or random labeling are not applicable to TSS microarrays. The development of new labeling methods was therefore important for proper usage of TSS microarrays (see *Fig. 3*, also available in the color section). In standard microarray protocols, poly(A)$^+$ or total RNA is amplified into complementary RNA (cRNA) using T7 polymerase-based methods. In the case of oligo(dT) priming, 5′ mRNA sequences tend to be under-represented compared with 3′ mRNA sequences, as oligo(dT) priming occurs on the 5′ end of the mRNA

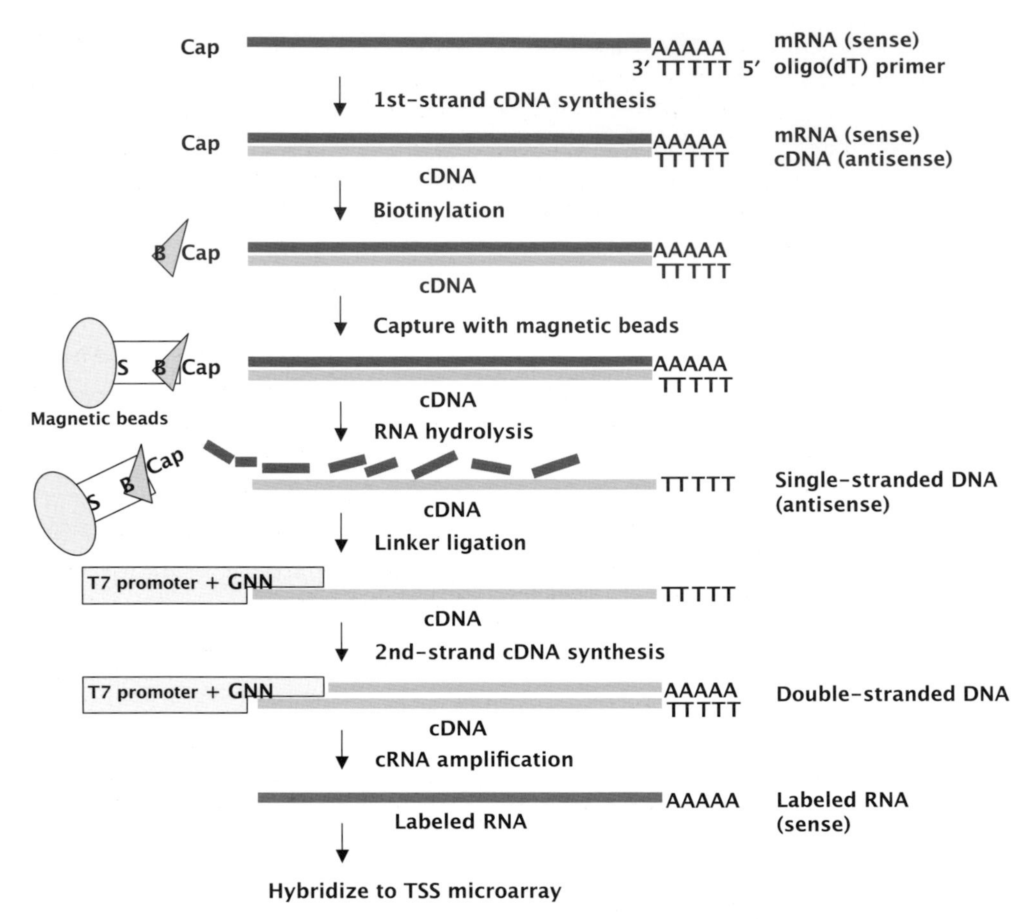

Figure 3. Schematic illustration of the protocol used for TSS microarray labeling (see page xxv for color version).
To avoid amplification biases and to ensure labeling from bona fide 5′ ends, 'cap-trapper' full-length cDNA selection was combined with cRNA amplification as a new labeling technique suitable for TSS microarray experiments. Cellular mRNAs (blue lines) are primed with oligo(dT) (TTTTT) and extended with reverse transcriptase to yield nascent cDNA (green lines). 'Cap trapping' with biotin (B)- and streptavidin (S)-conjugated magnetic beads highly enriches the number of full-length cDNAs. A primer sequence containing a T7 promoter sequence is annealed to the 5′ ends of the cDNAs, followed by second-strand (pink line) cDNA synthesis. The cRNA (red line) for hybridization to TSS microarrays is synthesized from the double-stranded cDNA templates using T7 RNA polymerase (see protocols for details).

and enzymatic extension becomes less efficient as the enzyme moves towards the 3′ end of the transcript. The presence of mRNA size bias can present difficulties in profiling expression transcripts if the oligonucleotides are not chosen carefully.

For gene expression applications, oligonucleotide sequences are typically designed to the 3′ ends of genes to avoid the loss of signal due to the under-representation of 5′ sequences. As the gene targets for TSS microarrays are designed to the 5′ end of transcripts, the cRNA labeling method needs to be modified to amplify the 5′ end more efficiently. To address this issue, so-called 'cap trapper' technology has been used. The cap trapper method was developed initially to address the difficulties involved in obtaining and selecting full-length cDNAs. The details of the cap trapper method are reviewed in detail in several key papers (19–21). In this strategy, full-length cDNAs are isolated from shorter products using a biotinylated cap structure to 'trap' the full-length cDNA. The full-length biotinylated cDNA is isolated using streptavidin-coated magnetic beads, which bind tightly to the biotinylated products. To maximize the extension efficiency of reverse transcriptase (RT), the RT reaction is heated to a high temperature to 'melt' any secondary structure in the mRNA that may inhibit the reverse transcription process. To avoid loss of function of the RT enzyme in the heating process, the disaccharide trehalose is added to the mixture. Trehalose is a stabilizing agent that enhances the thermostability of RT (22), and correspondingly the elongation efficiency is much improved in the presence of trehalose.

For TSS microarray experiments, cDNA synthesis is performed using total RNA and first-strand cDNA primer (oligo(dT)$_{12-18}$) with SuperScript II RT in the presence of trehalose and sorbitol. Subsequently, full-length cDNA products are selected using a biotinylated cap trapper, making it possible to investigate the 5′ promoter without the limitations of mRNA size bias introduced by standard extension procedures.

2.6 Fluorescent cRNA synthesis

In this labeling procedure, one sample is labeled with Cy3- or Cy5-labeled CTP. In the standard method, an oligonucleotide primer containing a string of T residues (poly(dT)) linked to a T7 polymerase promoter is annealed to the poly(A)$^+$ mRNA. Extension of the primed mRNAs by RT creates cDNA molecules that contain T7 promoters on the end complementary to the 3′ end of the mRNA. However, the requirement for oligo(dT) priming and the use of RT create cDNA molecules that are statistically deficient in 5′ sequences, and thus this synthesis and amplification bias are not appropriate for use with TSS microarrays.

One method that prevents the statistical under-representation of 5′ sequences involves a ligation-based approach described as follows. Following synthesis of single-stranded cDNA, a biotinylated linker containing a T7 polymerase promoter is ligated to the 5′ ends, double-stranded cDNA is made, and full-length products are selected using the biotinylated cap-trapper approach. Synthesis of cRNA via T7 RNA polymerase produces full-length cRNA products. The polymerase simultaneously incorporates Cy3- or Cy5-labeled CTP into cRNA; importantly, the resulting cRNA labeled using this method represents the opposite cRNA strand (sense strand) relative to the standard T7-based synthesis approach, which produces antisense cDNA products. In the TSS method, the labeled cRNA represents the sense strand because the primer containing the T7 polymerase promoter is annealed to the 5′ end of the first strand of cDNA, which is converted to a double-stranded product following second-strand synthesis. The successful hybridization of labeled sense cRNA to a microarray requires the microarray target sequences to contain antisense gene sequence information.

2.7 Columns; detection sensitivity in oligonucleotide microarrays

DNA microarray analysis, particularly the oligonucleotide microarray approach, has been used extensively to explore global gene expression patterns in many organisms, tissues, cells, and samples under myriad physiological conditions. Microarray density is sufficient to enable tens of thousands of genes to be measured quantitatively in a single hybridization. On the other hand, real-time RT-PCR is a highly sensitive and specific method that is widely used to explore gene expression although it is relatively low throughput; typically only several genes are measured with the method at one time. As we have established a medium-scale real-time RT-PCR system focusing on transcription factors (http://genome.gsc.riken.jp/qRT-PCR/) (23), we explored the sensitivity of detection on two microarray platforms with different oligonucleotide target lengths. Affymetrix (25mers) and Agilent (60mers) were compared with each other and verified using real-time RT-PCR data.

We first examined the correlation between microarray fluorescence signals and gene expression levels measured by real-time RT-PCR to explore the quantitative aspects of the two methods. For this purpose, we selected the real-time RT-PCR data from 818 genes that showed amplification efficiency via specific primer pairs close to the ideal value of 1.00; as such, the threshold cycle (Ct) values for the 818 genes most likely reflect true gene expression levels. The Ct value represents the first PCR cycle at which there is a detectable difference between the reporter signal and the baseline control. The Ct values determined by RT-PCR of transcription factor genes in adult mouse hippocampus were compared with microarray fluorescence signal intensities of the same genes using the Affymetrix and Agilent microarray platforms (see *Fig. 4*). As expected, microarray signals correlated quite well with RT-PCR up to Ct values of ~25 on both microarray platforms. The correlation coefficient values were

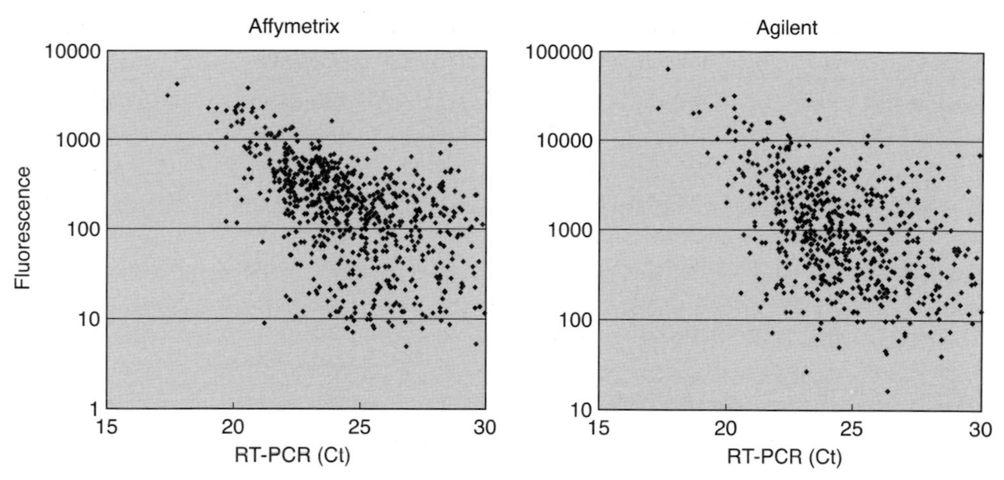

Figure 4. Scatter plots comparing microarray and RT-PCR gene expression data. Fluorescence intensities from Affymetrix and Agilent oligonucleotide microarray data were compared with the Ct values for selected genes in real-time RT-PCR experiments. Expression values from the Affymetrix microarray platform were taken from the SymAtlas database (http://symatlas.gnf.org/SymAtlas/) from the Genomics Institute of the Novartis Research Foundation (GNF), whereas the Agilent expression values were obtained from our own 'in-house' data set. Data points with RT-PCR Ct values greater than 30 were omitted from the analysis as such values are less reliable, as indicated by large variations in the reproducibility test.

0.63 and 0.53, respectively, for the Affymetrix and Agilent microarrays. For genes in which Ct values exceeded 25, microarray signals showed a greater divergence from the RT-PCR data, but typically concordance was within a factor of 2. Taken together, the data indicate good agreement between RT-PCR and microarray analysis, and indicate that both approaches provide quantitative gene expression information; moreover, both the Affymetrix and Agilent platforms showed very good performance in our studies.

One of the most important benchmarking criteria of a DNA microarray platform is the capacity to detect changes in gene expression levels for specific genes among different biological samples. This criterion bears directly on the hybridization specificity of the target sequences present at each location on the microarray. To this end, we examined whether cerebellum-specific genes, selected from our real-time RT-PCR data, were detectable using the Affymetrix 25mer oligonucleotide microarray platform (see *Table 1*). The genes consisted of 23 prominent cerebellum-specific genes whose Ct values were at least 4.0 cycles less than the average Ct values observed in cerebral cortex and hippocampus. As a practical matter, this means that the level of expression of the genes in the cerebellum were on average about tenfold higher than in the other tissues. Many of the genes we tested show differential expression in the cerebellum according to the primary scientific literature (data not shown).

Table 1. Detectivity data for selected cerebellum-specific transcription factor genes measured using RT-PCR and an Affymetrix oligonucleotide microarray

Gene symbol	RT-PCR (Ct)	GeneAtlas ID	Microarray specificity	Microarray fluorescence
Zic1	20.5	gnf1m01967_a_at	Yes	7246.0
Neurod1	20.8	gnf1m28985_a_at	Yes	6009.3
Zic2	21.5	gnf1m27330_at	Yes	332.2
Neurod2	21.7	gnf1m13274_at	Yes	542.4
St18	21.7	gnf1m13674_at	Yes	659.0
St18	21.7	gnf1m28100_at	Yes	85.3
Pax6	22.3	gnf1m12449_a_at	Yes	327.0
Pax6	22.3	gnf1m26555_at	Yes	38.4
Pax6	22.3	gnf1m21933_at	No	138.0
Hes3	22.5	gnf1m28557_at	Yes	575.0
E130309B19	22.9	gnf1m21556_at	Yes	210.1
E130309B19	22.9	gnf1m21557_a_at	No	214.3
Tcfap2b	23.2	gnf1m17826_at	Yes	331.0
Tcfap2b	23.2	gnf1m21186_at	No	322.2
Tcfap2b	23.2	gnf1m23564_at	No	41.6
En2	23.4	gnf1m02223_a_at	Yes	1586.8
Uncx4.1	24.1	gnf1m03244_at	Yes	383.7
Uncx4.1	24.1	gnf1m26901_at	Yes	150.4
Eomes	24.9	gnf1m29385_s_at	Yes	293.3
Eomes	24.9	gnf1m26813_at	No	91.6
Lhx1	25.3	gnf1m24326_a_at	Yes	368.0
Lhx1	25.3	gnf1m01379_s_at	Yes	177.0
Tcfap2a	25.3	gnf1m28633_s_at	Yes	180.8
Tcfap2a	25.3	gnf1m28151_at	Yes	143.4
Og9x	25.7	gnf1m01534_at	No	175.7
Ebf2	26.2	gnf1m13003_a_at	No	268.4
Ebf2	26.2	gnf1m14967_at	No	9.6
Pax2	26.4	gnf1m02606_a_at	No	7.9
Pax3	26.5	gnf1m01545_a_at	No	176.8
Pax3	26.5	gnf1m21278_at	No	122.0
Msx2	26.6	gnf1m26263_at	No	87.2
Msx2	26.6	gnf1m03176_a_at	No	155.9
Tead4	27.5	gnf1m13821_at	No	260.0
Barhl1	28.1	gnf1m00266_a_at	No	286.9
Nr0b2	28.2	gnf1m10720_a_at	No	19.9
Lhx5	29.9	gnf1m01380_a_at	No	280.1

Gene symbols are provided for select cerebellum-specific genes. The Ct values are those obtained from RT-PCR. Expression values for the Affymetrix microarray platform were taken from the SymAtlas database (http://symatlas.gnf.org/SymAtlas/) from the Genomics Institute of the Novartis Research Foundation (GNF) and GeneAtlas IDs are provided for the genes. Microarray specificity was scored as positive (yes) or negative (no) if the fluorescence signals for genes expressed in the cerebellum were greater than or equal to twofold higher than the signals for the same genes expressed in the cerebral cortex and hippocampus. Raw fluorescence intensity values for each gene in the cerebellum are also shown.

The Affymetrix data used in this study were obtained from the SymAtlas database (http://symatlas.gnf.org/SymAtlas/) from the Genomics Institute of the Novartis Research Foundation (GNF). The data in SymAtlas concurred with the RT-PCR data for genes with Ct values up to about 25, whilst expression specificity was not observed for genes with Ct values greater than 25 (see *Table 1*). Together with the previous estimate that a Ct value of 27 corresponds to approximately one mRNA copy per cell in our RT-PCR experiments (data not shown), the results suggest that the Affymetrix 25mer oligonucleotide microarray detectivity limit is approximately several mRNA copies per cell, which is consistent with a previous report that estimated the detectivity limit of a DNA microarray in one study to be one to ten mRNA copies per cell (23).

To estimate the reliability of the fluorescence intensity of the Affymetrix data, we analyzed the correlation between the ability of the microarray to specifically detect a twofold or greater change in specific genes between the cerebellum and cortex or hippocampus gene and raw fluorescence intensity values obtained by microarray (see *Table 1*). We found that a fluorescence value of 327 was the lower limit for complete correlation in the table, although the sample number was relatively small, and in some cases differential expression was detected for genes that gave much lower fluorescence values (see *Table 1*). The comparative data also revealed that the Affymetrix microarray concordantly detected differential expression for 19 out of 36 (53%) of the genes detected using RT-PCR (see *Table 1*).

2.8 Recommended protocols

Herein we present protocols for RNA purification, labeling, cDNA and amplified RNA synthesis, and related methods. We have used commercial kits for the purification of total RNA, but we have found that total RNA prepared from various tissues using a modification of a standard procedure and cetyl triammonium bromide (CTAB) precipitation is superior as it allows selective removal of polysaccharides (21, 24). This protocol has also been used to purify RNA from cultured cells. Contaminating polysaccharides inhibit the binding of full-length cDNAs to the streptavidin-coated magnetic beads, apparently by competing for streptavidin sites.

Protocol 1

Polysaccharide removal

Reagents

- CTAB solution (1% CTAB, 4 M urea, 50 mM Tris-HCl, pH 7.0, and 1 mM EDTA, pH 8.0)
- 7 M Guanidinium chloride
- 5 M NaCl
- 80 and 100% Ethanol

Method

1. Separate the polysaccharides from the RNA by selective precipitation with CTAB solution. Add 4 vols of CTAB solution and 1.3 vols of 5 M NaCl to the RNA sample.
2. Recover the RNA by centrifugation at 9500 ***g*** for 15 min at room temperature (20–25°C).
3. Discard the supernatant containing the contaminating polysaccharides and resuspend the RNA pellet in 4.0 ml of 7 M guanidinium chloride.
4. Add 8 ml of 100% ethanol to the RNA mixture. Mix the solution well and incubate for 1–2 h at –20°C to allow RNA precipitation.
5. Recover the RNA by centrifugation at 9500 ***g*** for 15 min at 4°C.
6. Wash the RNA pellet with 80% ethanol and recentrifuge at 9500 ***g*** for 5 min at 4°C.
7. Redissolve the RNA in 0.1–1 ml of H_2O to achieve the desired final concentration of 10 μg/μl.

Protocol 2

First-strand cDNA synthesis

Equipment and Reagents

- 0.5 ml PCR tubes
- Oligo(dT) primer (2 μg/μl)
- 2× GCI (LA Taq) buffer (TaKaRa)
- 10 mM dNTP solution
- Sorbitol/trehalose solution (100 μl of 4.9 M sorbitol and 50 μl of a saturated solution of trehalose)
- SuperScript II RT (200 units/μl; Invitrogen)
- 0.5 M EDTA (pH 8.0)
- Thermal cycler

Method

1. To synthesize first-strand cDNA, prepare the following reagents in a 0.5 ml PCR tube (tube A): 3.0 μl of 10 μg/μl total RNA (30 μg), 1.5 μl of 2 μg/μl oligo(dT) primer (3 μg), and 0.5 μl of dH_2O for a final volume of 5 μl.

2. Heat the tube A RNA and primer mixture to 65°C for 10 min to remove RNA secondary structure. Place on ice for 5 min to stabilize the RNA–oligo(dT) hybrids. During the 10 min incubation in this step, prepare the tube B reaction as described in step 3.
3. Prepare the following reagents in a 0.5 ml PCR tube (tube B): 20 µl of 2× GCI (IA Taq) buffer, 1.0 µl of 10 mM dNTP solution, 12 µl of sorbitol/trehalose solution, and 2 µl 200 units/µl SuperScript II RT.
4. Transfer the contents of tube B into tube A and transfer tube A (40 µl total volume) into a thermal cycler. Initiate the 'hot start' cycling program using the following parameters: 42°C for 30 min, 50°C for 10 min, 56°C for 10 min, and 4°C for 10 min. The reaction can be stored overnight at 4°C for convenience if necessary.
5. At the end of the thermal cycling program, add 1 µl of 0.5 M EDTA (pH 8.0) to a final concentration of 12.5 mM to inactivate the RT enzyme and halt the enzymatic reaction.

Protocol 3

Purification of first-strand cDNA by digestion with proteinase K

Reagents

- Proteinase K (20 µg/µl)
- Phenol : chloroform (1 : 1, v/v)
- 5 M NaCl
- 100% Ethanol

Method

1. To the 41 µl first-strand cDNA synthesis reaction (from *Protocol 2*), add 0.5 µl of 20 µg/µl proteinase K and incubate the reaction for 15 min at 45°C.
2. Add 40 µl of phenol : chloroform (1 : 1, v/v). Mix vigorously by vortexing to emulsify the organic and aqueous phases, and spin the sample by centrifugation (>12 000 *g*) for 5 min at 25°C to separate the phases. Transfer the top (aqueous) layer containing the cDNA–mRNA product to a fresh tube, being careful not to disturb the interface between the layers.
3. Add 2 µl of 5 M NaCl and 80 µl of 100% ethanol.
4. Centrifuge the mixture at 15 000 *g* at room temperature (20–25°C) for 20 min to pellet the cDNA–mRNA product. Remove and discard the supernatant, being careful not to disturb the pellet at the bottom of the tube. Wash the pellet twice by adding 200 µl of 80% ethanol, spinning for 1 min at 15 000 *g*, and discarding the supernatant. Dry the pellet completely using vacuum centrifugation.
5. Dissolve the dried cDNA–mRNA pellet in 44.7 µl of dH_2O. It is critical not to use Tris/EDTA or other buffers containing Tris or polyhydroxy groups that can be oxidized to diols. Buffers other than dH_2O will interfere with downstream steps.

Protocol 4

Oxidation and biotinylation of mRNA diol groups[a]

Equipment and Reagents

- 1 M Sodium acetate (pH 4.5)
- 250 mM $NaIO_4$ (prepared just prior to use)
- 80% Glycerol
- 80 and 100% Ethanol
- 10% SDS
- 5 M NaCl
- 1 M Sodium citrate (pH 6.1)
- 10 mM Biotin hydrazine (long-arm variant)
- Micropipette fitted with a fine tip

Method

1. Obtain the 44.7 µl cDNA–mRNA mixture from step 5 of *Protocol 3.* Add 3.3 µl of 1 M sodium acetate (pH 4.5) to a final concentration of 66 mM and mix well. Add 2 µl of 250 mM $NaIO_4$ (freshly prepared) to a final concentration of 10 mM and mix well.
2. Incubate the mixture on ice in the dark for 45 min to oxidize the RNA component of the cDNA–mRNA hybrids. Oxidation produces diol groups on the ribose moieties of the mRNA.
3. Stop the oxidation reaction by adding 1 µl of 80% glycerol. Mix well[b].
4. Precipitate the cDNA–mRNA mixture by adding 103 µl of 100% ethanol, 0.5 µl of 10% SDS, and 2.58 µl of 5 M NaCl. Mix well. Centrifuge at 15 000 ***g*** at room temperature (20–25°C) for 20 min to pellet the cDNA–mRNA hybrids.
5. Wash the pellet in 500 µl of 80% ethanol and centrifuge (20–25°C) at 15 000 ***g*** for 5 min. Carefully discard the supernatant.
6. Centrifuge again briefly and remove any remaining ethanol using a micropipette fitted with a fine tip. Dissolve the cDNA–mRNA hybrids in 50 µl of dH_2O.
7. Add 5 µl of 1 M sodium citrate (pH 6.1), 5 µl of 10% of SDS and 150 µl of a freshly prepared solution of 10 mM biotin hydrazine (long-arm variant) to the cDNA–mRNA hybrids. Mix the solution gently and incubate overnight (10–16 h) at room temperature (20–25°C).
8. After the biotinylation step (step 7), add 75 µl of 1 M sodium acetate (pH 6.1), 5 µl of 5 M NaCl, and 750 µl of 100% ethanol. Mix gently.
9. Recover the biotinylated cDNA–mRNA hybrids by centrifugation at 15 000 ***g*** for 20 min (20–25°C). Discard the supernatant and wash the pellet twice with 80% ethanol. Make certain to spin for 1 min (15 000 ***g***) after each wash to repellet the nucleic acids. After the second wash step, spin briefly and remove any residual ethanol using a micropipette fitted with a fine tip.
10. Dissolve the biotinylated cDNA–mRNA pellet in 180 µl of dH_2O.

Notes

[a]The RNA diol groups (cap structure and 3′ terminus) are oxidized with $NaIO_4$ and coupled with the biotin hydrazide long-arm variant biocytin hydrazide. Biotinylated cDNA–mRNA hybrids are then isolated using streptavidin-coated beads as described below.
[b]Glycerol reacts with $NaIO_4$ and quenches the reaction.

Protocol 5

Cap trapping and purification of full-length cDNAs

Equipment and Reagents

- 10× RNase I buffer (Promega)
- RNase I (10 units/μl)
- 10% SDS
- Proteinase K (20 μg/μl)
- Phenol : chloroform (1 : 1, v/v)
- Chloroform
- Yeast tRNA (10 μg/μl, DNA-free)
- 5 M NaCl
- 80 and 100% Ethanol
- Magnetic porous glass (MPG) beads
- Binding buffer (4.5 M NaCl, 50 mM EDTA, pH 8.0)
- Wash buffer 1 (0.3 M NaCl, 1 mM EDTA)
- Wash buffer 2 (0.4% SDS, 0.5 M sodium acetate, 20 mM Tris/HCl, pH 8.5, and 1 mM EDTA)
- Wash buffer 3 (0.5 M sodium acetate, 10 mM Tris/HCl, pH 8.5, and 1 mM EDTA)
- Elution buffer (50 mM NaOH, 5 mM EDTA)
- 1 M Tris/HCl (pH 7.0)
- Isopropanol
- MicroSpin S-400 HR columns (GE Healthcare)
- 5 M Ammonium acetate (pH 5.2)

Method

1. Obtain the 180 μl biotinylated cDNA–mRNA sample from step 10 of *Protocol 4*. Add 20 μl of 10× RNase I buffer and mix gently. Add 3 μl of 10 units/μl RNase I and mix gently.
2. Incubate the reaction mixture for 30 min at 37°C to degrade the mRNA strands and then place the sample on ice.
3. Add 4 μl of 10% SDS and 1 μl of 20 μg/μl proteinase K. Incubate the reaction mixture for 15 min at 45°C to degrade the RNase I enzyme.
4. After proteinase K treatment, extract the sample once with 200 μl of phenol : chloroform (1 : 1, v/v) and once with 200 μl of chloroform. Transfer the upper phase containing the nucleic acid to a new tube. Back-extract the organic phase with dH_2O and transfer the upper phase to a new tube.
5. Add 3.2 μl of 10 μg/μl yeast tRNA as carrier nucleic acid. Add 20 μl of 5 M NaCl and 800 μl of 100% ethanol. Recover the cDNA by centrifugation at 15 000 ***g*** for 20 min at 4°C.

Carefully decant and discard the supernatant. Respin briefly and remove residual ethanol using a micropipette fitted with a fine tip. Dissolve the cDNA pellet in 50 μl of dH_2O.

6. Pre-treat 300 μl of MPG beads (100 μl per 10 μg of starting total RNA) with 30 μg of 10 μg/μl DNA-free yeast tRNA. Incubate the beads for 30 min on ice with occasional mixing to keep the beads in solution.
7. Place the tube in a magnetic rack for 2–3 min to draw the magnetic beads to the bottom of the tube. Carefully remove and discard the supernatant and wash the beads three times with 300 μl of binding buffer. Suspend the magnetic beads in 300 μl of binding buffer.
8. Add 210 μl of the tRNA-treated MPG beads (step 7) to a fresh tube containing the 50 μl biotinylated first-strand cDNA sample (step 5). Incubate and mix the reaction for 10 min at 50°C using slow end-over-end rotation.
9. After the 10 min incubation, transfer the remaining 90 μl of magnetic beads to the reaction tube and incubate for an additional 20 min at 50°C.
10. Place the tube in a magnetic rack for 3 min to draw the beads to the bottom of the tube. Carefully remove and discard the supernatant and wash the beads with 300 μl per wash as follows: twice with binding buffer, once with wash buffer 1, three times with wash buffer 2, and twice with wash buffer 3.
11. After the final wash, elute the cDNA from the magnetic beads by adding 60 μl of elution buffer. Mix the beads gently by vortexing and then place the tube on a rotary wheel and mix for 5 min at 70°C.
12. Place the tube in a magnetic rack for 3 min to draw the beads to the bottom of the tube and transfer the supernatant containing the eluted cDNA to a fresh tube. Store the 60 μl cDNA sample on ice to prevent hybridization between the cDNA molecules and trace amounts of contaminating RNA that might be present.
13. Repeat the elution twice more, each time using 60 μl of of elution buffer.
14. Add 60 μl of 1 M Tris/HCl (pH 7.0) to each of the three tubes stored on ice and mix quickly to neutralize the pH.
15. Add 120 μl of phenol:chloroform (1:1, v/v) to each of the three tubes. Mix vigorously by vortexing. Centrifuge the three tubes in a microfuge for 5 min (15000 *g*) at room temperature (20–25°C) to separate the phases. Transfer the aqueous phase containing the cDNA (i.e. the upper layer) of each of the three tubes to three fresh tubes.
16. To each tube, add 24 μl of 5 M NaCl and 240 μl of isopropanol. Mix well by vortexing. Pellet the cDNAs by centrifugation in a microfuge (15000 *g*) for 20 min at 4°C. Decant and discard the supernatant, being careful not to disturb the cDNA pellet at the bottom of the tube. Spin each tube briefly to collect any remaining supernatant and remove using a micropipette fitted with a fine tip.
17. Add 1 ml of 80% ethanol to each of the three tubes and centrifuge in a microfuge (15000 *g*) for 5 min to repellet the cDNA. Decant and discard the supernatant, spin each tube briefly to collect any remaining supernatant, and remove using a micropipette fitted with a fine tip. This step removes residual salt and isopropanol from the cDNA pellets.
18. Dissolve each cDNA pellet in 50 μl of dH_2O.
19. Transfer each 50 μl cDNA sample onto a MicroSpin S-400 HR Column and centrifuge at 700 *g* for 2 min at room temperature (20–25°C). Apply 50 μl of dH_2O and repeat the centrifugation.

20. To each 100 μl of purified cDNA sample, add 100 μl of 5 M ammonium acetate (pH 5.2) and 250 μl of 100% ethanol. Centrifuge each sample in a microfuge (15 000 *g*) for 20 min to pellet the cDNA, and remove and discard the supernatant, being careful not to disturb the pellet.
21. Wash each pellet twice with 500 μl of 80% ethanol. Centrifuge in a microfuge (15 000 *g*) for 5 min at room temperature (20–25°C) after each wash. Spin each tube briefly to collect any remaining supernatant and remove residual ethanol using a micropipette fitted with a fine tip.
22. Dissolve the cDNA pellet in 2.5 μl of dH_2O.
23. Heat the 2.5 μl cDNA solution for 10 min at 65°C and place each tube on ice to prevent intra- and intermolecular annealing.

Protocol 6

Adding a priming site to the 5′ end of first-strand cDNA molecules

Equipment and Reagents

- 0.4 μg/μl Linker mixture[a]
- DNA Ligation Kit Solution II, Concatenation Buffer (TaKaRa)
- DNA Ligation Kit Solution I, Enzyme Solution (TaKaRa)
- 0.5 M EDTA
- 10% SDS
- Proteinase K (10 μg/μl)
- Phenol : chloroform (1 : 1, v/v)
- Chloroform
- Column buffer (10 mM Tris/HCl, pH 8.0, 1 mM EDTA, and 100 mM NaCl)
- MicroSpin S-400 HR column (GE Healthcare)
- 5 M NaCl
- 80 and 100% Ethanol

Method

1. To a tube, add 1.5 μl of linker mixture and 3.5 μl of dH_2O. Mix well, incubate at 37°C for 5 min, and transfer onto ice.
2. To the 5.0 μl linker solution on ice, add 2.5 μl of heat-treated cDNA (*Protocol 5*, step 23), 7.5 μl of DNA Ligation Kit Solution II, and 15 μl of DNA Ligation Kit Solution I.
3. Incubate the reaction overnight at 16°C to allow ligation to proceed.
4. Stop the reaction by adding 10 μl of H_2O, 1 μl of 0.5 M EDTA, 1 μl of 10% SDS and 1 μl of 10 μg/μl proteinase K. Incubate the solution for 15 min at 45°C.
5. Extract the solution once with phenol : chloroform (1 : 1, v/v) and once with chloroform. Back-extract the organic phases with 40 μl of column buffer.

6. Transfer the 83 µl of ligation sample onto a MicroSpin S-400 HR column and centrifuge at 700 *g* for 2 min at room temperature (20–25°C) to remove excess linkers and purify the ligation products.
7. To the 80 µl of purified ligation products, add 4 µl of 5 M NaCl and 160 µl of 100% ethanol. Centrifuge the mixture in a microfuge at full speed (15 000 *g*) for 10 min and remove and discard the supernatant.
8. Gently wash the ligated cDNA pellet twice with 800 µl of 80% ethanol, centrifuging in a microfuge at full speed (15 000 *g*) for 3 min after each wash to make sure that the ligated cDNA pellet remains adhered to the bottom of the tube.
9. Dissolve the ligated cDNA pellet in 18.5 µl of dH_2O.

Note

[a]The linker mixture is a 0.4 µg/µl equimolar solution of four oligonucleotides:
5′-ACTAATACGACTCACTATAGGNNN-3′, 5′-TGATTATGCTGAGTGATATCC-3′,
5′-ACTAATACGACTCACTATAGGGNN-3′, and 5′-TGATTATGCTGAGTGATATCC-3′.

Protocol 7

Synthesis of second-strand cDNA

Equipment and Reagents

- T7 promoter primer (1 µg/µl)
- 10× Reaction buffer (TaKaRa)
- 10 mM dNTP solution
- 0.1 M DTT
- 4 units/µl DNA polymerase I (TaKaRa)
- 0.5 M EDTA
- 10% SDS
- Proteinase K (10 µg/µl)
- Phenol : chloroform (1 : 1, v/v)
- Chloroform
- MicroSpin S-200 HR column (GE Healthcare)
- 5 M NaCl
- 80 and 100% Ethanol

Method

1. To the 18.5 µl of ligated cDNA sample (*Protocol 6*, step 9), add 0.4 µl of 1 µg/µl T7 promoter primer.
2. Incubate for 10 min at 65°C to denature the secondary structure and for 5 min on ice to allow annealing of the primer to the ligated cDNA templates.
3. Add the following to the primed cDNA template mixture: 3 µl of 10× reaction buffer, 1.0 µl of 10 mM dNTP solution, 2.3 µl of 0.1 M DTT, and 4.5 µl of 4 units/µl DNA polymerase. Incubate at 16°C for 2 h.

4. Stop the reaction by adding 10 μl of dH_2O, 1 μl of 0.5 M EDTA, 1 μl of 10% SDS and 1 μl of 10 μg/μl proteinase K. Incubate the solution for 15 min at 45°C.
5. Extract the double-stranded cDNA sample once with phenol : chloroform (1 : 1, v/v) and once with chloroform. Back-extract the organic phases with 40 μl of dH_2O. Combine the aqueous samples to give 80 μl of total sample volume.
6. Transfer the 80 μl of double-stranded cDNA sample onto a MicroSpin S-200 HR column and centrifuge at 700 *g* for 2 min at room temperature (20–25°C) to remove contaminants.
7. To the 80 μl of purified double-stranded cDNA sample, add 2 μl of 5.0 M NaCl and 160 μl of 100% ethanol. Centrifuge the sample in a microfuge at full speed (15 000 *g*) for 20 min at room temperature (20–25°C) and remove and discard the supernatant.
8. Very gently wash the pellet twice with 500 μl of 80% ethanol. Centrifuge the sample in a microfuge at full speed (15 000 *g*) for 5 min at room temperature (20–25°C) after each wash to repellet the double-stranded cDNA at the bottom of the tube. Remove any residual ethanol using a micropipette fitted with a fine tip.
9. Dissolve the double-stranded cDNA in 7.31 μl of dH_2O.

Protocol 8

cRNA amplification from double-stranded cDNA templates

Reagents

- CUGA 7 *in vitro* transcription kit (Nippongene)
- 5× Reaction buffer (Nippongene)
- 100 mM each ATP, UTP, GTP, and CTP (Nippongene)
- CUGA T7 RNA polymerase (Nippongene)
- 10 mM Cy3–CTP or Cy5–CTP (Perkin Elmer)
- RNeasy kit (Qiagen)

Method

1. To the 7.31 μl double-stranded cDNA sample (*Protocol 7*, step 9), add 4 μl of 5× reaction buffer, 1.5 μl each of 100 mM ATP, UTP, and GTP, and 1.31 μl of 100 mM CTP. Mix well by gentle vortexing.
2. Add 1 μl of CUGA T7 RNA polymerase and mix gently by vortexing.
3. Add 1.88 μl of 10 mM of Cy3–CTP or Cy5–CTP and mix gently by vortexing. Incubate for 2 h at 37°C to allow cRNA synthesis.
4. Purify the cRNA using an RNeasy kit[a,b] according to the Agilent manual.

Notes

[a]Use the RNeasy kit (Qiagen) according to the instructions provided by Agilent in their user manual (http://www.chem.agilent.com/).
[b]Microarray hybridization and scanning procedures are performed according to the instructions of Agilent in their user manual (http://www.chem.agilent.com/).

3. TROUBLESHOOTING

The protocols for mRNA isolation and labeling for TSS microarrays have been highly optimized. This section briefly identifies some of the potential pitfalls of the above protocols and provides suggestions for avoiding the more challenging experimental aspects.

Protocols 1–5 and *8* require extensive manipulation of cellular RNA, which is enzymatically and chemically unstable. For the best results, it is imperative to use a fastidious experimental technique for every RNA-oriented protocol. Gloves should be worn at all times and all reagents, solutions, and laboratory hardware should be RNase-free. RNA degradation usually manifests in the form of reduced RNA or cDNA yields or truncated product length. If either is observed, samples, kit components, and the laboratory should be tested for RNA contamination and decontaminated where necessary. When working with RNA, extremes of temperature and prolonged periods at room temperature should also be avoided. RNA samples should be stored at –80°C and thawed just prior to use.

A second potential pitfall pertains to the fact that many of the protocols require precipitation and centrifugation of the nucleic acid products, resulting in DNA and RNA pellets that can be fragile and dislodged if care is not taken during handling. In particular, when washing DNA and RNA pellets, care should be taken to decant the supernatants carefully. In some cases, pellets will be visible with the naked eye and should be observed throughout the washing steps to ensure that they do not become dislodged and discarded. For the best results, pellets should also be as free as possible from contaminating ethanol and isopropanol used for the precipitations and washes. Residual alcohol can be removed using a fine pipette tip or by vacuum centrifugation.

Acknowledgements

We thank Yuki Tsujimura and Noriko Niniomiya for their technical assistance. This study was supported courtesy of MEXT Research Grants for the RIKEN Genome Exploration Research Project, for the Advanced and Innovational Research Program in Life Science and for National Projects on Genome Network Analysis to Y.H.

Note added in proof

After finishing this chapter, further developments in the field of labeling methods have emerged. The basic concept and the technical developments are the same as in this manuscript, with some improvements. The improved method was shown to be a more effective RNA labeling method and subsequently we adjusted our custom arrays (TSS arrays) to this and tested it thoroughly. These additional data are now published (**Katayama S, Kanamori-Katayama M, Yamaguchi K, Carninci P, Hayashizaki Y.** (2007) *Genome Biol.* **8**: R42).

4. REFERENCES

★ 1. **Okazaki Y, Furuno M, Kasukawa T, *et al.*** (2002) *Nature*, **420**, 563–573. – *An important publication providing a comprehensive annotation of the mouse transcriptome.*

★★★ 2. **Carninci P, Kasukawa T, Katayama S, *et al.*** (2005) *Science*, **309**, 1559–1563. – *This study provides a comprehensive view of transcription start and stop sites in the mouse genome.*

3. **Katayama S, Tomaru Y, Kasukawa T, *et al.*** (2005) *Science*, **309**, 1564–1566.

4. **Suzuki Y, Yamashita R, Sugano S & Nakai K** (2004) *Nucleic Acids Res.* **32**, D78–D81.

5. **Bajic VB, Tan SL, Suzuki Y & Sugano S** (2004) *Nat. Biotechnol.* **22**, 1467–1473.

★★ 6. **Shiraki T, Kondo S, Katayama S, *et al.*** (2003) *Proc. Natl. Acad. Sci. U.S.A.* **100**, 15776–15781. – *First publication describing CAGE.*

7. **Kornblihtt AR** (2005) *Curr. Opin. Cell Biol.* **17**, 262–268.

8. **Landry JR, Mager DL & Wilhelm BT** (2003) *Trends Genet.* **19**, 640–648.

9. **Fossat N, Courtois V, Chatelain G, Brun G & Lamonerie T** (2005) *Dev. Dyn.* **233**, 154–160.

10. **Walsh NC, Cahill M, Carninci P, *et al.*** (2003) *Gene*, **307**, 111–123.

11. **Benson DA, Karsch-Mizrachi I, Lipman DJ, Ostell J & Wheeler DL** (2005) *Nucleic Acids Res.* **33**, D34–D38.

12. **Pruitt KD, Tatusova T & Maglott DR** (2005) *Nucleic Acids Res.* **33**, D501–D04.

13. **Hubbard T, Andrews D, Caccamo M, *et al.*** (2005) *Nucleic Acids Res.* **33**, D447–D453.

14. **Harbers M & Carninci P** (2005) *Nat. Methods*, **2**, 495–502.

15. **Waterston RH, Lindblad-Toh K, Birney E, *et al.*** (2002) *Nature*, **420**, 520–562.

16. **Kent WJ** (2002) *Genome Res.* **12**, 656–664.

17. **Florea L, Hartzell G, Zhang Z, Rubin GM & Miller W** (1998) *Genome Res.* **8**, 967–974.

18. **McGinnis S & Madden TL** (2004) *Nucleic Acids Res.* **32**, W20–W25.

19. **Carninci P, Kvam C, Kitamura A, *et al.*** (1996) *Genomics*, **37**, 327–336.

20. **Carninci P, Westover A, Nishiyama Y, *et al.*** (1997) *DNA Res.* **4**, 61–66.

★★ 21. **Carninci P & Hayashizaki Y** (1999) *Methods Enzymol.* **303**, 19–44. – *An important publication describing the experimental basis for the biotinylated cap-trapper approach.*

22. **Carninci P, Nishiyama Y, Westover A, *et al.*** (1998) *Proc. Natl. Acad. Sci. U.S.A.* **95**, 520–524.

23. **Kane MD, Jatkoe TA, Stumpf CR, Lu J, Thomas JD & Madore SJ** (2000) *Nucleic Acids Res.* **28**, 4552–4557.

★ 24. **Suzuki H, Okunishi R, Hashizume W, *et al.*** (2004) *FEBS Lett.* **573**, 214–218. – *This key paper describes a medium-scale real-time RT-PCR approach to quantitatively measuring gene expression.*

CHAPTER 6

Methods for increasing the utility of microarray data

Kazuro Shimokawa, Rimantas Kodzius, Yonehiro Matsumura, and Yoshihide Hayashizaki

1. INTRODUCTION

The use of DNA microarrays for comprehensive and quantitative expression measurements has progressed rapidly (1). In terms of cost per measurement, this technique is vastly superior to other methods such as Northern blotting or quantitative reverse transcriptase polymerase chain reaction (QRT-PCR) (2, 3). However, the output values of DNA microarrays are not always highly reliable or accurate compared with other techniques, and the output data sometimes consist of measurements of relative expression (treated sample vs untreated) rather than absolute expression values as desired. In effect, some measurements from some laboratories do not represent absolute expression values (such as the number of transcripts) and as such are experimentally deficient. Here, we explore computational approaches to microarray data accuracy, and present two principally different methods that address the problem of accuracy and absolute value, allowing researchers to increase the utility of microarray data.

The first approach addresses one problem in some microarray data: the absence of accurate measurements. Spot reliability evaluation score for DNA microarrays (SRED) (4) offers a reliability value for each spot in the microarray. SRED does not require an entire microarray to assess the reliability, but rather analyzes the reliability of individual spots of the microarray. The calculation of a reliability index can be used for different microarray systems, which will facilitate the analysis of multiple microarray datasets from different experimental platforms. To address the second problem that some microarray datasets fail to reflect the number of mRNA molecules sufficiently in a given sample (i.e. fail to provide absolute expression levels), additional methods are required. We provide a

DNA Microarrays: *Methods Express* (M. Schena, ed.)

new method for converting microarray data to absolute expression values with the use of external data (5) such as expressed sequence tags (ESTs) and cap analysis of gene expression (CAGE) tags (6, 7).

2. METHODS AND APPROACHES

2.1 Spot reliability evaluation

The abundance of small-scale microarray experiments using various experimental platforms has yielded large quantities of data and rendered database deposition of results an important process for the entire research community. Expression data are usually associated with supplementary data such as RNA sample information, experimental methods used, and so forth. When storing expression data in databases, the concept of storing accompanying detailed experimental conditions is important. Access to such information and the subsequent integration of the microarray measurement data are necessary for detailed analysis of any microarray data set. The Microarray Gene Expression Data (MGED) Society have defined a standard known as the Minimum Information About a Microarray Experiment (MIAME) (8). MIAME is a document that outlines the minimum information that should be reported with a microarray experiment to ensure its unambiguous interpretation and reproduction. When submitting articles that include microarray experiments, a database submission of the microarray data adherent to this standard is typically a strict requirement. In effect, the number of publicly available datasets has increased dramatically in recent years and the need for tools and methods to compare and integrate these data sets has grown equally quickly.

In this context, the SRED reliability index was developed to offer the end user an intelligible metric by which to estimate the accuracy of expression values for each spot in a DNA microarray. This method estimates the accuracy of expression data by applying multivariate analysis, using multiple data sources originating from the actual microarray spot measurements. The SRED score is calculated principally using multivariate discriminant analysis. The analysis parameters are chosen by comparing the difference between expression values derived from the DNA microarray experiment and corresponding measurements obtained by QRT-PCR; consequently, SRED is calculated for each spot (gene) in the DNA microarray not validated by QRT-PCR.

An implicit assumption when calculating SRED scores is that QRT-PCR gives accurate reference expression values. This assumption is likely to be true for most genes. QRT-PCR is a highly sensitive and specific method that is widely used to measure gene expression and utilizes several distinct chemical reactions for detection (2, 3). For increased accuracy, primers for QRT-PCR can be selected from the RTPrimerDB public database (9) using the following electronic link (http://medgen.ugent.be/rtprimerdb/).

One way to represent spot intensity values from a DNA microarray is as the expression ratio of two samples such as $\log_2(Cy3/Cy5)$, where the Cy3 and Cy5 values represent the measured intensities of each sample. SRED requires that the corresponding values of QRT-PCR be calculated in the same form. The difference, R, between microarray and QRT-PCR is given by the following equation:

$$R(t_a, c_a, t_p, c_p) = |\log_2(t_a/c_a) - \log_2(t_p/c_p)| \quad (1)$$

In Equation (1), t_a is the intensity of microarray test data, c_a is the corresponding control value of the spot, t_p is the intensity of QRT-PCR sample, and c_p is the corresponding QRT-PCR control intensity.

It is important to consider the variance in the measurements between different technologies. For instance, we have found previously that the standard deviation of expression ratios derived from QRT-PCR was 1.8 times larger than that of microarray experiments (4). It is therefore necessary to normalize these two values when calculating the difference. The normalized difference, R_n, is calculated as in Equation (2):

$$R_n(t_a, c_a, t_p, c_p) = \left| \frac{|\log_2(t_a/c_a) - \overline{\log_2(t_a/c_a)}|}{\sigma_a} - \frac{|\log_2(t_p/c_p) - \overline{\log_2(t_p/c_p)}|}{\sigma_p} \right| \quad (2)$$

In Equation (2), σ_a is the standard deviation of the microarray data, σ_p is the standard deviation of the QRT-PCR data, t_a is the intensity of microarray test data, c_a is the corresponding control value of the spot, t_p is the intensity of QRT-PCR sample, and c_p is the corresponding QRT-PCR control intensity. As an aside, we do not consider the nonlinearity of the expression ratio between the techniques using this approach.

As mentioned above, SRED is calculated using multivariate discriminant analysis. The SRED score is defined as the probability $P(t)$ that the difference between the relative expression values of each spot (gene) determined using the DNA microarray or QRT-PCR is less than a factor of 2 (this condition is synonymous with $R_n \leq 1$). The reliability of spot intensities is evaluated using parameters obtained from the microarray experiment itself. We considered nine possible different parameters from the microarray experiment (summarized in *Table 1*). The most suitable combination of parameters for predicting the experimental result was decided by applying multivariate discriminant analysis to all of the possible combinations. In our case, a discriminant function using two parameters, 7 and 8 (see *Table 1*), proved to show the greatest efficiency and sensitivity when the threshold value was set to $R_n = 1$ (i.e. a twofold difference between microarray spot intensity and QRT-PCR expression value) (4).

2.2 Theoretical background of the method

In the process of defining the SRED scores, we employed a combination of two approaches to improve predictive accuracy: Mahalanobis distances and Bayesian methods. In general, discriminant function analysis is used to determine variables that discriminate between two or more naturally occurring groups (usually called source groups). Conceptually, the distance

Table 1. Candidate parameters for discriminant analysis obtained from DNA microarray data

Parameter	Category	Description
1	Signal intensity of each microarray spot	Average of replicate signal intensities of channel 1 (e.g. Cy3) and channel 2 (e.g. Cy5)
2		Minimum signal intensity of channel 1 (e.g. Cy3) and channel 2 (e.g. Cy5)
3	Difference between the signal intensity and background intensity of each microarray spot	Average of the difference between the signal intensity and the background intensity of channel 1 (e.g. Cy3) and channel 2 (e.g. Cy5)
4		Minimum difference between the signal intensity and the background intensity of channel 1 (e.g. Cy3) and channel 2 (e.g. Cy5)
5	Ratio of signal intensity to backvground intensity of each spot	Average of the ratio of the signal intensity to the background intensity of channel 1 (e.g. Cy3) and channel 2 (e.g. Cy5)
6		Minimum value of the ratio of the signal intensity to the background intensity of channel 1 (e.g. Cy3) and channel 2 (e.g. Cy5)
7		Reproducibility of the measurement value for each spot calculated as the difference between replicate experiments: $\log_2(t_{array1}/c_{array1}) - \log_2(t_{array2}/c_{array2})$, where t_{array1} is the intensity of test data from microarray 1, c_{array1} is the corresponding control value of the spot from microarray 1, t_{array2} is the intensity of test data from microarray 2, and c_{array2} is the corresponding control value of the spot from microarray 2
8		Relative expression value of each gene calculated as: $\log_2(t_{array}/c_{array})$
9		Overall reproducibility of the microarray (Pearson's correlation coefficient of duplicate experiments)

between each sample and the center of every source group is computed in the multi-dimensional space described by the data properties, which in this case are parameters 7 and 8 in *Table 1*, with the sample belonging to the closest source group. In this case, we used discriminant analysis to determine which (source) group ($R_n \leq 1$ or $R_n > 1$) a spot (sample) most likely belongs to. If Mahalanobis distances are used, the probability of the sample belonging to any source group can be calculated, as probabilities are inversely proportional to the Mahalanobis distances. In our case, Mahalanobis distances (a nonlinear distance model) perform better than the linear distance model. *Fig. 1* shows a graphical example of discriminant analysis using Mahalanobis distances.

In order to incorporate prior knowledge of the distribution of spots in the two groups, we use a Bayesian approach. Using the QRT-PCR and cDNA microarray test data as in Equation (2), we can obtain a prior probability, $P(t)$, describing how likely a microarray spot is to be within twice the value of the corresponding QRT-PCR measurement. The term

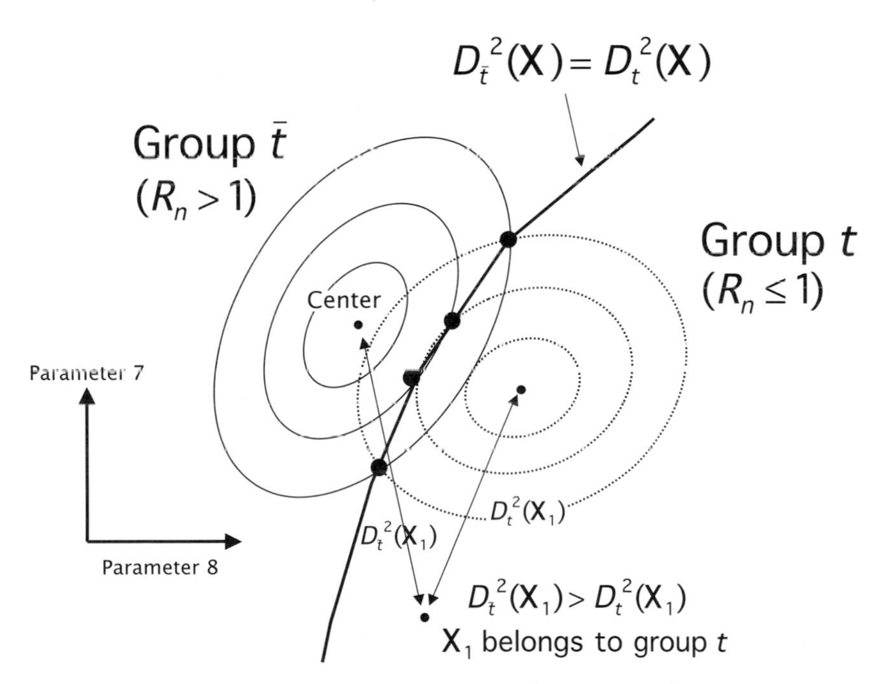

Figure 1. Graphical visualization of discriminant analysis using Mahalanobis distance. It is possible to determine the grouping for the spot data X_1 by calculating the distance between X_1 and the center of each group in the space set using the chosen parameters. In the implementation described in this chapter, the distance is used for the subsequent calculation of the posterior probability that X_1 belongs to a certain group (see Equation 3).

t denotes a subset of the group of the spot that satisfies $R_n \leq 1$, and $\bar{t}$ corresponds to $R_n > 1$. The prior probability enables us to calculate a posterior probability $P(t|\mathbf{X})$: the probability that a spot that has the parameter vector $\mathbf{X}$ satisfies $R_n \leq 1$. The posterior probability will be the final SRED score (the spot reliability). For the final calculation of SRED scores, we combine both methods to obtain Equation (3):

$$\begin{cases} d_t^2(\mathbf{X}) = (\mathbf{X} - \mathbf{m}_t)' S_p^{-1} (\mathbf{X} - \mathbf{m}_t) \\ D_t^2(\mathbf{X}) = d_t^2(\mathbf{X}) - 2\ln(q_t) \\ p(t \mid \mathbf{X}) = \exp(-\frac{D_t^2(\mathbf{X})}{2}) / \sum_u \exp(-\frac{D_u^2(\mathbf{X})}{2}) \end{cases} \quad (3)$$

In Equation (3), $\mathbf{X}$ is a vector of values corresponding to the parameters in *Table 1*, t is a subset of the group of test data for which the difference of the expression intensities of microarray and QRT-PCR is less than twofold, $d_t^2(\mathbf{X})$ is the squared Mahalanobis distance from $\mathbf{X}$ to group t, m_t is a vector containing variable means in group t, S_p is the pooled covariance matrix in group t, $D_t^2(\mathbf{X})$ is the generalized squared distance from $\mathbf{X}$ to group t, q_t is the prior probability of membership in group t, and $p(t|\mathbf{X})$ is the posterior probability of an observation $\mathbf{X}$ belonging to group t, $u = \{t, \bar{t}\}$.

Both Mahalanobis distance and the Bayesian approach are well-known methods that have been used previously for multivariate discriminant analysis. For a deeper understanding, we refer interested readers to more comprehensive publications on these subjects (e.g. 10, 11). For convenience, statistical software packages such as R (12) or SYSTAT (Systat Software Inc.) are recommended for use in calculating SRED scores. In one implementation, SRED scores were assigned to approximately 1 500 000 spots in the Riken Expression Microarray Database (READ) (13). It is important to emphasize that the parameters used for the SRED calculation are specific to this microarray system. However, it is possible to apply SRED to other microarray systems by readjusting the parameters.

2.3 Calculating absolute expression values

Most microarray experiments utilize either single- or dual-color labeling and detection approaches. Both methods have advantages and disadvantages. The use of dual-color labeling produces two probe mixtures having distinct labels (e.g. Cy3 and Cy5), allowing the measurement of expression ratios reliably by competitive hybridization. In the two-color approaches, one probe mixture typically serves as reference and is derived, for example, from all of the transcripts represented on the microarray. Some measurements that provide relative expression levels between two samples (two-probe mixtures) may not provide absolute expression values. The use of single-color labeling eliminates differences in hybridization efficiency seen in most dual-color approaches. Differences in hybridization efficiency lead to a loss of quantification and artifacts in ratiometric data. Single-color approaches also generally allow a simpler experimental set-up and represent the predominant approach used by commercial microarray providers.

Single-color labeling methods primarily allow direct measurement of absolute expression values using pre-calibration or exogenous controls (14, 15). However, these efforts can be limited by insufficient information available on the reference data. These limitations can be partially overcome by integration with external data obtained by different experimental methods such as EST sequencing (16), serial analysis of gene expression (SAGE) (17) or the novel CAGE method (6, 7). The tags produced by these methods can be used to provide absolute expression values for every sample used on a DNA microarray, with the units represented in transcripts per million (t.p.m.). By way of example, we present here the calculation of absolute expression values for mouse transcripts in READ using quantitative CAGE and EST tag data (18). The READ database contains expression information for 50 mouse tissues, where dual-labeled relative gene expression levels are shown using the expression levels obtained from mouse embryo E17.5 mRNA as the reference sample (19). E1.5 mRNA is derived from whole body, mixed-sex mouse embryo tissue taken at mouse embryonic day 17.5. Using the absolute expression values of the E17.5 mRNA sample, the READ values of the mRNA samples from the 50 tissues can be converted into absolute values.

Both CAGE and EST data are independent of microarray data and have different data properties. CAGE and EST sequencing technologies involve the sequencing and mapping of transcripts (tags) to the genome. In order to link external EST and CAGE data to the cDNA targets used for microarray analysis, we used the FANTOM representative transcript set (RTS) based on RIKEN cDNAs and associated transcriptional unit (TU) definitions, as described previously (18). Briefly, EST and CAGE tag sequences were mapped to the mouse genome and then linked to unique TUs by identifying the closest TU within a 10 kb window. The cDNA microarray targets are generally based on RIKEN cDNA clones and also have an annotated TU.

The cDNA library made from E17.5 mRNA contains 49 806 5′-ESTs grouped into 7164 unique TUs by RTS. In this way, the correspondence between sequenced tags and READ clone IDs used for the microarray analysis was established. With this pre-processing, each sequence tag and microarray spot is annotated with a corresponding TU identifier. It is then possible to count the number of tags per TU and multiply those by the corresponding READ expression value to obtain the conversion to absolute t.p.m. values as shown in Equation (4):

$$S_{Array\_relative}(TU_x) = \log_2 \left[\frac{S_{1Array}(TU_x)}{S_{2Array}(TU_x)}\right]$$
$$S_{1Array\_TPM}(TU_x) = S_{2CAGE\_TPM}(TU_x) \cdot 2^{S_{Array\_Relative}(TU_x)} \quad (4)$$

In Equation (4), $S_{1Array\_TPM}(TU_x)$ is the t.p.m. that corresponds to a specific TU_x in sample 1, $S_{2CAGE\_TPM}(TU_x)$ is the CAGE or EST expression value that corresponds to a specific TU_x in sample 2, and $S_{Array\_relative}(TU_x)$ is the relative expression value in each microarray spot that corresponds to a specific TU_x. In the case of absolute expression for READ (5), the relative expression values are obtained from the READ database. Sample 2 is always the mouse E17.5 library and sample 1 is one of the mRNA samples from the 50 tissues used in READ.

2.4 Confirming absolute expression data conversions

Confirmation of the approach presented in section 2.3 is described in detail in a previous publication (5). A short description of the outcome is included here as it provides additional insights. Once relative microarray expression values are converted to absolute values, it is possible to compare the converted dataset directly with other externally obtained data including CAGE, EST, or SAGE data not used in the conversion procedure. This can be used to verify the conversion efficiency and accuracy.

To confirm the converted absolute expression values, publicly available expression data from SAGE and EST databases were used as a control set for direct comparison as described previously (5). *Table 2* shows the library source, total tags, TUs, and the correlation between the converted READ data and the control data set. We noticed that as the number of tags contained in the libraries increases, a higher correlation can be observed between the libraries (see *Fig. 2*). For example, the CAGE cerebellum library has the highest number of tags (327 178) and the highest correlation of READ absolute values (0.699). Thus, the number of CAGE and EST tags used in sample 2 (see Equation 3) is important for both the accuracy of the absolute data and the detection of rare transcripts. To improve the accuracy of the absolute expression values, TUs with few tags should be ignored before applying Equation (3). However, as the system may fail to detect rare transcripts because of this operation, this is a trade-off between specificity and sensitivity. In our case, we ignored TUs that had less than three tags mapped.

Table 2. Summary of the library and database sources, tags, TUs, and the correlation between the converted READ data and the control data set

Data were reproduced from (5) with permission.

Library or database source	Total entries[a]	Mapped entries[b]	Number of TUs[c]	Number of shared tags[d]	Correlation coefficient[e]
READ database	57931	57931	22406	22406	–
CAGE E17.5	86555	36284	7507	6229	–
CAGE brain	42349	10273	3617	3023	0.689
CAGE cerebellum	327178	123387	14227	10874	0.699
SAGE kidney	12154	1168	205	172	0.562
EST E17.5	49806	47493	7164	6284	–
EST placenta	5347	3727	1049	936	0.592
EST cerebellum	6409	4667	2266	2062	0.529
E17.5 standard	136361	83777	10675	8845	–
RTS	42690	42690	42690	22406	–

[a]Total number of tags or sequences.
[b]Total number of tags or sequences mapped to a representative transcript set (RTS) transcription unit (TU).
[c]Total number of TUs is the number of TUs obtained after grouping into RTSs.
[d]Total number of shared tags indicates the number of sequence tags that are shared with the RIKEN Expression Microarray Database (READ) TUs.
[e]Correlation coefficient indicates the correlation coefficient calculated from the READ absolute values and the corresponding library used for validation.

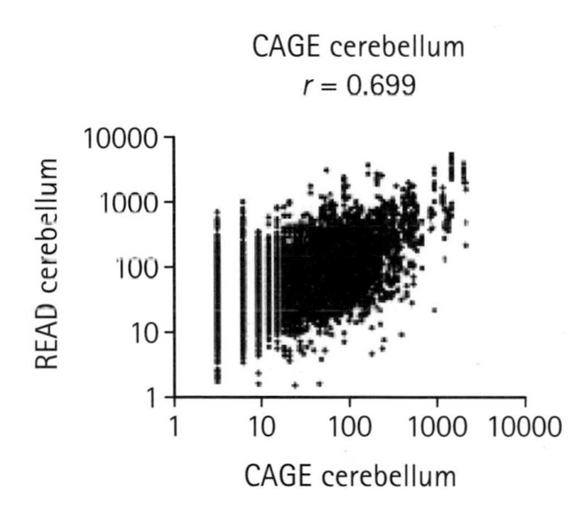

Figure 2. Scatter plot and unique transcript distributions for the E17.5 and cerebellum libraries.
Given the high correlation, transcript expression data obtained by DNA microarray, absolute READ, and CAGE methods can be compared directly. Data were reproduced from (5) with permission.

2.5 Conclusions

It is likely that many important biological observations are hidden and as yet undiscovered in the wealth of data available from microarray experiments. The need for tools to enhance the value of data collections is considerable, particularly methods that can standardize data and make data sets comparable. This chapter discusses two such methods. These can be used for both critical reanalysis of older sets, and new dataset comparison and standardization, as well for comparisons of microarray data sets with data from other experimental techniques. These computational tools allow viewing of the data from multiple angles to gain new biological insights. The concepts presented here have been shown to work satisfactorily (4, 5), and can be improved further, both from a theoretical standpoint (e.g. improved algorithms) and by incorporating new types of data.

2.6 Recommended protocols

Protocol 1

Outline of the calculation of SRED scores for microarray data

Equipment

- Personal computer running Windows 2000 or newer version with at least an Intel Pentium IV CPU, 3.6 GHz processor, and 800 MHz front side bus

Method

1. For each spot or gene element in the microarray, collect all of the parameter data listed in *Table 1*.
2. Prepare as many QRT-PCR control measurements corresponding to as many microarray spots as possible. In the READ example, 133 QRT-PCR controls were used.
3. Place each microarray spot for which there is a corresponding QRT-PCR measurement into one of two groups according to Equation (2), $R_n \leq 1$ or $R_n > 1$.
4. Find the most suitable combination of parameters to explain the experimental results (i.e. which group a given microarray spot belongs to) by applying discriminant analysis.
5. Calculate SRED scores for all of the microarray spots using the parameters obtained in step 4 and Equation (3).

Protocol 2

Outline of the calculation of absolute expression values

Equipment

- Personal computer running Windows 2000 or newer version with at least an Intel Pentium IV CPU, 3.6 GHz processor, and 800 MHz front side bus

Method

1. Establish a bioinformatic link between external, absolute expression data such as CAGE, EST, or SAGE tags and each microarray spot or element. This was done using genome mapping in the example provided in section 2.
2. Given the annotations established in step 1, calculate absolute expression values using Equation (4).
3. (Optional) Ideally, an external data set should be used for confirmation, especially if the absolute values will be used to evaluate many different microarrays.

3. TROUBLESHOOTING

The concepts and methods presented here have been shown to work well on bona fide microarray data (4, 5). *Protocols 1* and *2* are intended as guides on how to apply the approaches. Several potential difficulties can be experienced in the course of applying the methods described. Provided below are some helpful troubleshooting tips.

Our experience is that superior microarray data produce superior results. Endeavor to input the highest-quality microarray data possible, taking care to manufacture and hybridize the microarrays using the most rigorous scientific procedures. Poorly printed microarrays and low-quality samples will produce inferior raw data, which will negatively affect the downstream computational processes. Superior robotics, printing technology, surface chemistry, target and probe preparation, and other molecular aspects will produce superior data for analysis.

In advance of calculating SRED values, it is advantageous to obtain as many QRT-PCR control measurements as possible. If fewer than 50 control measurements are used and the data seem dubious, it makes sense to obtain additional QRT-PCR measurements. The parameter data for calculating SRED scores can be improved further by using improved algorithms and by adding new types of external data to the analyses.

Acknowledgements

We would like to thank Albin Sandelin, and Ann Karlsson for help with editing. This work was supported (in part) by a grant from the Genome Network Project from the Ministry of Education, Culture, Sports, Science and Technology, Japan. R.K. was supported courtesy of an FP5 INCO2 to JAPAN fellowship from the European Union.

4. REFERENCES

1. **Duyk GM** (2002) *Nat. Genet.* **32** (Suppl.), 465–468.
2. **Giulietti A, Overbergh L, Valckx D, Decallonne B, Bouillon R & Mathieu C** (2001) *Methods,* **25**, 386–401.
3. **Bustin SA** (2002) *J. Mol. Endocrinol.* **29**, 23–39.
4. ★★★ **Matsumura Y, Shimokawa K, Hayashizaki Y, Ikeo K, Tateno Y & Kawai J** (2005) *Gene,* **350**, 149–160. – *Original publication describing SRED.*
5. ★★★ **Kodzius R, Matsumura Y, Kasukawa T, *et al.*** (2004) *FEBS Lett.* **559**, 22–26. – *Original publication describing the calculation of absolute expression values.*
6. **Shiraki T, Kondo S, Katayama S, *et al.*** (2003) *Proc. Natl. Acad. Sci. U.S.A.* **100**, 15776–15781.
7. **Kodzius R, Kojima M, Nakamura M, *et al.*** (2006) *Nat. Methods,* **3**, 211–222.
8. **Ball C, Brazma A, Causton H, *et al.*** (2004) *Microbiology,* **150**, 3522–3524.
9. **Pattyn F, Speleman F, De Paepe A & Vandesompele J** (2003) *Nucleic Acids Res.* **31**, 122–123.
10. **SAS/STAT User's Guide** (2000) STATS Publishing Inc.
11. **Huberty C** (1994) *Applied Discriminant Analysis.* Wiley Interscience, New York
12. **Ihaka R & Gentleman R** (1996) *J. Comput. Graph. Stat.* **5**, 299–314.
13. ★★ **Bono H, Kasukawa T, Hayashizaki Y & Okazaki Y** (2002) *Nucleic Acids Res.* **30**, 211–213. – *The cDNA microarray database used to calculate SRED.*
14. **Dudley AM, Aach J, Steffen MA & Church GM** (2002) *Proc. Natl. Acad. Sci. U.S.A.* **99**, 7554–7559.
15. **Carter MG, Sharov AA, VanBuren V, *et al.*** (2005) *Genome Biol.* **6**, R61.
16. **Banfi S, Guffanti A & Borsani G** (1998) *Trends Genet.* **14**, 80–81.
17. **Velculescu VE, Zhang L, Vogelstein B & Kinzler KW** (1995) *Science,* **270**, 484–487.
18. ★★ **Kasukawa T, Katayama S, Kawaji H, Suzuki H, Hume DA & Hayashizaki Y** (2004) *Genomics,* **84**, 913–921. – *The preferred protocol for connecting cDNA microarray data to CAGE data.*
19. **Miki R, Kadota K, Bono H, *et al.*** (2001) *Proc. Natl. Acad. Sci. U.S.A.* **98**, 2199–2204.

CHAPTER 7

Key features of bacterial artificial chromosome microarray production and use

Timon P.H. Buys, Ian M. Wilson, Bradley P. Coe, William W. Lockwood, Jonathan J. Davies, Raj Chari, Ronald J. DeLeeuw, Ashleen Shadeo, Calum MacAulay, and Wan L. Lam

1. INTRODUCTION

Somatic DNA copy number alterations are hallmarks of cancer, leading to disruptions in the expression of oncogenes and/or tumor suppressor genes, whilst constitutional DNA copy number variations have been associated with developmental disorders (1, 2). Identification of these alterations will impact disease susceptibility characterization, disease subclassification, and treatment suitability. It will also lead to the development of novel prognostic and diagnostic markers and provide new therapeutic targets.

1.1 Surveys of DNA copy number changes

Comprehensive analysis of genetic alterations requires high-resolution techniques (3–5). Initial microarray comparative genomic hybridization (CGH) experiments were performed on cDNA microarrays (6); however, cDNA targets lack introns that are present in genomic probe mixtures, resulting in relatively low signal-to-noise ratios. Throughout both this chapter and the rest of this book, the term 'target' refers to elements displayed on the microarray, whilst the term 'probe' refers to the labeled sample applied to the microarray, maintaining the traditional usage of the terms. Oligonucleotide-based platforms, such as those used for single-nucleotide polymorphism (SNP) analysis and representative

DNA Microarrays: *Methods Express* (M. Schena, ed.)

oligonucleotide microarray analysis, are powerful means of assessing DNA copy number integrity (7, 8). However, even with >100 000 loci represented on SNP microarrays, only a subset of loci will be informative. Also, there is evidence that the use of genomic reduction techniques (e.g. whole genome sampling) and polymerase chain reaction (PCR) amplification steps that are typically used for oligonucleotide microarray analysis contribute to experimental variability, bias, and loss of feature details (4, 9).

Microarrays comprised of bacterial artificial chromosomes (BACs) offer another means of high-throughput DNA copy number analysis. These large-insert-clone microarrays can be obtained from several sources (see *Table 1* for useful web links). BAC microarrays have much lower DNA sample input requirements than other platforms, require no amplification steps, and are more sensitive at detecting single-copy changes than other platforms. Whilst microarrays with smaller probe targets will ultimately provide higher resolution, current BAC microarray platforms combine high resolution with high sensitivity to give the most reliable results.

1.2 BAC microarrays

BAC microarrays can be subdivided into two main categories: disease-specific (or region-specific) and genome-wide. These categories can be further divided into low and high resolution, depending on the number of clones used to span a given genomic region.

A number of microarrays exist for examining specific diseases or specific chromosomal regions (10–32). Sources of academic BACs as well as manufacturers of commercially available BAC microarray products are provided (see *Table 1*). All of these platforms offer reliable interrogation of copy number for their respective targets, and the focused designs have proven useful for both cancer and constitutional disease studies. However, they are of limited use to researchers investigating novel loci. For effective design, selection of regions for focused microarrays requires *a priori* knowledge of disease-driving genomic alterations.

Gene discovery research is best served by the use of CGH microarrays that provide unbiased coverage of the entire human genome. These include interval marker-based microarrays such as the Spectral Chip 2600 (Spectral Genomics) and the HumMicroarray (University of California, San Francisco), which cover the human genome at a density of approximately one BAC clone per Mb (33, 34); see *Table 1*). The term 'marker-based' refers to those BAC platforms consisting of targets that sample the genome at various intervals. Due to the relationship between genome coverage

Table 1. Selected resources for BAC microarray users

Description	Type of resource	Universal resource locator (URL)
CHORI website (Roswell Park Clone Library Information)	BAC resource	http://bacpac.chori.org
Caltech Genome Research Laboratory (CTD Library Information)	BAC resource	http://informa.bio.caltech.edu
Finger Print Contig (FPC) Database		http://www.agcol.arizona.edu/software/fpc
UCSC Genome Mapping		http://genome.ucsc.edu
Ensembl		http://www.ensembl.org/index.html
University of California, San Francisco – Comprehensive Cancer Center	Academic BAC microarray provider	http://cancer.ucsf.edu/array/services.php
British Columbia Cancer Research Centre BAC Array Group (SeeGH software)		http://www.bccrc.ca/cg/ArrayCGH_Group.html, http://www.arraycgh.ca
Fred Hutchison Cancer Research Centre		http://www.fhcrc.org
Agilent	Commercial reagents, microarrays, and scanners	http://www.agilent.com
Amersham	Commercial reagents and microarrays	http://www.amersham.com
API	Commercial scanners	http://www.api.com
Corning	Commercial slides	http://www.corning.com
Genetix	Commercial printers, microplates, and reagents	http://www.genetix.com
Invitrogen	Commercial reagents	http://www.invitrogen.com
Perkin Elmer	Commercial microarrayers, scanners, and reagents	http://www.perkinelmer.com
Roche	Commercial reagents	http://www.roche-applied-science.com
Schott US	Commercial slides	http://www.us.schott.com/english/index.html
TeleChem International (ArrayIt)	Commercial microarrayers, scanners, pins, tools, and reagents	http://www.arrayit.com
University Health Network	Commercial microarrays	http://www.microarray.ca
BioRad	Commercial scanners and microarrayers	http://www.biorad.com
Vysis	Reagents, arrays, and scanners	http://www.vysis.com
Spectral Genomics	Commercial reagents, microarrays, and scanners	http://www.spectralgenomics.com
Genomic Solutions	Commercial scanners, microarrayers, and reagents	http://www.genomicsolutions.com
Signature Genomics	Commercial microarrays	http://www.signaturegenomics.com

and the probability of alteration detection, marker-based methods will not typically detect genomic changes smaller than their average clone spacing, and, as such, small alterations may not be detected and alteration boundaries may not be finely mapped (4). Additionally, poor-quality DNA samples such as those extracted from paraffin-embedded tissue tend to increase the noise observed in experiments, further reducing the resolution (35). To detect subtle genomic alterations, the submegabase-resolution tiling set (SMRT) CGH microarray was developed (36). This microarray contains >32 000 overlapping BAC clones and thus there is no need to infer alteration status among clones. SMRT microarray CGH enables the fine mapping of breakpoints to within a single BAC clone and the detection of genomic alterations as small as 40–80 kb.

1.3 Platform choice

Currently available microarrays may not meet the precise needs of a particular genotyping research project, leaving laboratories to produce their own BAC microarray platforms. What follows is a description of the steps required for BAC microarray production and experimental use, including specific details based on our experience manufacturing a high-density SMRT microarray. We provide protocols and troubleshooting tips for microarray production, sample preparation, labeling, hybridization, scanning of hybridized microarrays, and analysis of experimental data.

2. METHODS AND APPROACHES

2.1 Preparation of BAC clones for printing

Ease of use and the ability to carry large inserts (50–200 kb) have rendered pBACe3.6 the modular vector of choice to date for constructing the majority of the BAC libraries (37). Several human BAC clone libraries are currently available; the most commonly cited include Roswell Park Cancer Institute (RPCI) and Caltech (CTD) (38; see *Table 1*). The SMRT microarray contains 32 433 BAC clones selected from the RPCI-11, RPCI-13, and CTD-D libraries, producing a 1.5-fold overlapping coverage of the human genome (36, 39).

Clone identity can be verified by multiple means. *Hin*dIII digestion 'fingerprints' of BAC clones can be compared with the physical map of the human genome using the fingerprint contig BAC fingerprint database (40, 41). Gel electrophoresis banding patterns will be unique to each clone, whilst multiple digests from a series of clones can be used to reveal clone relationships with respect to chromosomal position, providing a less expensive and less time-consuming alternative to other approaches used to generate contiguous clones ('contigs'). At the time of SMRT microarray construction, we used this approach because the sequence of the human genome was not available and 'fingerprint contigs' could be used to bridge sequence gaps (39). Two ways to confirm amplified fragment pool clone identity include fluorescence *in situ* hybridization and DNA sequencing. High-throughput amplified fragment pool sequencing as described by Watson *et al.* (42) is more efficient for large clone sets such as the SMRT microarray BAC content.

As BACs utilize low-copy-number vectors, DNA yield per bacterial cell is low. Additionally, DNA isolation from primary bacterial cells does not afford a high-throughput means of generating renewable target material, particularly because optimal microarray printing requires 0.5–1.0 μg/μl BAC DNA concentration. For efficient amplification of the large clone set needed for the SMRT microarray, we developed a high-throughput PCR-based strategy that allows the generation of large amounts of target DNA from low-yield BAC preparations. The two most common methods of BAC clone amplification are degenerate-oligonucleotide-primer PCR (DOP-PCR) and linker-mediated PCR (LM-PCR). We elected to use LM-PCR as it offers a linear, unbiased amplification strategy in which linkers with primer sites are ligated to all fragments obtained by 'four-cutter' restriction enzyme digestion (43–45). The LM-PCR protocol used for SMRT microarray production (42) is provided (see *Fig. 1*, available in the color section, and *Protocol 1*). A major advantage of this method is the generation of source LM-PCR #1 product, which can be stored and used to generate additional LM-PCR #2 product as needed (*see Protocol 1*). Both DOP-PCR and LM-PCR can be employed to modify fragments in a manner that promotes efficient coupling to specific slide or substrate surface chemistries.

A variety of printing buffers or 'spotting solutions' are available commercially including Micro Spotting Plus (TeleChem International) and Pronto (Corning), as well as 'home-made' buffers based on sodium phosphate and 3× saline sodium citrate (SSC). Our experience with Micro Spotting Plus and SSC-based buffers has been positive. One important consideration is that evaporation is a problem with all spotting solutions, although evaporation can be greatly reduced using elevated humidity and temperature control. The nonproprietary solutions (i.e. those with known reagents) are advantageous because they allow reconstitution using the appropriate solvents. In most cases, the addition of dH_2O followed by thorough mixing is an effective means of replenishing sample volume lost because of evaporation.

2.2 Surface chemistry

There is a variety of slide and substrate types available for BAC microarray printing. Two common surface chemistries utilize silane reagents to provide glass surfaces containing reactive amine groups that bind DNA by electrostatic interactions, and reactive aldehyde groups that bind in a covalent manner to DNA containing a 5′ amino modification. Whilst aldehyde-treated slides are not appropriate for binding native BAC DNA, they are ideal for PCR-based products amplified using amino-modified primers. Other characteristics that are important in the selection of microarray slides and substrates are surface flatness, hydrophobicity, reactive group uniformity, and optical properties of the glass. Surface flatness, hydrophobicity, and reactive group uniformity are of the utmost importance as these parameters determine spot diameter and morphology. Glass with high optical purity (i.e. low intrinsic autofluorescence) is helpful in reducing background fluorescence.

A major consideration when selecting a microarray slide or substrate type is the extent to which the surface can be blocked or inactivated as a means of reducing background on the unprinted and nonreacted portions of the surface. Whilst certain amino-silane protocols require a pre-hybridization step to block the still-active slide chemistry, aldehyde-treated slides can be inactivated after printing (see section 2.4). Aldehyde-treated slides are currently used for SMRT microarray production for the following reasons: (i) the 5′ amino-modified BAC DNA binds in a highly stable covalent manner; (ii) the high hydrophobicity produces small spot size; and (iii) the ability to inactivate unprinted regions of the slide provides the best overall signal-to-noise ratios. We have found that slide batches can vary among certain manufacturers and therefore recommend testing each batch carefully prior to committing to full-scale production (see section 3).

2.3 Microarray printing and processing

For the manufacture of microarrays from PCR products, we have found that machines capable of using either microarray printing pins containing an exterior sample channel or solid pins perform best (see section 2.3.5). Many instrument makers provide high-quality microarrayers, although the specifications for throughput, accuracy, sample handling, and ease of use vary from robot to robot. When deciding on the microarrayer that best suits your needs, there are a number of key points to consider.

2.3.1 Microplate handling

The ability to handle multiple microplates through an automated stacking mechanism is essential for reducing the labor involved in managing extended print runs. Plate stackers with humidity control are highly recommended as they offer a means of reducing microplate sample evaporation during extended print runs (see section 2.3.3).

2.3.2 Number of slides or substrates

Depending on the application, a user might require a high-throughput microarrayer with a 'nest' capacity of hundreds of slides or a compact personal system that holds a dozen. Microarrayer prices increase proportionately with size of the slide bed (i.e. nest or platen), which in turn determines the number of slides that can be printed in a single session. As the cost of labor is constant irrespective of slide capacity, the labor cost per slide generally decreases as the slide capacity increases because high-throughput systems can be operated more quickly and with a larger number of pins than small personal systems. When choosing a microarrayer, it is critical to select a system that is sufficient both at the time of purchase and in case of future expansion. Printing speed is determined by a number of factors, with the number of pins, motion control speed, and wash/dry station efficiency having the largest impact on total printing time. For most high-density microarrays, a 48-pin printhead represents the most efficient option for printing from 384-well microplates, as a 4 × 12 matrix of 48 pins at 4.5 mm centers loads exactly eight groups of samples per 384-well microplate and prints into a convenient 18 × 54 mm area on a standard microarray slide or substrate (25 × 76 mm).

2.3.3 Air filtration and humidity control

Humidity control is essential for preventing evaporation of printing material from the microplates and for controlling spot morphology. The recommended relative humidity range is 45–55% for most applications. Control of air purity is also important for high-density microarrays as dust particles can compromise the quality of the microscopic printed features (~100 µm) and clog the channels in the printing pins (46, 47). For this reason, high-efficiency particulate air (HEPA) filters are common both on microarrayers and in laboratories (e.g. clean rooms) in which they are used.

2.3.4 Levelness of platen

The amount of time a pin spends in contact with a slide (dwell time) as well as the impact speed can have an effect on feature size. The best printing results are obtained using a relatively short dwell time (e.g. 25–50 ms), which can be achieved by programming the microarrayer to make minimal contact with the top surface of the slide. With a short 'up–down' printing stroke, dwell time is minimized and so is spot diameter. In this context, small differences in platen height or slide thickness (e.g. 50 μm) can cause a noticeable difference in printing quality and spot diameter. Unfortunately, many microarrayers exhibit deck variations in excess of 100 μm across the surface of the bed. Variations in platen height can be mitigated by individually calibrating each slide location, although it is advisable to check the manufacturer's specifications and to assay for variability with test prints prior to purchasing an expensive system.

2.3.5 Printing pins

The choice of printing pins is critical in determining spot diameter and therefore maximum printing density. Higher printing densities allow a greater number of spots (e.g. BAC clones) to be printed on a single microarray slide, and the ability to reduce the surface area of the printed microarray. This in turn allows the use of smaller volumes of hybridization solution per microarray and the printing of multiple microarrays per slide.

There are two major styles of pin suitable for the production of BAC microarrays. Solid pins are simple, highly durable pins that print one spot per loading or 'dip' into the printing solution. Microarray printing pins contain a slit in the tip of the pin that acts as a spotting fluid reservoir, drawing up liquid from the printing plate by capillary action. Microarray printing pins are capable of producing as many as 100–200 spots per loading, but tend to be less durable than solid pins. For any high-throughput or high-density printing, microarray printing pins are the best option, as they substantially reduce printing times and overall cost.

The second major factor that determines spot size is the pin tip diameter. Both solid and microarray printing pins produce spots slightly larger than the diameter of the pin tip, but the effect is most pronounced in microarray printing pins. Pins with smaller tip diameters produce smaller spots and thus allow high-density printing, but sacrifice some durability and printing speed compared with pins with larger diameters. For pins capable of the highest printing densities, the impact speed needs to be reduced to extend pin longevity and minimize spot size. Recent advances in ultrahard alloys afford pins that are ten times more durable than the current stainless steel microarray printing pins, although the cost is higher (see http://arrayit.com/Products/Printing/Pro/pro.html).

Microarray printing pins require careful use and maintenance to preserve their performance. Microarrayers are configured with high-performance pin-cleaning stations that include sonicators and sophisticated wash and dry stations to prevent sample build-up on the pins. Improperly cleaned pins will demonstrate reduced print efficiency or lose the ability to print entirely, as contaminants cause sample channel clogging and a loss of sample flow from the channel onto the tip. Clogged pins will produce microarrays with missing spots, which either require 'spot repair' or reprinting. It is important to follow the recommended cleaning protocols provided by the manufacturer and to make certain that pins are cleaned properly both during and after every major print run.

We have used two major brands of pin with good results (see *Table 1*). We find that TeleChem International manufactures high-quality pins that allow the printing of small features (90 μm spot diameter with optimization) and consistent high-density microarrays. Genetix brand pins are constructed of durable stainless steel and perform well for the production of feature sizes in the order of 100 μm; however, Genetix pins generally do not match TeleChem Stealth or 946 pins in terms of the feature (spot) size parameter (46, 47).

2.3.6 Microplates

The choice and handling of microarray source plates is critically important in terms of optimizing the quality and number of microarrays obtained from a given set of samples. Premium commercial plates such as those provided by TeleChem International have been optimized for well geometry and low sample volume (5–10 μl) capabilities, and are manufactured to the tight tolerances required for automated plate handling. Plate lids are critical for preventing sample evaporation while the plates are in the plate stacker. The use of foil microplate sealing tape immediately after use and subsequent to freezing prevents evaporation and sample degradation, and increases the number of print runs attainable from a single preparation of solution. Plate choice is sometimes linked to the microarray printer used, although other options are possible under specific circumstances.

Although most high-end microarrayers include an integrated plate stacker that allows automated handling of a large number of plates, it is still important to minimize the amount of time that the plates are thawed and unsealed. In our experience, limited evaporation is observed if plates are left unsealed with the lids on for no more than 1.5 h at 50% humidity. If the plate stacker offers independent humidity control, a humidity setting of >50% inside the stacker as compared with the microarrayer can be used to prevent sample evaporation without impacting on spot morphology, which tends to diminish if humidity levels exceed 50%.

2.3.7 Optimizing printing parameters

After purchasing a microarrayer, preparing the samples, and selecting a slide type, there are several optimizations that need to be performed to produce consistently high-quality microarrays. It is well known that humidity has a drastic effect on spot morphology. We find that a setting of 50–60% relative humidity consistently produces the best results. Additionally, the *z*-axis printing speed and the dwell time (affected by the pin overstroke) are important considerations. Ultimately, it is important to emphasize that there are no fixed settings that will work optimally for every printing environment, microarrayer, and sample type. Thus, it is of vital importance to optimize each printing parameter using test printing prior to commencing a large manufacturing run (47).

2.3.8 Microarray processing

The goals of slide processing after BAC printing are twofold: (i) to remove BAC DNA not coupled to the surface; and (ii) to inactivate reactive groups not coupled to BAC DNA molecules. Removing unbound DNA is essential as this material can migrate beneath the cover slip during hybridization and cause poor spot morphology and high background. Inactivation of reactive groups reduces background by preventing coupling of fluorescent molecules in the probe solution to the surface. SMRT microarrays are typically printed on aldehyde slides, and for this surface chemistry, reactive group inactivation with a strong reducing agent (e.g. sodium borohydride ($NaBH_4$)) is an effective means of reducing background.

2.4 BAC microarray probe preparation

This section discusses the important aspects of biological samples and probe preparation and labeling for BAC microarrays.

2.4.1 Biological samples

The SMRT microarray is currently the most high-throughput and comprehensive means of profiling genomic alterations in archival samples. Archival samples are typically fixed with formalin, embedded in paraffin, and stored in hospital archives. SMRT microarrays require minimal input DNA and have the ability to withstand non-target-cell DNA contamination (4, 35). Currently, DNA from archived specimens cannot be applied to other high-throughput platforms such as the Affymetrix 100K SNP microarray (48). The main considerations for assessing the applicability of a sample for a BAC microarray experiment are the relative heterogeneity of the contributing cell population, and the quantity and quality of DNA.

If tissue heterogeneity is a concern, manual or laser-assisted microdissection allows the targeting of specific cells within a tissue cross-section (49–51). Even with an increased ability to tolerate heterogeneity, selection of a subpopulation of cells will produce clearer genomic profiles and facilitate more effective BAC microarray analysis (35). Whilst a number of other platforms employ genome sampling and amplification steps, the low DNA input requirement of BAC microarrays generally removes this need (4). Currently, BAC microarray researchers report input DNA requirements in the range of 300–3000 ng (52), although our experience with the SMRT microarray is that 50–400 ng of DNA per slide suffices. Eliminating the requirement of an amplification step is advantageous, as factors such as template length, secondary structure, and GC content can contribute to DNA segment misrepresentation and analysis bias (53, 54).

DNA samples derived from fresh, frozen, and archival specimens can be analyzed by BAC microarray CGH. Relative to formalin fixation (which can degrade nucleic acids), freezing a sample preserves DNA quality better and allows concurrent isolation of RNA and proteins to assess gene expression. However, for clinical samples, fixed tissues produce more effective histological references, facilitating the selection of a desired subpopulation of cells in a heterogeneous sample. In addition, archived samples may be associated with a wealth of clinical data (e.g. outcome), making them extremely valuable in retrospective analyses. In our experience, regardless of the source material, the DNA extraction protocol can also have a significant impact on DNA quality and yield. For more factors affecting DNA sample quality, see section 3.

2.4.2 Reference samples

In terms of reference samples for competitive hybridization, different samples will satisfy user needs depending on the experiment (see *Fig. 2*). Issues that need to be considered in selecting the reference sample are gender (matched vs mismatched to the test sample), composition (individual vs pooled from multiple sources), and source (allogenic vs autogenic). A single reference type should be used for an entire experimental set. As reference DNA will contain copy number variations (i.e. natural polymorphisms), these need to be taken into account during experimental analysis. Significantly, further characterization of copy number variations using a tiling-set microarray will build on existing work and serve as an invaluable baseline to identify population variance as well as susceptibility loci driving various constitutional diseases (2, 55–57).

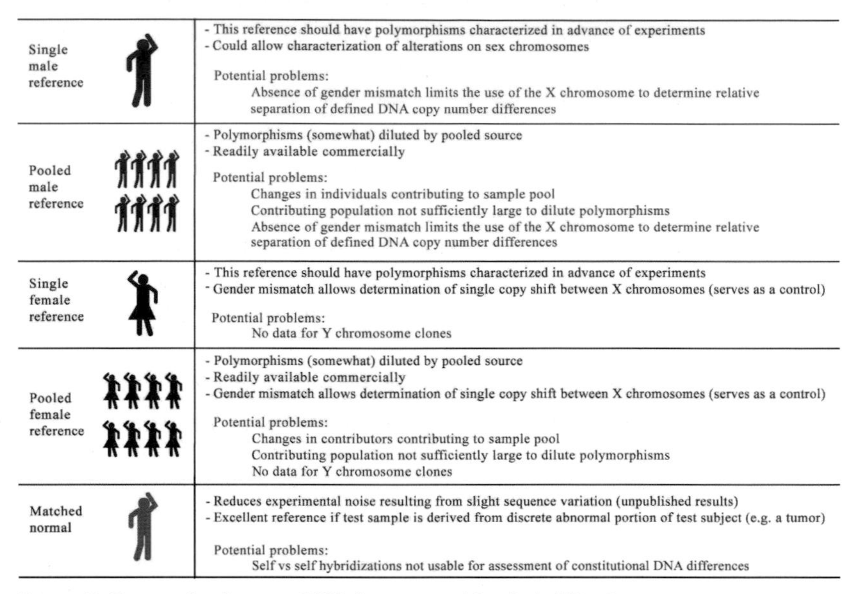

Reference	Features
Single male reference	- This reference should have polymorphisms characterized in advance of experiments - Could allow characterization of alterations on sex chromosomes Potential problems: Absence of gender mismatch limits the use of the X chromosome to determine relative separation of defined DNA copy number differences
Pooled male reference	- Polymorphisms (somewhat) diluted by pooled source - Readily available commercially Potential problems: Changes in individuals contributing to sample pool Contributing population not sufficiently large to dilute polymorphisms Absence of gender mismatch limits the use of the X chromosome to determine relative separation of defined DNA copy number differences
Single female reference	- This reference should have polymorphisms characterized in advance of experiments - Gender mismatch allows determination of single copy shift between X chromosomes (serves as a control) Potential problems: No data for Y chromosome clones
Pooled female reference	- Polymorphisms (somewhat) diluted by pooled source - Readily available commercially - Gender mismatch allows determination of single copy shift between X chromosomes (serves as a control) Potential problems: Changes in contributors contributing to sample pool Contributing population not sufficiently large to dilute polymorphisms No data for Y chromosome clones
Matched normal	- Reduces experimental noise resulting from slight sequence variation (unpublished results) - Excellent reference if test sample is derived from discrete abnormal portion of test subject (e.g. a tumor) Potential problems: Self vs self hybridizations not usable for assessment of constitutional DNA differences

Figure 2. Types of reference DNA for competitive hybridization.

2.4.3 Probe generation and labeling

There are numerous methods that can be used to label the sample and reference DNAs differentially with fluorescent nucleotides (e.g. Cy3- and Cy5-dCTP) for use in BAC microarray CGH. The ability to detect single-copy chromosomal gains and losses accurately depends greatly on the ability to generate high-quality labeled probes (58, 59). Many factors influencing labeling, such as the incorporation efficiency, the spectral separation of fluorophores, and the nonbiased amplification of DNA, must be considered to obtain maximum sensitivity and accuracy in detecting copy number alterations.

Conceptually, the two main approaches to DNA labeling are whole genome and representational (see *Fig. 3*, available in the color section). Whole genome labeling is the primary method used for BAC microarray CGH as it creates probe molecules from all of the genomic material present in the sample without complexity reduction. Genome representation, also known as 'genome sampling', is a complexity reduction approach used to reduce cross-hybridization when using an oligonucleotide platform (60). In contrast to whole genome labeling, the representational approach enriches for short DNA fragments by performing a restriction enzyme digest followed by LM-PCR amplification. Complexity reduction is not needed in BAC microarray CGH because of the high degree of hybridization specificity (4, 9, 61).

The most common whole genome probe generation approach is random primer DNA labeling, originally developed by Feinberg and Vogelstein (58, 59, 62, 63). This procedure facilitates incorporation of cyanine–dNTP nucleotide analogs into template DNA and requires between 50 and 3000 ng of starting material per reaction (52, and unpublished data). Cyanine dyes have proven useful because of their detection sensitivity and efficiency as polymerase substrates (64). Random primer labeling uses a mixture of all possible sequences of short primers (usually hexamers or octamers), which hybridize to the template and provide starting points for DNA synthesis (see *Fig. 3*). Random primers hybridize extensively along the template DNA, ensuring priming along the entire length of the template. Klenow, the large fragment of DNA polymerase I, synthesizes the complementary strand starting from the 3′-OH of the primer and incorporating labeled nucleotides along the template. Klenow has the capacity to displace strands on the template DNA and therefore fragments can be generated that are larger than the template regions defined by adjacent primers. Random priming leads to greater than or equal to fourfold nonbiased linear amplification of the starting material so that larger amounts of probe are generated for the hybridization step, an attractive feature when using clinical samples with low DNA yield (58, 64).

Whole genome methods employed to label DNA for microarray CGH experiments fall into two categories, direct and indirect (see *Fig. 3*) (65–68). With direct labeling, tagged nucleotides are incorporated directly into probe molecules generated from template DNA, which simplifies the labeling procedure. Indirect labeling requires a secondary dye-coupling step and is not widely used in microarray CGH experiments at this time.

Troubleshooting tips for labeling experiments are given in section 3. Please note that spectrophotometer readings only measure the amount of cyanine-labeled dNTPs incorporated into the probe and not the emission of the dyes; therefore fluorometric readings may reflect the activity of the dyes more accurately (69). Also, please note that a major concern when using cyanine dyes is their sensitivity to environmental agents. Extended exposure to light or high levels of atmospheric ozone have been shown to affect the fluorescence of these dyes and should be avoided as much as possible during microarray experimentation (70, 71).

Whenever using probes or targets obtained from genomic DNA, it is important to 'block' repetitive sequences in the hybridization reaction to maximize sequence-specific fluorescence ratios. This is typically achieved by adding Cot-1 DNA to the hybridization mixture (72–74).

Cot-1 is species-specific genomic DNA that has been enriched for repetitive sequences through sonication and controlled denaturation and reannealing. The binding of the Cot-1 DNA to the repetitive elements present in the probe mixture prevents binding to the target elements on the microarray. Cot-1 is typically used in excess of the probe DNA (e.g. 3 : 1 molar ratio) to ensure adequate blocking, with the final concentration ranging from 1 to 20 μg/μl in 15–110 μl of hybridization buffer (75). Labeled probes are precipitated in the presence of Cot-1, suspended in hybridization buffer, and then typically incubated to allow blocking prior to hybridization. In our experience, Cot-1 DNA varies considerably from manufacturer to manufacturer, and even from lot to lot. We strongly recommend that test microarray experiments and rigorous quality control standards be applied prior to purchasing any large amount of Cot-1 DNA (see section 3).

2.5 BAC microarray hybridization and washing

This section outlines the hybridization buffers and wash solutions used for BAC microarrays.

2.5.1 Hybridization buffers

Hybridization buffers are generally formulated to reduce the melting point of DNA and maintain stringency at lower hybridization temperatures. Formamide or urea can be used to reduce the melting point of DNA, with formamide being more common and generally used at a 50% concentration (76). However, urea-based hybridization solutions are desirable as they are less toxic than their formamide counterparts (77). Every 1% increase in formamide concentration reduces the melting point of DNA by approximately 0.7°C (76). In the presence of 10% dextran sulfate or polyethylene glycol, the hybridization rate is increased by about tenfold (78–80), as these 'volume excluders' effectively increase the probe concentration. The buffer 2× SSC is common in microarray CGH hybridization buffers (75). Detergents are also used frequently, with sodium dodecyl sulfate (SDS) being the most common. For exact concentrations used by different groups, see (75). Commercially obtained hybridization buffers can offer a higher level of consistency than 'home-made' buffers; we have had success with the DIG Easy product from Roche (see *Table 1*).

The addition of other macromolecules such as yeast tRNA and sheared herring sperm DNA to a hybridization solution is also common practice. This is done in an effort to minimize nonspecific binding to both the slide surface and the printed target DNA sequences. Yeast tRNA is used by various groups at final concentrations ranging from 50 ng/μl to 10 μg/μl and sheared herring sperm DNA at final concentrations ranging from 50 ng/μl to 400 ng/μl in hybridization volumes of 15–110 μl (75).

2.5.2 Hybridization reactions

In addition to buffer composition and hybridization temperature, there are other factors to consider before setting up a hybridization experiment. These include whether or not a pre-hybridization step is necessary, the optimal buffer volume and probe concentration, and the type of hybridization to be performed.

A pre-hybridization step can be used to block repetitive DNA sequences present in microarray targets, thus reducing nonspecific binding of probe DNA (81). In addition, pre-hybridization blocks reactive groups on slides that were not chemically inactivated during pre-processing (e.g. amine-coated slides; see section 2.2). For aldehyde slides that are chemically inactivated, this step is not necessary as pre-treatment with $NaBH_4$ reduces the reactive surface to alcohol groups that are unable to bind DNA.

All that is required to perform a hybridization reaction is static addition of the probe mixture onto the microarray surface. Performing hybridizations under slide cover slips allows the use of reduced probe concentrations and smaller volumes, important features when test material is limited, as with most clinical specimens (see *Fig. 4*). One potential concern with the cover slip approach is reduced probe diffusion, which reduces the chances of productive hybridization events between complementary probe strands and spotted targets on the slide surface (82). Automated hybridization apparatuses that use air bladders, rocking, sonic waves, and other means to agitate the hybridization mixture and increase probe diffusion reportedly give increased sensitivity, although they ultimately require a greater probe volume to work effectively and are best used for experiments where test DNA is not limiting (75, 83).

'Sandwich' hybridizations offer one way to control hybridization differences between microarrays of a multi-slide set (84). Briefly, this entails placing one microarray on top of the other with the spotted surfaces facing each other. The benefit of this approach is that both slides are exposed to identical experimental conditions. However, drawbacks to this technique include the fact that microarrays seem to be more prone to drying out and that some hybridization chambers are not deep enough to accommodate stacked slides. ArrayIt offers an extra-deep hybridization cassette that caters for sandwich hybridizations (see *Table 1*).

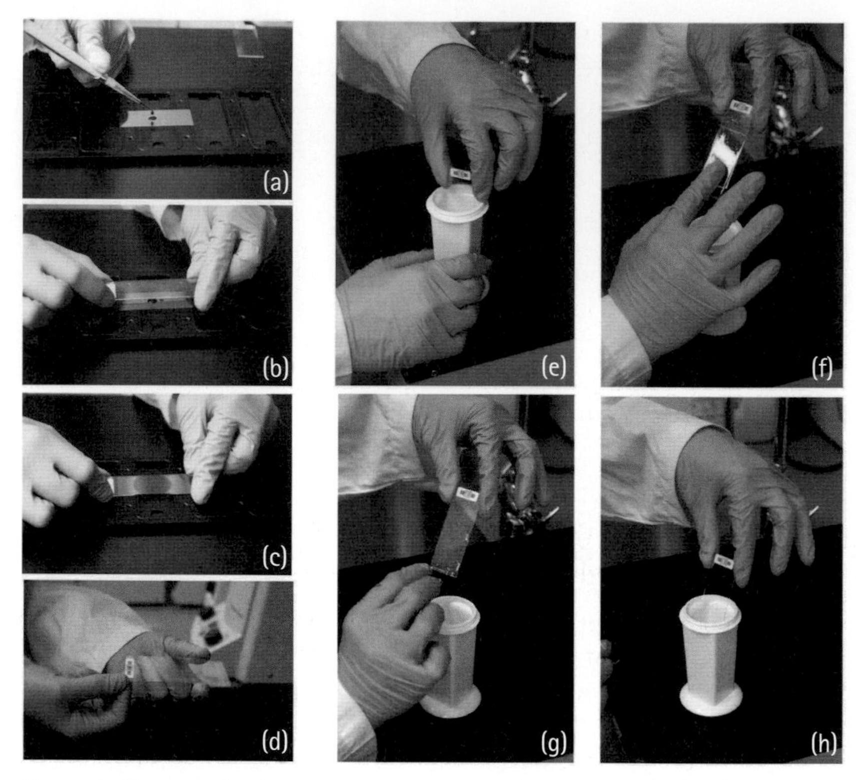

Figure 4. Photographs depicting microarray hybridization set-up (a–d) and post-hybridization washing steps (e–h).
(a) Deposit the hybridization solution onto the cover slip. (b) Orient the cover slip perpendicular to the hybridization cassette. (c) Place the microarray slide directly onto the cover slip, taking care to ensure that the hybridization solution covers the entire printed area, which can be marked by etching the back of the slide with a diamond pencil to help with alignment. (d) The hybridization solution should cover the entire microarray. If the solution does not reach the edges of the cover slip, more solution is required. Care must be taken not to introduce bubbles between the cover slip and the microarray, as bubbles will prevent hybridization between the printed targets and the probe molecules in solution. The microarray/cover slip ensemble may be placed into a hybridization chamber to allow temperature-controlled hybridization. (e) Place the hybridized microarray into wash solution for approximately 30 s. (f) Remove the microarray from the wash solution and slide the cover slip gently downwards so that at least 5 mm of the cover slip extends beyond the microarray slide. The cover slip should fall off the slide easily. If not, additional soaking time will remedy the problem. (g) Use the protruding edge of the cover slip to lift it off the slide. (h) Place the slide into the wash solution and begin the washing process. All of the steps above (a–h), including the cover slip application and removal steps, should be performed quickly to prevent drying of the probe solution.

2.5.3 Post-hybridization washes

Once hybridization is complete, the slides must be washed to remove probe molecules that are either adhering or bound nonspecifically to the slide surface (see *Protocol 5*). Several buffers and components are commonly used for this purpose including formamide, SSC, and SDS (75). Cyanine dyes have been shown to be particularly sensitive to atmospheric ozone degradation during the washing and subsequent drying steps (71). Thus, ozone levels should be controlled to obtain the best-quality microarray CGH results (see section 3).

2.6 BAC microarray scanning and experimental analysis

The section describes the latest approaches to BAC microarray scanning and analysis of the data.

2.6.1 Scanning hybridized slides

Post-hybridization image acquisition is a critical step in the production of microarray CGH data. BAC microarray CGH-slides rarely exhibit the large dynamic range seen with gene expression microarrays. This is due to the relatively low ratios observed for single-copy changes, which account for the greatest number of alterations in the average cancer sample (85). Additionally, high probe complexity and Cot-1 blocking can result in microarray slides that exhibit much lower peak fluorescence than a typical gene expression microarray. Due to these factors, it is much more important to use a scanner that exhibits low noise and high sensitivity than one that boasts a very large dynamic range.

The two primary scanning technologies available today utilize either a laser and photomultiplier tube configuration or a white light source and charge-coupled device (CCD) sensor in combination, each with excitation and emission filters (86). In our experience, current CCD-based systems offer the best low-intensity performance; however, the need to replace the excitation light source frequently, as well as slow scan times, limit their effectiveness in a high-throughput setting. Laser-based scanners are the more traditional microarray imaging system and can vary drastically in performance. ArrayIt has recently developed an ultrahigh-sensitivity scanner based on 'cool excitation' technology (see *Table 1*), but we have not tested this system for BAC microarray analysis.

All scanners require adjustment, through controller software, of the imaging sensors and excitation sources to achieve optimal images that lie within the linear range of the scanner. CCD-based scanners are arguably the easiest to pre-configure as they exhibit very wide linear ranges and all configuration settings behave in a linear manner (i.e. doubling the exposure time results in twice the intensity). Laser-based scanners require a more complex adjustment where both the laser power and photomultiplier tube sensitivity can be adjusted and neither represents a linear scale. However, this problem is remedied using the automatic adjustment algorithms incorporated into many newer scanners, allowing true automated image acquisition.

2.6.2 Analysis of BAC hybridization results

After the microarray is scanned, there is a series of steps that needs to be performed before sample DNA copy number can be assessed. These include image processing, data filtering and visualization, and statistical analysis. Furthermore, prior to visualization and analysis, normalization of the data may also be required (see section 3).

2.6.3 Image processing and data filtering

After scanning the microarray, the signal intensity from each spot must be quantified. Specifically, the intensity values of the two fluorophores must be calculated and, depending on which dyes are coupled to which sample, the appropriate $\log_2$ ratio (taken as the ratio of the test sample intensity divided by the reference sample intensity) needs to be calculated. There are many programs available that can perform image quantification, some of which are bundled with the scanners, whilst others are available as stand-alone applications (see *Table 1*).

After the image has been processed and ratios have been calculated, one can inspect the image and intensity values manually looking for a small percentage of spots of suboptimal quality. Moreover, with any DNA microarray there will be a small percentage of spots that do not contain usable information. Abnormal shape morphology, low intensity values, low signal-to-noise ratio, and contamination (e.g. dust particles) are potential factors leading to exclusion of such suboptimal spots.

2.6.4 Normalization

Normalization addresses broader experimental trends that cannot be resolved through simple spot exclusion. It is required to account for the various CGH biases that produce suboptimal results. Differential dye effects (e.g. Cy3 and Cy5 detection or incorporation differences), variation in scanning parameters between experiments, spatial effects associated with spot locations, slide intensity gradients, and differences in the amount of starting material between the test and reference sample are among the factors that normalization addresses.

There are two broad levels of normalization that can be performed: global and local. Most methods of normalization can be used in either a global or a local context. Two of the most common methods are scaling normalization and locally weighted scatter-plot smoothing. Scaling normalization is usually performed by transforming each of the individual intensities such that, for example, the mean or median intensity for each channel in each microarray or over the whole set of microarrays is identical across all experiments. Locally weighted scatter-plot smoothing is a normalization approach that uses a locally based least-squares-weighted regression model to fit subsets of data until all data points have been evaluated under the regression function (87, 88). It is a locally based method because evaluation of a particular data point uses its neighboring points as well.

Many successful normalization approaches for expression microarray data are currently being applied to other DNA microarrays. However, a normalization framework was recently developed in the context of BAC DNA microarrays that targets three areas of potential bias: spot position, clone origin, and intensity data (see *Fig. 5*, also available in the color section) (89). As can be seen in the post-normalization SeeGH karyogram, programs such as this greatly reduce the systematic biases in the process that are usually perceived as noise. This facilitates clearer assessment of DNA copy number. Continuing development of normalization methods for DNA copy number data will improve microarray CGH data analysis.

2.6.5 Visualization and statistical analysis

After the elimination of poor-quality spots and transformation of the data present in the high-quality spots, the final steps are assessing the statistical significance of the computed ratios and visually representing the data to highlight DNA copy number alterations. Currently, there are freely available software tools that employ various techniques to assess statistical significance (5, 90). Software choice is ultimately made based on user needs. Whereas clinical usage requires a limited robust form of software that allows only selected types of data interrogation, exploratory scientific research requires a more sophisticated and comprehensive application (5, 91–93). SeeGH software (see *Table 1*) is currently the only program that correctly displays tiled clone data (91). Future applications will enhance the power of tiled clone platforms by providing analysis approaches that capitalize on the degree of overlap and the polymorphic status of the microarray elements.

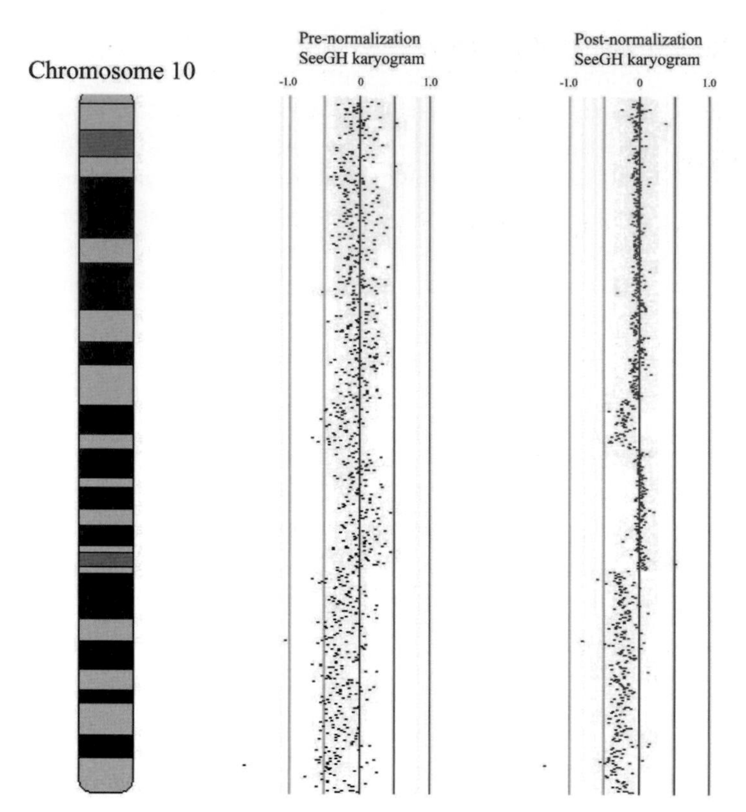

Figure 5. Normalization methods for BAC microarray data analysis (see page xxviii for color version).
The effects of microarray CGH data normalization for competitive hybridization of a cancer cell line against its drug-resistant derivative. Normalized and non-normalized $\log_2$ signal intensity ratios were plotted using SeeGH software. Clones with standard deviations among the triplicate spots of >0.09 or a signal-to-noise ratio of >3 were filtered from the analysis. Chromosome arm 10q is represented on the left. Vertical lines denote $\log_2$ signal ratios from −1 to 1 with copy number increases to the right (red lines) and decreases to the left (green lines) of zero (purple line). A $\log_2$ signal ratio of zero represents equivalent copy number among the hybridized samples. Each black dot represents a single BAC clone. Normalization was performed using a custom normalization program with the parameters set as a default (89).

2.7 Recommended protocols

Protocol 1

Generation of BAC clones by LM-PCR

Equipment and Reagents

- BAC DNA
- *Taq* polymerase and buffer (5 units/μl; Promega)
- T4 DNA ligase (400 units/μl; New England Biolabs)
- *Mse*I (10 units/μl; New England Biolabs)
- 10× *Mse*I buffer (New England Biolabs)
- dNTPs (10 mM each; Promega)
- 25 mM $MgCl_2$ (Promega)
- *Mse*I long primer (10 μM, 5′-AGTGGGATTCCGCATGCTAGT-3′; Alpha DNA)
- *Mse*I short primer (10 μM, 5′-TAACTAGCATCG-3′; Alpha DNA)
- 3.0 M Sodium acetate (pH 5.5)
- 3× SSC buffer
- 96-Well microplates
- Incubator (16°C)
- Thermocycler
- 100% Ethanol
- Centrifuge

Method

1. Transfer 50 ng of BAC DNA from each clone to a 96-well microplate.
2. Add 5 units of *Mse*I restriction enzyme and the appropriate restriction enzyme buffer for a total reaction volume of 40 μl.
3. Digest for 8 h at 37°C and then heat-inactivate the *Mse*I by incubating at 65°C for 10 min.
4. Transfer 4.0 μl of digested product to a new 96-well microplate.
5. Set up the following linker reaction containing 4.0 μl of digested DNA (from step 4), 0.8 μl of *Mse*I long primer and 0.8 μl *Mse*I short primer. Pre-anneal the linker reaction for 5 min at room temperature.
6. Add 80 CEL (cohesive end ligation) units of T4 DNA ligase and the appropriate buffer for a final reaction volume of 40 μl. Incubate the mixture for 12–16 h at 16°C.
7. Remove 2.5 μl of the ligation product for use in a 50 μl PCR (termed LM-PCR #1).
8. Set up LM-PCR #1 by combining 2.5 μl of linker-ligated DNA (from step 7), 16 μl of $MgCl_2$, 5 μl of each dNTP, 2 μl of *Mse*I long primer (amino-modified if using aldehyde slides), 5 units of *Taq* polymerase and the appropriate buffer for a final reaction volume of 50 μl.
9. Amplify the LM-PCR #1 mixture using the following cycling parameters: 1 cycle for 3 min at 95°C; 30 cycles of 1 min at 95°C, 1 min at 55°C and 3 min at 72°C; and 1 cycle for 10 min at 72°C.
10. Remove 0.25 μl of LM-PCR #1 to use in a second 50 μl PCR (termed LM-PCR #2).

11. Set up and amplify LM-PCR #2 exactly as for LM-PCR #1 except that the number of PCR cycles is increased from 30 to 35.
12. Once LM-PCR #2 is complete, add 5 µl of 3 M sodium acetate (pH 5.5) and 137.5 µl of 100% ethanol. Mix well and incubate on ice for 30 min to precipitate the LM-PCR #2 product.
13. Pellet the PCR product by spinning at 4°C for 30 min in a microplate centrifuge at 2750 *g* (if using 96-well microplates).
14. After centrifugation, remove and discard the supernatant by inverting the microplate onto a paper towel or suitable laboratory absorbent material.
15. Dissolve the BAC DNA pellet in 100 µl of 3× SSC buffer or another suitable print buffer. The typical DNA yield is 40–50 µg, so a final volume of 100 µl would provide a final DNA concentration of 0.4–0.5 µg/µl for printing.

Protocol 2

Post-production processing of BAC microarray aldehyde slides[a]

Equipment and Reagents

- 0.1% SDS
- $NaBH_4$
- 100% Ethanol
- 10× Phosphate-buffered saline (PBS)
- Centrifuge
- Water bath
- Slide holder

Method

1. After printing, incubate the microarrays in the microarrayer at ambient humidity for ≥2 h to allow drying. This drives the dehydration reaction that couples the DNA to the aldehyde surface.
2. Remove the printed microarrays from the microarrayer and place into wash stations[b].
3. Wash the microarrays in 0.1% SDS twice for 2 min each.
4. Rinse the microarrays in dH_2O for 1 min.
5. Transfer the microarrays to 1× PBS containing 0.5% $NaBH_4$ and 20% ethanol.
6. Agitate the microarrays by mixing for 3 min.
7. Rinse in dH_2O for 1 min.
8. Place the microarrays into boiling water for 30 s.
9. Rinse at room temperature in dH_2O for 1 min.
10. Dry the microarrays by centrifugation in 50 ml conical tubes with no lids for 5 min (700 *g*)[c].
11. Store in a desiccated, vacuum-sealed bag in the dark until required for hybridization.

Notes

[a]This protocol was adapted from Schott Nexterion (http://www.schott.com/nexterion).
[b]ArrayIt offers an affordable high-throughput wash station that works well for this purpose (http://www.arrayit.com).
[c]ArrayIt offers an affordable microarray high-speed centrifuge that works well for this purpose (http://www.arrayit.com).

Protocol 3

Fluorescent probe labeling for BAC microarrays[a]

Equipment and Reagents

- 1 mM Cy3-dCTP and Cy5-dCTP (Perkin Elmer)
- Klenow (9 units/µl; Promega)
- Klenow buffer (10×; Promega)
- 10× Random-priming dNTP mix containing 2 mM each of dATP, dGTP, and dTTP, and 1.2 mM dCTP (Invitrogen)
- Random octamers (150 µg/µl stock; Alpha DNA)
- Incubator or water bath set to 37°C
- ProbeQuant Sephadex G-50 columns (Amersham) or YM-30 Microcons (Millipore)
- Spectrophotometer

Method

1. Prepare separate reference and test sample reactions by combining the following components in a microfuge tube: 1.0 µl of DNA (50–400 ng), 0.75 µl of random octamers, 2.5 µl of Klenow buffer, and 10.5 µl of dH_2O.
2. Heat the reactions for 10 min at 100°C to denature the DNA.
3. Transfer the reactions immediately onto ice for 2 min.
4. Add 3.75 µl of 10× random-priming dNTP mix.
5. Add 2 µl of Cy3-dCTP to either the reference or the test DNA sample. Add 2 µl of Cy5-dCTP to the other DNA sample. Add 2.5 µl of Klenow enzyme and mix gently. The final reaction volume should be 25 µl.
6. Incubate the reactions for 18–36 h at 37°C.
7. Remove unincorporated cyanine dyes using a G-50 column or Microcon YM-30 column.
8. Combine the Cy3 and Cy5 labeling reactions. The total volume of the sample should be 50 µl.
9. Assess the incorporation values using a spectrophotometer[b].

Notes

[a]This protocol is adapted from (36).
[b]Incorporation values below 3.0 pmol/µl for 50 µl combined labeling reactions in either channel have shown variable results. Incorporation values of 8–25 pmol/µl are preferred. The Cy3 fluor typically shows higher values than Cy5.

Protocol 4

Probe preparation and hybridization

Equipment and Reagents

- Cot-1 DNA (1 μg/μl; Roche/Invitrogen)
- 3.0 M Sodium acetate (pH 5.5)
- 100% Ethanol
- 1× Hybridization buffer containing 4.617 g of DIG Easy Hyb granules (Roche) in 10 ml of dH_2O
- Sheared herring sperm DNA (20 μg/μl; Invitrogen)
- Yeast tRNA (10 μg/μl; Calbiochem)
- Hybridization cassettes (TeleChem International, ArrayIt)
- Heating block (capable of heating to 45 and 85°C)

Method

1. To the 50 μl labeled probe sample (*Protocol 3*, step 9), add 100 μl of Cot-1 DNA.
2. Add 15 μl of 3 M sodium acetate (pH 5.5) and 412.5 μl of 100% ethanol.
3. Pellet the labeled probe material by centrifugation in a microfuge for 15 min at full speed (15 000 *g*). The DNA pellet will appear purple in color after centrifugation.
4. Resuspend the DNA pellet in 45 μl of hybridization solution prepared as follows: 36 μl of 1× DIG Easy hybridization buffer, 4.5 μl of sheared herring sperm DNA, and 4.5 μl of yeast tRNA.
5. Denature the probe solution by incubating at 85°C for 10 min.
6. Incubate the denatured probe solution at 45°C for 1 h to allow Cot-1 annealing.
7. Place 45 μl of probe solution onto the cover slip (see *Fig. 4a*).
8. Gently lower the microarray slide onto the cover slip to allow a thin layer of probe solution to form between the cover slip and the microarray slide (see *Fig. 4b* and *c*)
9. Place the slide into a hybridization cassette pre-warmed to 45°C[a].
10. Incubate for 36–40 h at 45°C.

Note

[a]Add an amount of dH_2O to the hybridization cassette necessary to achieve 100% humidity. A 10 μl volume works well for TeleChem ArrayIt hybridization cassettes.

Protocol 5

Microarray washing and scanning

Equipment and Reagents

- Wash solution 1 (0.1× SSC, 0.1% SDS)
- Wash solution 2 (0.1× SSC)
- Coplin jar or slide staining boxes
- (Optional) High-throughput wash station (TeleChem International, ArrayIt)

Method

1. Pre-warm wash solution 1 to 45°C.
2. Remove the cover slip from the microarray slide (see *Fig. 4*) and place the slide immediately in wash solution 1.
3. Perform five sequential 5 min washes with agitation. Washes can be performed in a Coplin jar, slide-staining box, or high-throughput wash station.
4. Rinse the slides five times for 1 min each in wash solution 2.
5. Dry the slides with an oil-free air stream or by centrifugation in 50 ml conical tubes (no caps) at 700 *g* for 5 min.
6. Store the slides in the dark until ready for scanning, as signal intensities will diminish over time[a].

Note

[a]The authors strongly recommend that BAC microarray production and experiments be performed in a dedicated room where ozone, light exposure, and humidity can be carefully controlled.

3. TROUBLESHOOTING

Target spot size is impacted by numerous factors and careful control is essential for producing microarrays of outstanding quality. The hydrophobicity of the slide chemistry as well as the wetting properties of the spotting solution (including viscosity due to DNA concentration) drastically affect spot size. Choosing a more hydrophobic substrate or a more viscous or 'lower spreading' spotting solution will yield smaller spots. The humidity at which printing is performed affects the wetting properties of the substrate. Optimizing spotter humidity to between 50 and 60% will produce the best results. The impact speed as well as the amount of time the printing pins make contact with the substrate can change spot size. Reducing the pin overstroke distance and setting the *z*-axis speed to the manufacturer's recommended settings will reduce spot size. The size of the printing pin tip is the most significant factor in determining the final spot size. Refer to the commercial pin supplier specifications to select the appropriate pin tip size for the application. Post-spotting treatment of microarrays can also affect spot diameter and therefore careful reference to slide manufacturer protocols is needed.

Missing target spots on the microarray is also detrimental to microarray analysis and several factors typically affect printing efficiency. Clogged or damaged pins can lead to poor printing efficiency. The pin channels may develop a contaminating coating over time caused by out-gassing or other fumes present in the microarrayer or laboratory that reduces pin loading and printing efficiency. Pins should be cleaned according to the instructions of the manufacturer and replaced as necessary. The printhead or holder allows the printing pins to move vertically during the 'up–down' motion of the printing cycle and users should check that the pins move freely and do not stick in the printhead. 'Sticking' pins are one of the most common sources of missing target spots. The microplates should contain a sample volume (e.g. 5–10 μl) sufficient to allow complete pin loading, and the microarrayer should be calibrated properly to allow the pins to contact the bottom of the microplate for proper sample loading. Ensure that the pin washing system is functioning properly during the print run, as failure to clean the printing pins properly can cause a significant decrease in spot quality over time.

Careful control over DNA sample purity is also essential for optimal BAC microarray results. Contaminating RNA in DNA samples will artificially elevate the absorbance at 260 nm, leading to erroneously high DNA concentrations that can negatively impact downstream reactions. Treatment of the DNA sample with RNase A will degrade the RNA and allow its removing by ethanol precipitation. DNA samples treated with RNase A should be extracted with organic solvents (e.g. phenol, chloroform) to remove the ribonuclease. Contaminating proteins will reduce the absorbance ratio (A_{260}/A_{280}), affecting concentration assessment and potentially impacting downstream reactions. Re-extracting a DNA sample with organic solvents will remove unwanted proteins. Residual organic solvents will also affect the A_{260}/A_{280} ratio and downstream reactions. When extracting with organic solvents, be careful not to disrupt the aqueous/organic interface. Ether extraction followed by evaporation at an elevated temperature can be used to remove contaminating phenol and chloroform.

High sample salt concentration can impact downstream enzyme function. Even for samples with anticipated low DNA yields, applying multiple 70% ethanol washes of the pellet during DNA extraction is an effective means of reducing salt concentration. If the average DNA fragment size is too small, subsequent enzyme reactions may not be effective. If there is sufficient material, run a small amount of sample against size standards to determine the size range of the sample DNA. Alternatively, low DNA yield quantification methods such as randomly amplified polymorphic DNA PCR (92) can be used.

Probe labeling efficiency can be another challenging aspect of BAC microarray analysis. The amount of probe generated can be reduced if the DNA is not completely denatured (i.e. single stranded), if the Klenow enzyme is inhibited by sample contaminants, or if the template DNA is degraded. Sites of DNA damage, such as thymidine dimers and abasic sites caused by various DNA damaging agents, may not be passable by the Klenow enzyme, and such samples may only produce short probes. To avoid this shortcoming, probes should be generated only from test and reference samples that meet the minimum DNA size and quality standards. Poor dye incorporation efficiency may be caused by DNA contaminants (e.g. ethanol, phenol, chloroform, proteins), poor-quality cyanine dyes, suboptimal nucleotide ratios, or poor coupling of cyanine to the labeling nucleotide. Ensure that test and reference DNA samples are free of contaminants and that the dye quality is high. Nucleotide ratios may need to be determined empirically for specific protocols.

Post-scanning troubleshooting is also an essential aspect of microarray analysis. Weak microarray signals may result from several different sources. Overly aggressive post-hybridization washing can remove bound probe and produce weaker signals. This problem can be countered by reducing the washing stringency (i.e. decreasing washing temperature or washing time). Dim signals can also be due to insufficient CCD exposure times or scanner sensitivity settings, both easily remedied by changing the instrument settings. Undertaking microarray experiments in a controlled environment may also mitigate weak signals owing to environmental factors that degrade cyanine dyes such as excess light or ozone. Dim images can also result if the target DNA sample concentration is too low, due to contaminants in the sample DNA (leading to erroneous quantification), or due to the excessive removal of target DNA during post-printing treatment. Close monitoring of these factors will resolve this issue. Slide signals can also be overly bright if there is inefficient blocking of repeats via the Cot-1 DNA. If the Cot-1 DNA is of low quality, or if too little Cot-1 DNA is used, repeats will not be blocked effectively, and the amount of probe binding to the microarray will be greatly increased. To counter this, ensure that a sufficient amount of Cot-1 DNA is used.

Slide intensity gradients can result from numerous factors. If the solution on the microarray dries out during hybridization, a gradient can arise. To prevent this, ensure that the hybridization chambers have sufficient humidity and that an adequate volume of probe solution is applied to the microarray. If performing the hybridization under a cover slip, ensure that the microarray slide does not make direct contact with the sides of the hybridization chamber, as this will result in probe solution being drawn out from under the cover slip, reduced probe volume, and increased drying. Gradients can also result during the microarray washing steps, causing unbound probe to be removed unevenly leading to the formation of gradients. This can be addressed by ensuring that washing solutions cover the entire slide and that the washing solution is agitated vigorously during the washing process.

Dust or precipitated probe can cause speckles in scanned images. Rinsing in 0.1× SSC or drying with compressed air may remove these. High image background, caused by inefficient slide washing or incomplete slide pre-treatment, can reduce the amount of usable data obtained from a microarray. Increasing wash stringency may improve this. Alternatively, if incomplete reduction of aldehyde to alcohol groups on the slide surface is the problem, the reducing agent used in the slide processing may need to be replaced.

Certain aspects of post-visualization may also require troubleshooting. Occasionally, completely unrelated samples may yield identical genomic profiles. Such profiles may be caused by several factors including degraded test DNA samples or samples of limited abundance. 'Noisy' hybridizations can be an artifact of a dim overall hybridization signal. Low overall signal intensities will require increased scanner sensitivity (increased time or gain), which can elevate background in the scanned image. Poor DNA quality may also contribute to noise. Ensure that the test and reference DNA are of high purity and high molecular mass. Noise in hybridizations may also be the result of a poor representational amplification being used. Low signal intensity ratios may be present due to poor suppression of repeats by Cot-1 DNA. It is best to evaluate Cot-1 DNA using a sample with known alterations. Heterogeneity of a sample (e.g. normal cell contamination in isolated tumor cells) may also cause a reduction in apparent fluorescent ratios. To avoid this, ensure that rigorous microdissection is undertaken.

Acknowledgements

The authors wish to acknowledge the contributions of all members of the Lam laboratory who offered critical insights into the production of the BAC microarrays. Specifically, we would like to thank Spencer K. Watson and Dr Cathie Garnis for their careful proofreading and thoughtful discussions.

4. REFERENCES

1. **Hanahan D & Weinberg RA** (2000) *Cell*, **100**, 57–70.
2. **de Vries BB, Pfundt R, Leisink M, *et al.*** (2005) *Am. J. Hum. Genet.* **77**, 606–616.
3. **Garnis C, Buys TPH & Lam WL** (2004) *Mol. Cancer*, **3**, 9.

★ 4. **Davies JJ, Wilson IM & Lam WL** (2005) *Chromosome Res.* **13**, 237–248. – *Current review of existing microarray CGH technologies and their application in a cancer research context.*

5. **Lockwood WW, Chari R, Chi B & Lam WL** (2005) *Eur. J. Hum. Genet.* **14**, 139–148.
6. **Pollack JR, Perou CM, Alizadeh AA, *et al.*** (1999) *Nat. Genet.* **23**, 41–46.
7. **Lucito R, Healy J, Alexander J, *et al.*** (2003) *Genome Res.* **13**, 2291–2305.
8. **Matsuzaki H, Dong S, Loi H, *et al.*** (2004) *Nat. Methods*, **1**, 109–111.
9. **Bignell GR, Huang J, Greshock J, *et al.*** (2004) *Genome Res.* **14**, 287–295.
10. **Greshock J, Naylor TL & Margolin A** (2004) *Genome Res.* **14**, 179–187.
11. **Kohlhammer H, Schwaenen C, Wessendorf S, *et al.*** (2004) *Blood*, **104**, 795–801.
12. **Massion PP, Kuo WL, Stokoe D, *et al.*** (2002) *Cancer Res.* **62**, 3636–3640.
13. **Mantripragada KK, Buckley PG, Jarbo C, Menzel U & Dumanski JP** (2003) *J. Mol. Med.* **81**, 443–451.
14. **Nessling M, Richter K, Schwaenen C, *et al.*** (2005) *Cancer Res.* **65**, 439–447.
15. **Roerig P, Nessling M, Radlwimmer B, *et al.*** (2005) *Int. J. Cancer.* **117**, 95–103.
16. **Cheung SW, Shaw CA, Yu W, *et al.*** (2005) *Genet. Med.* **7**, 422–432.
17. **Garnis C, Campbell J, Zhang L, Rosin MP & Lam WL** (2004) *Oral Oncol.* **40**, 511–519.
18. **Schwaenen C, Nessling M & Wessendorf S** (2004) *Proc. Natl. Acad. Sci. U.S.A.* **101**, 1039–1044.
19. **Coe BP, Henderson LJ, Garnis C, *et al.*** (2005) *Genes Chromosomes Cancer*, **42**, 308–313.
20. **Garnis C, Baldwin C, Zhang L, Rosin MP & Lam WL** (2003) *Cancer Res.* **63**, 8582–8585.
21. **Garnis C, Campbell J, Davies JJ, *et al.*** (2005) *Hum. Mol. Genet.* **14**, 475–482.
22. **Garnis C, Coe B, Henderson LJ, *et al.*** (2004) *Chest*, **125**, 104S–105S.
23. **Garnis C, Davies JJ, Buys TP, *et al.*** (2005) *Oncogene*, **24**, 4806–4812.
24. **Garnis C, MacAulay C, Lam S & Lam W** (2004) *Lung Cancer*, **44**, 403–404.
25. **Henderson LJ, Coe BP, Lee EH, *et al.*** (2005) *Br. J. Cancer*, **92**, 1553–1560.
26. **Kameoka Y, Tagawa H, Tsuzuki S, *et al.*** (2004) *Oncogene*, **23**, 9148–9154.
27. **von Duin M, van Marion R, Watson JE, *et al.*** (2005) *Cytometry A*, **63**, 10–19.
28. **Buckley PG, Mantripragada KK, Benetkiewicz M, *et al.*** (2002) *Hum. Mol. Genet.* **11**, 3221–3229.
29. **Davison EJ, Tarpey PS, Fiegler H, Tomlinson IP & Carter NP** (2005) *Genes Chromosomes Cancer*, **44**, 384–391.
30. **Zafarana G, Grygalewicz B, Gillis AJ, *et al.*** (2003) *Oncogene*, **22**, 7695–7701.
31. **Redon R, Rio M, Gregory SG, *et al.*** (2005) *J. Med. Genet.* **42**, 166–171.
32. **Solomon NM, Ross SA, Morgan T, *et al.*** (2004) *J. Med. Genet.* **41**, 669–678.
33. **Snijders AM, Nowak N, Segraves R, *et al.*** (2001) *Nat. Genet.* **29**, 263–264.

★★ 34. **Pinkel D, Segraves R, Sudar D, *et al.*** (1998) *Nat. Genet.* **20**, 207–211. – *The first application of microarray CGH in a cancer context.*

★★★ 35. **Garnis C, Coe BP, Lam SL, MacAulay C & Lam WL** (2005) *Genomics*, **85**, 790–793. – *Paper demonstrating that increased resolution allows greater alteration detection sensitivity, invaluable when analyzing heterogeneous clinical specimens.*

★★★ 36. **Ishkanian AS, Malloff CA, Watson SK, *et al.*** (2004) *Nat. Genet.* **36**, 299–303. – *The first report of a whole genome tiling set microarray.*

37. **Frengen E, Weichenhan D, Zhao B, *et al.*** (1999) *Genomics*, **58**, 250–253.
38. **Osoegawa K, Mammoser AG, Wu C, *et al.*** (2001) *Genome Res.* **11**, 483–496.

39. **Krzywinski M, Bosdet I, Smailus D, *et al.*** (2004) *Nucleic Acids Res.* **32**, 3651–3660.
40. **Marra MA, Kucaba TA, Dietrich NL, *et al.*** (1997) *Genome Res.* **7**, 1072–1084.
★★★ 41. **McPherson JD, Marra M, Hillier L, *et al.*** (2001) *Nature*, **409**, 934–941. – *Presents a physical map of entire human genome.*
42. **Watson SK, DeLeeuw RJ, Ishkanian AS, Malloff CA & Lam WL** (2004) *BMC Genomics*, **5**, 6.
★ 43. **Pfeifer G, Steigerwald S, Mueller P, Wold B & Riggs A** (1989) *Science*, **246**, 810–813. – *The initial report of LM-PCR approaches.*
★ 44. **Telenius H, Carter NP, Bebb CE, *et al.*** (1992) *Genomics*, **13**, 718–725. – *The first report of DOP-PCR.*
45. **Fiegler H, Carr P, Douglas EJ, *et al.*** (2003) *Genes Chromosomes Cancer*, **36**, 361–374.
★ 46. **Hegde P, Qi R, Abernathy K, *et al.*** (2000) *Biotechniques*, **29**, 548–550, 552–54, 556 passim. – *A thorough description of microarray printing.*
★ 47. **McQuain MK, Seale K, Peek J, Levy S & Haselton FR** (2003) *Anal. Biochem.* **320**, 281–291. – *A thorough description of the use of microarray printing pins in microarray production.*
48. **Conrad W** (2005) In *Affymetrix Microarray Bulletin*, pp. 1–24. Affymetrix, Inc.
49. **Emmert-Buck MR, Bonner RF, Smith PD, *et al.*** (1996) *Science*, **274**, 998–1001.
50. **Bonner RF, Emmert-Buck M, Cole K, *et al.*** (1997) *Science*, **278**, 1481–1483.
51. **Rekhter MD & Chen J** (2001) *Cell Biochem. Biophys.* **35**, 103–113.
52. **Pinkel D & Albertson DG** (2005) *Annu. Rev. Genomics Hum. Genet.* **6**, 331–354.
53. **Hughes S, Arneson N, Done S & Squire J** (2005) *Prog. Biophys. Mol. Biol.* **88**, 173–189.
54. **Lasken RS & Egholm M** (2003) *Trends Biotechnol.* **21**, 531–535.
55. **Iafrate A, Feuk L & Rivera M** (2004) *Nat. Genet.* **36**, 949–951.
56. **Sebat J, Lakshmi B, Troge J, *et al.*** (2004) *Science*, **305**, 525–528.
57. **Sharp AJ, Locke DP, McGrath SD, *et al.*** (2005) *Am. J. Hum. Genet.* **77**, 78–88.
58. **Lieu PT, Jozsi P, Gilles P & Peterson T** (2005) *J. Biomol. Tech.* **16**, 104–111.
59. **Tsubosa Y, Sugihara H, Mukaisho K, *et al.*** (2005) *Cancer Genet. Cytogenet.* **158**, 156–166.
60. **Lisitsyn N & Wigler M** (1993) *Science*, **259**, 946–951.
61. **Lucito R, West J, Reiner A, *et al.*** (2000) *Genome Res.* **10**, 1726–1736.
★★★ 62. **Feinberg AP & Vogelstein B** (1983) *Anal. Biochem.* **132**, 6–13. – *Original report of the random priming reaction.*
63. **Feinberg AP & Vogelstein B** (1984) *Anal. Biochem.* **137**, 266–267.
64. **Pinkel D & Albertson DG** (2005) *Nat. Genet.* 37 (Suppl.), S11–S7.
65. **Richter A, Schwager C, Hentze S, *et al.*** (2002) *BioTechniques*, **33**, 620–628, 630.
66. **Yu J, Othman MI, Farjo R, *et al.*** (2002) *Mol. Vis.* **8**, 130–137.
67. **Xiang CC, Kozhich OA, Chen M, *et al.*** (2002) *Nat. Biotechnol.* **20**, 738–742.
★★★ 68. **Solinas-Toldo S, Lampel S, Stilgenbauer S, *et al.*** (1997) *Genes Chromosomes Cancer*, **20**, 399–407. – *The first use of printed large insert clones instead of metaphase chromosome spreads as targets in microarray CGH experiments.*
69. **Yu H, Chao J, Patek D, *et al.*** (1994) *Nucleic Acids Res.* **22**, 3226–32.
70. **Petrescu AD, Payne HR, Boedecker A, *et al.*** (2003) *J. Biol. Chem.* **278**, 51813–51824.
71. **Fare TL, Coffey EM, Dai H, *et al.*** (2003) *Anal. Chem.* **75**, 4672–4675.
72. **Britten RJ & Kohne DE** (1968) *Science*, **161**, 529–540.
73. **Marx KA, Allen JR & Hearst JE** (1976) *Biochim. Biophys. Acta*, **425**, 129–147.
74. **Schrock E, du Manoir S, Veldman T, *et al.*** (1996) *Science*, **273**, 494–497.
★★ 75. **Carter NP, Fiegler H & Piper J** (2002) *Cytometry*, **49**, 43–48. – *This publication presents protocols from various established microarray CGH groups.*
76. **Casey J & Davidson N** (1977) *Nucleic Acids Res.* **4**, 1539–1552.

77. **Simard C, Lemieux R & Cote S** (2001) *Electrophoresis,* **22**, 2679–2683.
78. **Wahl GM, Stern M & Stark GR** (1979) *Proc. Natl. Acad. Sci. U.S.A.* **76**, 3683–3687.
79. **Renz M & Kurz C** (1984) *Nucleic Acids Res.* **12**, 3435–3444.
80. **Amasino RM** (1986) *Anal. Biochem.* **152**, 304–307.
81. **Southern EM** (1975) *J. Mol. Biol.* **94**, 51–69.
82. **Borden JR, Paredes CJ & Papoutsakis E** (2005) *Biophys. J.* **89**, 3277–3284.
83. **Adey NB, Lei M, Howard MT, *et al.*** (2002) *Anal. Chem.* **74**, 6413–6417.
84. **Ting AC, Lee SF & Wang K** (2003) *BioTechniques,* **35**, 808–10.
85. **Hyman E, Kauraniemi P, Hautaniemi S, *et al.*** (2002) *Cancer Res.* **62**, 6240–6245.
86. **Burgess JK** (2001) *Clin. Exp. Pharmacol. Physiol.* **28**, 321–328.
87. **Cleveland WS** (1979) *J. Amer. Stat. Assoc.* **74**, 829–836.
88. **Quackenbush J** (2002) *Nat. Genet.* **32**, 496–501.
89. **Khojasteh M, Lam WL, Ward RK & MacAulay C** (2005) *BMC Bioinformatics,* **6**, 274.
90. **Lai WR, Johnson MD, Kucherlapati R & Park PJ** (2005) *Bioinformatics,* **21**, 3763–3770.
91. **Chi B, DeLeeuw RJ, Coe BP, MacAulay C & Lam WL** (2004) *BMC Bioinformatics,* **5**, 13.
92. **Margolin AA, Greshock J, Naylor TL, *et al.*** (2005) *Bioinformatics,* **21**, 3308–3311.
93. **Siwoski A, Ishkanian A, Garnis C, *et al.*** (2002) *Modern Path.* **15**, 889–892.

CHAPTER 8
Epigenetic analysis of cellular immortalization

Aviva Levine Fridman, Scott A. Tainsky, and Michael A. Tainsky

1. INTRODUCTION

Carcinogenesis is a multi-step process ensuing from the accumulation of mutations in tumor-suppressor genes and oncogenes that confer growth advantages and/or genomic instability on the cell. One of the critical steps in this process is immortalization, a process in which cells must acquire an infinite lifespan. In the absence of immortalization, a cell may undergo malignant transformation but cannot proliferate indefinitely. Frequently, the loss of tumor-suppressive genes contributes to cellular immortalization. This often occurs by multiple mechanisms, including loss of heterozygosity and/or silencing of these genes by methylation of CpG islands in their promoters (1). CpG islands, which represent about 1% of the human genome, are 200–2000 bp stretches of CpG clusters. More than 60% of human genes contain a CpG island in their promoter (2). Typically, CpG islands are unmethylated. However, when they become methylated, either as a consequence of aberrant hypermethylation during immortalization or during normal cellular processes including developmental imprinting, X-chromosome inactivation, and tissue-specific gene expression, the genes become transcriptionally inactive. This process is referred to as epigenetic silencing (1, 3–5).

Treatment with an inhibitor of DNA methyltransferase such as 5-aza-deoxycytidine (5-aza-dC) can restore expression of genes aberrantly silenced in immortal cells (6, 7). Normal cells in culture grow until they reach a growth crisis and then cease to proliferate. In our laboratory, we identify genes and gene pathways that are regulated epigenetically during immortalization by comparing pre-crisis, mortal cells with immortal cells, and immortal cells before and after treatment with 5-aza-dC, using high-throughput microarray technology. The methods for

DNA Microarrays: *Methods Express* (M. Schena, ed.)

identifying gene expression changes using high-throughput expression profiling are well established and reliably measure up- and downregulated genes. Using standard methods, genes whose expression decreases during immortalization and increases after 5-aza-dC treatment are identified as putative epigenetically downregulated genes. Bioinformatics programs, such as METHPRIMER (http://www.urogene.org/methprimer/index1.html), are then used to determine whether there are CpG islands in the promoters of such genes.

In general, when researchers identify genes whose expression increases after 5-aza-dC treatment and have a computational CpG island, they assume that the gene is regulated epigenetically and do not perform additional verification. We have found that this is not a valid assumption. First, DNA methyltransferase inhibitors, such as 5-aza-dC, are known to affect DNA in ways other than inhibiting DNA methyltransferase. Secondly, a gene with a CpG island may appear to respond to a DNA methyltransferase inhibitor in a manner consistent with methylation, but in actuality the methylation status of the CpG island within its promoter may not change. Our experience indicates that it is important to verify that a gene is regulated epigenetically by performing additional experiments, such as methylation-specific PCR, PyroMeth, MethylLight, combined bisulfite restriction analysis, or bisulfite sequencing (reviewed in 8). Of these methods, the detection of DNA methylation using bisulfite sequencing is the most reliable.

Researchers also use microarray data to determine the significant pathways that are regulated epigenetically in their particular experiments. There are several bioinformatics programs that are available for identifying the most significant pathways given a set of genes. These programs are based on gene ontology associations and/or known pathways such as those found in the Kyoto Encyclopedia of Genes and Genomes (KEGG) (9). Website links and descriptions of several gene ontology tools can be found at http://www.geneontology.org/GO.tools.shtml#in_house. In these analyses, researchers generally compare a list of genes whose expression is altered with the entire set of genes on a microarray chip. Using the entire set of genes on a microarray may not provide the optimal reference set, as not all of the genes on a given chip are necessarily relevant. Some genes, for example, may not be regulated due to cell-type specificity of gene expression. Using genes that never change in a given system as the reference set of genes should improve the quality of the pathway analysis. In this chapter, we outline how to identify and confirm epigenetically regulated genes and the pathways in which these genes participate using microarray data.

2. METHODS AND APPROACHES

2.1 Methods for detecting CpG island methylation

There are several methods, including methylation-sensitive arbitrary primed PCR, methylated CpG island amplification, and restriction landmark genomic scanning, that have been employed to identify methylated genes (8). The limitations of these methods include a requirement for high-quality DNA and being restricted by methylation-specific restriction sites. To identify novel epigenetically regulated genes, we and others have performed large-scale screenings using microarray analysis comparing cells before and after treatment with DNA methyltransferase inhibitors, specifically 5-aza-dC (7, 10). These studies make the assumption that a gene is regulated epigenetically if it is upregulated by 5-aza-dC and has a CpG island, as assessed bioinformatically. However, this assumption is not valid. We find that there is little difference in the fraction of genes containing a significant CpG island that are regulated epigenetically during immortalization compared with control sets of genes such as those that decrease during immortalization without an increase in expression after 5-aza-dC treatment or a randomly selected set of genes (see *Protocol 2* and *Table 1*). Interestingly, a difference was observed when we compared the number of genes in these three sets with >80% CpG island content. There was a higher fraction of epigenetically regulated genes with this very high CpG content. However, CpG content is not an indicator of whether a

Table 1. Comparison of the percentage of genes with a CpG island as a function of gene category

Percentage of gene with X% or greater CpG island content from −1000 to +500 bp from the transcription start site	Genes in each group with X% or greater CpG island[a] content		
	Set of 14 genes that decrease during immortalization and increase after 5-aza-dC treatment	Set of 16 genes that decrease after immortalization and show no change after 5-aza-dC treatment	60 genes chosen at random
>5% CpG island	78%	94%	82%
>40% CpG island	71%	75%	53%
>50% CpG island	36%	37%	42%
>60% CpG island	21%	31%	28%
>70% CpG island	21%	25%	10%
>80% CpG island	21%	0%	7%

[a]A CpG island (or CG island) is a short sequence of DNA in which the frequency of CG nucleotides is higher than in other regions.

CpG island in a gene will be methylated (11). Furthermore, we found that some genes, such as STAT1, that have a CpG island in their promoter and fit the profile of an epigenetically regulated gene (downregulated after immortalization and upregulated in immortal cells by 5-aza-dC treatment), do not contain CpG island methylation in their promoters (12). Therefore, it is necessary to perform an assay that specifically detects whether or not CpG islands are methylated.

Many of the CpG methylation detection techniques are a variation of methylation-specific PCR. In general, the advantages of the methylation-specific PCR assays are that they are high-throughput, sensitive, and quantitative. The disadvantage of these assays is that they are highly dependent on the efficiency of bisulfite conversion (8). If bisulfite conversion is not complete, then there may be unmethylated cytosines that are not converted to uracil, which may give the appearance that the gene is methylated when in fact it is unmethylated (13). To avoid these problems, bisulfite sequencing should be used for the initial analysis of gene methylation. Bisulifte sequencing is the most comprehensive method of detecting CpG island methylation and does not rely on the analysis of short regions of each sequence. The disadvantage of the bisulfite sequencing method over other methylation detection methods is that it is not as sensitive at detecting differences in methylation if only a small percentage of the DNA molecules (less than about 10%) are differentially methylated. Bisulfite sequencing is also more time-consuming and in general more expensive to perform.

Following analysis of microarray data for genes that are up- and downregulated, there are several bioinformatics approaches available to determine the pathways that are regulated in a particular system. These software methods all depend on gene ontology annotation (14) or KEGG (15, 16), which map biological pathways to identify the significant pathways for a particular set of up- and downregulated genes. A major drawback of these programs is that not all of the known genes are annotated or mapped to a pathway, and therefore the process is inherently incomplete. If bioinformatics programs alone are used to identify pathways, then all of the possible pathways that are disrupted/regulated for a particular experimental condition may not be identified. Furthermore, frequently when these programs are used they commonly compare the list of up- and downregulated genes with the entire microarray chip. The problem in using the entire microarray dataset as the reference is that some of the genes on the chip may not change under any circumstance in the tissue in question. This could result in an inflated significance of some pathways, whilst other pathways appear insignificant (see *Table 2, Protocol 7*, and section 3). To avoid this problem, it may be important to identify the genes that could

Table 2. Comparison of the GOMINER results when the entire microarray chip is used as the reference list compared with using the 'ever changed' genes as the reference

GO category[b]	Entire HGU95Av2[a] used as the reference list			'Ever changed' genes on HGU95Av2 used as the reference list		
	Total genes	Changed genes	FDR[c]	Total genes	Changed genes	FDR[c]
GO:0050875 cellular physiological process	4987	210	0.00	2246	210	0.32
GO:0051243 negative regulation of cellular physiological process	335	24	0.02	179	24	0.08
GO:0016043 cell organization and biogenesis	509	33	0.02	261	33	0.09
GO:0030832 regulation of actin filament length	10	3	0.03	7	3	0.06
GO:0043118 negative regulation of physiological process	355	24	0.03	183	24	0.08
GO:0008361 regulation of cell size	86	9	0.03	49	9	0.08
GO:0016049 cell growth	86	9	0.03	49	9	0.08
GO:0001558 regulation of cell growth	71	8	0.03	41	8	0.08
GO:0050794 regulation of cellular process	1616	79	0.03	741	79	0.09
GO:0051244 regulation of cellular physiological process	1535	76	0.03	709	76	0.09
GO:0000910 cytokinesis	107	10	0.03	69	10	0.15
GO:0051301 cell division	107	10	0.03	69	10	0.15
GO:0050789 regulation of biological process	1798	86	0.04	804	86	0.08
GO:0009082 branched chain family amino acid biosynthesis	3	2	0.04	3	2	0.08
GO:0006026 aminoglycan catabolism	4	2	0.04	3	2	0.08
GO:0006027 glycosaminoglycan catabolism	4	2	0.04	3	2	0.08
GO:0006271 DNA strand elongation	4	2	0.04	3	2	0.08
GO:0050791 regulation of physiological process	1647	79	0.04	746	79	0.11
GO:0007275 development	1180	55	0.25	419	55	0.01
GO:0009653 morphogenesis	823	38	0.36	285	38	0.02

[a]Affymetrix Human Genome U95 Set Annotation Data (HGU95Av2).
[b]This table shows only the gene ontology categories with a value of $P < 0.05$ when one reference list is used and $P \geq 0.05$ when the other reference list is used. The electronic link is http://discover.nci.nih.gov/gominer/hi-thruput-defs.jsp#GO.
[c]FDR, false discovery rate.

change for a particular cell line or tissue and use this as the reference for the bioinformatics programs.

2.2 Recommended protocols

Protocol 1 describes the treatment of cells, in this case immortal cells, with 5-aza-dC. Microarray chip and data analysis were performed on results obtained from RNA extracted from the parental pre-crisis cells, and untreated and 5-aza-dC-treated immortal cells. However, these protocols are beyond the scope of this chapter and therefore are not described. Genes that decreased during immortalization and increased after 5-aza-dC treatment were considered to be epigenetically regulated. *Protocol 2* is used to determine whether there are CpG islands in the promoters of these genes. DNA extraction (*Protocol 3*) followed by bisulfite sequencing (*Protocols 4* and *5*) is performed to confirm methylation of the promoter during immortalization and demethylation following 5-aza-dC treatment. Generally speaking, we find that there are two critical factors for obtaining good microarray data. First, microarray analysis should be performed using only high-quality RNA. Secondly, multiple microarray chips should be analyzed for each treatment group and the RNA for each of the analyses should be from independent experiments.

Protocol 1

Treatment of cells with 5-aza-dC (6, 7)

Reagents

- 5-Aza-dC (Sigma)
- Appropriate cell culture medium
- RNeasy kit (Qiagen)

Method

1. On day 1, seed 3×10^5 cells per plate in the appropriate medium[a,b].
2. On days 2, 4, and 6, replace the medium in the plates with medium containing 1 μM 5-aza-dC[c].
3. On day 7, replace the medium on the plates with medium that does not contain 5-aza-dC.
4. On day 8, harvest total RNA from the cells using an RNeasy kit or a comparable method.

Notes

[a]It is necessary to seed two to three times as many plates of cells that contain 5-aza-dC to obtain similar numbers of cells as control plates following 5-aza-dC treatment.
[b]To prepare a 1 mM stock solution of 5-aza-dC, dissolve in tissue culture sterile water and prepare fresh on the day of use.
[c]An alternative program that can be used is CPGPLOT (http://www.ebi.ac.uk/emboss/cpgplot/). Both METHPRIMER and CPGPLOT analyze windows of 100 bp with each subsequent window shifting by 1 bp from the previous window. To determine whether there is a CpG island for each window, these programs calculate the observed ratio of C + G to CpG, and a minimum average percentage of G + C; the default values used for these parameters are >0.6 and >50, respectively (17).

Protocol 2

Identification of CpG islands in the promoter region of genes

Equipment

- Computer with Internet access
- METHPRIMER (http://www.urogene.org/methprimer/index1.html)[a]

Method

1. In a gene of interest, identify −1000 to +500 bp of the transcription start site using a program of choice. We use the UCSC GENOME BROWSER (http://genome.ucsc.edu)[b] with good success.
2. Analyze the 1500 bp sequence using a bioinformatics program that identifies CpG islands. We use METHPRIMER with good success to identify CpG islands (18).

Notes

[a]An alternative program that can be used is CPGPLOT (http://www.ebi.ac.uk/emboss/cpgplot/). Both METHPRIMER and CPGPLOT analyze windows of 100 bp, with each subsequent window shifting by 1 bp from the previous window. To determine whether there is a CpG island for each window, these programs calculate the observed ratio of C + G to CpG and the minimum average percentage of G + C; the default values used for these parameters are >0.6 and >50, respectively (17).
[b]Another program that can be used to identify the promoter of a gene is to be found at the Advanced Biomedical Computer Center at the National Cancer Institute (http://grid.abcc.ncifcrf.gov/promoters.php).

Protocol 3

DNA extraction

Reagents

- DNA extraction kit, such as Gentra Puregene Cell Kit (Qiagen)

Method

1. Follow the protocol provided in the Gentra Puregene Cell Kit.

Protocol 4

Genomic DNA bisulfite treatment and purification (19, 20)

Equipment and Reagents

- Spectrophotometer
- Thermocycler
- Heat block, heated water bath, or thermocycler
- Microcentrifuge
- 0.5 ml Microfuge tubes
- 0.2 ml PCR tubes
- 20-Gauge needle
- Genomic DNA
- 3 M NaOH
- 10 mM Hydroquinone (make fresh on day of experiment)
- 3.6 M Sodium bisulfite with pH adjusted to 5.0 using 3 M NaOH (make fresh on day of experiment)
- 100% Ethanol
- 70% Ethanol
- 3 M Sodium acetate (pH 5.2)
- Genomic DNA purification kit (e.g. Wizard Genomic DNA purification kit; Promega)

Method

1. Determine the concentration of the genomic DNA using a spectrophotometer.
2. Resuspend 3 μg of DNA in dH_2O to a final volume of 54 μl[a].
3. Shear the DNA by passing it through a 20-gauge needle ten times.
4. To a 0.2 ml PCR tube, add 18 μl of sheared DNA (1 μg) and 2 μl of 3 M NaOH. Incubate the mixture for 15 min at 37°C to denature the DNA.
5. To the denatured DNA, add 112.8 μl of freshly prepared 3.6 M sodium bisulfite (pH 5.0) and 7.2 μl of freshly prepared 10 mM hydroquinone.
6. Incubate the DNA samples using the following thermocycler program: 20 cycles of 30 s at 95°C and 15 min at 50°C; then 1 cycle holding at 4°C.
7. Desalt and purify the DNA using a Wizard Genomic DNA purification kit following the protocol provided by the manufacturer.

8. Add 5.56 μl of 3 M NaOH to the sample and incubate at 37°C for 15 min to desulfonate the DNA sample.
9. To the desulfonated genomic DNA, add 6.2 μl of 3 M sodium acetate (pH 5.2) and 154 μl 100% ethanol. Incubate the samples at −70°C for 10 min or −20°C for 60 min.
10. Centrifuge the samples in a microfuge for 15 min at full speed (15 000 *g*).
11. Decant and discard the ethanol, being careful not to disturb the DNA pellet.
12. Wash the pellet once with 1 ml of 70% ethanol.
13. Centrifuge the samples in a microfuge for 5 min at maximum speed (15 000 *g*).
14. Decant and discard the supernatant, being careful not to disturb the DNA pellet.
15. Dry the pellet by vacuum centrifugation and then resuspend the DNA in 50 μl of sterile dH_2O.

Note

[a]We recommend shearing a larger volume of DNA than the required 18 μl (step 4), as some is lost during shearing.

Protocol 5

Bisulfite sequencing (19)

Equipment and Reagents

- Spectrophotometer
- Thermocycler
- 0.2 ml PCR tubes
- 10× High Fidelity PCR Buffer (Invitrogen)
- 50 mM $MgSO_4$
- dNTP mix (10 mM each dATP, dTTP, dGTP, and dCTP)
- 10 μM Forward primer-1
- 10 μM Reverse primer-1
- 10 μM Forward primer-2
- 10 μM Reverse primer-2
- 5 units/μl Platinum *Taq* High Fidelity DNA polymerase
- 1.5% Agarose gel containing 0.5 μg/ml ethidium bromide
- QIAquick Gel Extraction kit (Qiagen)
- pCR2.1-TOPO TA Cloning kit (Invitrogen)
- QIAprep Spin Miniprep kit (Qiagen)
- Luria–Bertani (LB) medium containing 50 μg/ml ampicillin
- *Eco*RI restriction enzyme
- M13 reverse primer: 5′-CAGGAAACAGCTATGAC-3′

Method

1. Design primers for PCR-based amplification[a] of the bisulfite-treated DNA using METHPRIMER (http://www.urogene.org/methprimer/index1.html)[a].

2. In a 0.2 ml PCR tube for the first PCR, combine the following reagents for a 25 μl total volume[b]: 2.5 μl of 10× High Fidelity Buffer, 1 μl of 50 mM $MgSO_4$, 0.5 μl of 10 mM dNTP mix, 0.5 μl of 10 μM Forward primer-1, 0.5 μl of 10 μM Reverse primer-1, 9.8 μl of sterile dH_20, 0.2 μl of 5 units/μl Platinum *Taq* High Fidelity DNA polymerase, and 10 μl of DNA (from *Protocol 4*, step 15).
3. Amplify the PCR mixture using the following program: 1 cycle of 2 min at 94°C; 35 cycles of 30 s at 94°C, 30 s at 55°C[c], and 1 min at 68°C; 1 cycle of 5 min at 68°C; and hold at 4°C.
4. For the second PCR, combine the following reagents in a 0.2 ml PCR tube for a 100 μl reaction[d]: 10 μl of 10× High Fidelity PCR Buffer, 4 μl of 50 mM $MgSO_4$, 2 μl of 10 mM dNTP mix, 2 μl of 10 μM Forward primer-2, 2 μl of 10 μM Reverse primer-2, 55.6 μl of sterile dH_20, 0.4 μl of 5 units/μl Platinum *Taq* High Fidelity DNA polymerase, and 20 μl of PCR-amplified DNA from the previous step (step 3).
5. Run the amplified products on a 1.5% preparatory agarose gel.
6. Extract the DNA from the gel using a QIAquick Gel Extraction kit.
7. Elute the DNA with 30 μl of elution buffer.
8. Clone the PCR products into the pCR2.1-TOPO TA cloning vector. Add 4 μl of gel-extracted PCR product from step 7 to 1 μl of cloning vector and mix by pipetting. For the remaining cloning steps, follow the protocol provided in the cloning kit.
9. Pick white colonies and inoculate into 2 ml of LB medium containing 50 μg/ml ampicillin (one colony per tube). Grow overnight in a shaker at 37°C.
10. Extract plasmid DNA from the bacterial cells using a QIAprep Spin Miniprep kit. Elute the plasmid DNA with 50 μl of elution buffer.
11. Digest the plasmid DNA with *Eco*RI and run the digested DNA on a 1.5% agarose gel. Check for the fragment of the appropriate size.
12. Perform DNA sequencing on 5 μl of plasmid DNA using the M13 reverse primer.
13. Analyze the sequencing results for CpG methylation using a program such as METHTOOLS (http://genome.imb-jena.de/methtools/) (21).

Notes

[a]As a consequence of the bisulfite treatment, the DNA becomes fragmented and it is therefore difficult to amplify sequences that are longer than 600 bp. The ideal length of a sequence to amplify is about 300 bp.

[b]A master mix can be made that contains all of the reagents except the DNA. Mix and aliquot 15 μl of master mix into each 0.2 ml PCR tube, then add the DNA (*Protocol 4*, step 15) and mix by pipetting up and down.

[c]The annealing temperature will depend on the sequence being amplified and must be determined empirically.

[d]A master mix can be made that contains all of the reagents except the DNA. Mix and aliquot 15 μl of master mix into each 0.2 ml PCR tube. Add the DNA from *Protocol 4*, step 15, and mix by pipetting up and down.

Protocols 6 and *7* describe gene ontology classification of epigenetically regulated genes using 'ever changed' genes, or those genes whose expression changes under any condition, as the reference set of genes, as opposed to the entire microarray chip. Determining the gene ontology categories that have a statistically significant number of up- and downregulated genes facilitates the identification of pathways that are dysregulated during immortalization.

Protocol 6

Identification of 'ever changed' genes[a]

Equipment

- Microarray data
- Software program for microarray analysis

Method

1. Identify the 'ever changed' genes, which are those genes on the microarray whose expression changes under one or more physiological conditions.
2. Identify the genes on the microarray whose expression does not change and would not be expected to change under any relevant physiological condition.
3. Eliminate the genes in the reference list whose expression does not change. Eliminating such genes in the reference list used in the bioinformatic and statistical analyses of the microarray data should increase the accuracy of the results.

Note

[a]This protocol will depend on the microarray analysis software used.

Protocol 7

Gene ontology analysis of genes using GOMINER

Equipment

- A text file containing an 'ever changed'[a] gene list (see *Protocol 6*)
- A text file containing a list of genes that are up- and downregulated[b]
- Computer with Internet access
- GOMINER (http://discover.nci.nih.gov/gominer/)

Method

1. Upload files for analysis at the following web address: http://discover.nci.nih.gov/gominer/GoCommandWebInterface.jsp.
2. Receive the results via e-mail.

3. Compare GOMINER results when the entire microarray chip is used as the reference list of genes versus when only the 'ever changed' genes are used as the reference list (see *Table 3*)[c].

Table 3. Comparison of GOMINER results when the entire microarray chip is used as the reference list of genes compared with using the 'ever changed' genes as the reference

	Number of significant gene ontology categories ($P < 0.05$)	Significant ($P < 0.05$) gene ontology categories unique to the reference list used	Number of gene ontology categories ($P \geq 0.05$)
Entire HGU95Av2 used as reference list	65	18	690
'Ever changed' genes from HGU95Av2 used as a reference list	49	2	806
Common to both	47	NA	688

NA, Not applicable.

Notes

[a]An example list can be found at http://discover.nci.nih.gov/gominer/files/cvid/input/total.txt.
[b]An example list can be found at http://discover.nci.nih.gov/gominer/files/cvid/input/cvidchanged.txt. Note it is not necessary to include the second column in the list that indicates genes increased by 1 or decreased by −1.
[c]Nearly 30% (18 out of 65) of the gene ontology categories identified using the entire microarray chip as the reference are potentially false positives, as these categories are not identified when the 'ever changed' genes are used as the reference. Furthermore, when our criteria 2 gene ontology categories are used, these categories were not identified as significant, even though they were identified as significant when the entire microarray chip was used as the reference. These data illustrate the significance that the reference list can have on the bioinformatics results. *Table 2* gives the gene ontology categories that were significant for only one of the reference lists used in GOMINER.

3. TROUBLESHOOTING

It is important to confirm that the region −1000 and +500 (i.e. 1000 bp upstream and 500 bp downstream) of the transcription start site has been correctly identified. The bioinformatics programs used to identify regions surrounding the transcription start site are not always accurate. For instance, the transcription start site might not be correctly identified for genes with splice variants. Care must therefore be taken when identifying transcription start sites.

Bisulfite sequencing is optimally performed using a nested PCR scheme, which helps to increase the quantity of DNA (20). Nested PCR also reduces the nonspecific binding of primers and hence the amplification of unmethylated DNA where C residues in the CpG islands were not converted to U residues in the bisulfite sequencing process. This is important if a technique such as methylation-specific PCR is performed in which nonspecific binding can result in an overestimation of methylated DNA (13).

We highly recommend using control methylated DNA and unmethylated DNA to ensure that the bisulfite treatment of DNA is being performed correctly. Bisulfite sequencing should be performed on at least ten colonies from at least two independent PCR amplification experiments. This will reduce the possibility of PCR disproportionately amplifying the rare products, if there are only a small percentage of methylated or unmethylated genes.

The gene identifier (i.e. gene symbol, microarray target identification, unigene number, etc.) used in the bioinformatic analysis can affect the results. We have found that there are differences in the results depending on the identifier used. Primarily, this is a consequence of the number of a particular identifier that a bioinformatics program recognizes. Our recommendation is to analyze microarray data using at least two different gene identifiers (i.e. target identification, gene symbol, and Entrez GeneID number).

There are differences among bioinformatics programs in the number of genes they recognize when given the same list of genes. Our recommendation is to analyze microarray data with at least two different bioinformatics programs and then compare the results. It is important to be aware of the statistics that the bioinformatics programs use to identify significant gene ontology categories. Assessing microarray data using the false discovery rate, which corrects for the problem of multiple comparisons, can reduce false positives.

Gene ontology analysis of microarray data is only a tool to help elucidate the pathways and processes that are regulated in a particular system. The gene ontology database is not complete though. Fifteen percent of the unique genes on the particular microarray used in this study were not annotated. Furthermore, even if a gene is annotated, there may be other categories under which the gene should have been listed.

4. REFERENCES

1. **Baylin SB & Herman JG** (2000) *Trends Genet.* **16**, 168–174.
2. **Antequera F** (2003) *Cell. Mol. Life Sci.* **60**, 1647–1658.

3. **Esteller M, Corn PG, Baylin SB & Herman JG** (2001) *Cancer Res.* **61**, 3225–3229.
4. **Feinberg AP, Cui H & Ohlsson R** (2002) *Semin. Cancer Biol.* **12**, 389–398.
5. **Baylin SB, Esteller M, Rountree MR, *et al.*** (2001) *Hum. Mol. Genet.* **10**, 687–692.
6. **Vogt M, Haggblom C, Yeargin J, Christiansen-Weber T & Haas M** (1998) *Cell Growth Differ.* **9**, 139–146.
7. ★★ **Kulaeva OI, Draghici S, Tang L, *et al.*** (2003) *Oncogene*, **22**, 4118–4127. – *Presents the protocol used in this chapter for identifying epigenetically regulated genes.*
8. ★★★ **Cottrell SE** (2004) *Clin. Biochem.* **37**, 595–604. – *Excellent review of the methods available for identifying CpG island methylation.*
9. **Harris MA, Clark J, Ireland A, *et al.*** (2004) *Nucleic Acids Res.* **32**, D258–D261.
10. **Liang G, Gonzales FA, Jones PA, Orntoft TF & Thykjaer T** (2002) *Cancer Res.* **62**, 961–966.
11. **Feltus FA, Lee EK, Costello JF, Plass C & Vertino PM** (2003) *Proc. Natl. Acad. Sci. U.S.A.* **100**, 12253–12258.
12. **Tang L, Roberts PC, Kraniak JM, Li Q & Tainsky MA** (2006) *J. Interferon Cytokine Res.* **26**, 14–26.
13. **Sasaki M, Anast J, Bassett W, *et al.*** (2003) *Biochem. Biophys. Res. Commun.* **309**, 305–309.
14. ★ **Ashburner M, Ball CA, Blake JA, *et al.*** (2000) *Nat. Genet.* **25**, 25–29. – *Provides a description of the Gene Ontology Database.*
15. **Kanehisa M** (1997) *Trends Genet.* **13**, 375–376.
16. **Kanehisa M & Goto S** (2000) *Nucleic Acids Res.* **28**, 27–30.
17. **Gardiner-Garden M & Frommer M** (1987) *J. Mol. Biol.* **196**, 261–282.
18. **Li LC & Dahiya R** (2002) *Bioinformatics*, **18**, 1427–1431.
19. ★ **Rein T, Zorbas H & DePamphilis ml** (1997) *Mol. Cell. Biol.* **17**, 416–426. – *The bisulfite treatment and sequencing protocols described in this chapter are based on this publication.*
20. ★★ **Warnecke PM, Stirzaker C, Song J, *et al.*** (2002) *Methods*, **27**, 101–107. – *Discussion of the artifacts that can occur during bisulfite treatment and DNA sequencing.*
21. **Grunau C, Schattevoy R, Mache N & Rosenthal A** (2000) *Nucleic Acids Res.* **28**, 1053–1058.

CHAPTER 9

Microarray comparative genomic hybridization

Simon Hughes, Richard Houlston, and Jeremy A. Squire

1. INTRODUCTION

The genetic changes associated with cancer frequently involve loss of specific chromosomal regions that may contribute to inactivation of tumor suppressor genes, or amplification of key regions possibly increasing the level of expression of oncogenes (1, 2). Screening for alterations in gene expression generally requires high-quality RNA that is not always available from the available tissue samples. In contrast, obtaining DNA from the same tissue for comparative genomic hybridization (CGH) analysis is less problematic and can permit the detection of alterations at the genomic level that can be indicative of modifications of gene expression.

The advent of CGH has contributed greatly to our understanding of chromosomal changes, by making all of the chromosomal changes associated with tumor development and progression detectable in a single experiment (3, 4). Furthermore, it has greatly simplified the study of many tumors for which obtaining metaphase chromosomes or interphase nuclei is technically challenging.

CGH as first described by Kallioniemi *et al.* (5) used chromosomal targets derived from a normal cell population to determine the cytogenetic composition of the tumor of interest. Even when optimal, chromosomal CGH has a resolving power of only approximately 10 Mb for a simple loss or gain, increasing to approximately 2 Mb for a high-copy-number gain (5). This low resolving power of conventional CGH means that small focal changes have a high probability of being missed. The advent of microarray CGH technology has resulted in a switch from chromosome targets to DNA sequence targets, which allows the analysis of chromosomal changes at a far higher resolution.

DNA Microarrays: *Methods Express* (M. Schena, ed.)

2. METHODS AND APPROACHES

The advantage of 'microarray CGH' or 'matrix CGH' is that it enables the researcher to measure DNA copy number changes accurately at high resolution and to use *in silico* analysis to accurately map these changes directly to chromosomal locations. For bacterial artificial chromosome (BAC) microarrays, this resolution can be less than 100 kb (6), and for cDNA microarrays, this can be on a gene-by-gene basis, whilst for oligonucleotide microarrays, oligonucleotides can be designed for any specific region of the genome that is of particular interest. Moreover, the most recent oligonucleotide microarray configurations can offer the benefit of both DNA copy number assessment and allelotyping determined based on single nucleotide polymorphisms ('SNP chips') (7). High-density SNP microarrays such as the Affymetrix GeneChip systems enable the genotyping of thousands of SNPs to determine allelic origins of DNA imbalances and to define loss of heterozygosity haplotypes at high resolution in one assay.

The approach used for the different microarray platforms, with the exception of the Affymetrix microarrays, is essentially the same (displayed graphically in *Fig. 1*, also available in the color section). Equal quantities of tumor and a normal reference genomic DNA are labeled either directly

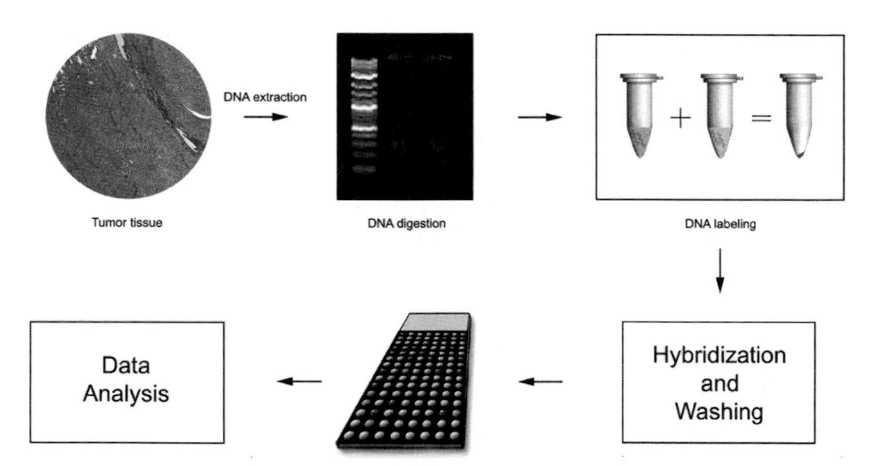

Figure 1. Graphical representation of the CGH technique (see page xxix for color version).
Test genomic DNA is first extracted from tumor tissue and control DNA is obtained from either commercial sources or from normal tissue. The DNA is digested using a restriction enzyme prior to labeling either directly or indirectly with different fluorescent dyes. The differentially labeled DNAs are combined and hybridized to a microarray of DNA sequences. Labeled tumor DNA competes with labeled normal reference DNA and the ratio of fluorescence for the two dyes can then be used to determine chromosomal loss or gain.

or indirectly with different fluorescent dyes; the DNAs are then mixed and hybridized in equal amounts to immobilized targets on microscope slides. Labeled tumor DNA competes with differentially labeled and equimolar concentrations of normal reference DNA for hybridization. The ratio of fluorescence for the two dyes at the target is then used to map out DNA copy number changes between the two DNA samples. Following hybridization, the interpretation of fluorescent signals requires sophisticated image analysis equipment and downstream analysis software. Although not dealt with in this chapter, there are several analysis programs now available that provide graphical representations of chromosomal changes. The available programs are relatively straightforward and can be mastered quite quickly.

This chapter will provide an overview of some of the CGH methods that are currently used in cytogenetics and genomics laboratories, and as an example their usefulness for genome-wide analysis of chromosomal imbalances in cancers will be described.

2.1 Recommended protocols

Protocol 1

Extraction of high-molecular-mass DNA from cells and tissue

Equipment and Reagents

- Tissue culture cells or fresh or frozen tissue/tissue sections
- Phosphate buffered saline (PBS) (pH 7.0; Sigma)
- 10× Trypsin (Sigma)
- 15 ml Polypropylene centrifuge tube (Nunc)
- Proteinase K buffer: 5 ml of 1 M Tris/HCl (pH 7.0; Sigma), 0.2 ml of 500 mM EDTA (pH 8.0; Sigma), 0.5 ml of Tween 20 (Sigma), and sterile water up to 100 ml
- Proteinase K (14 mg/ml; New England Biolabs)
- Sterile scalpel blade
- Buffer-saturated phenol (Sigma)
- Chloroform : isoamyl alcohol (24 : 1; Sigma)
- Isopropanol (Sigma)
- 70% Ethanol (Sigma)
- Sterile water (DNase/RNase-free; Promega)

Method

1. Grow cells to 75–90% confluency. Remove the medium and wash cells twice with 1× PBS. For a tissue sample, chop 10–100 mg of tissue into small pieces using a scalpel blade.
2. Dislodge the cells from the bottom of the flask by treatment with 10 ml of 1× trypsin (diluted 1 : 10 with 1× PBS) for 5 min at 37°C.

3. Transfer the suspended cells to a 15 ml polypropylene centrifuge tube and pellet by centrifugation (200 *g* for 5 min). Discard the supernatant. For a tissue sample, transfer the tissue to a 15 ml polypropylene tube.
4. Resuspend the cell pellet in 2 ml of proteinase K solution (100 μg/ml proteinase K in proteinase K buffer). Incubate overnight at 37°C with constant gentle shaking. For a tissue sample, add 2 ml of proteinase K solution (100 μg/ml proteinase K in proteinase K buffer) and incubate for 24–48 h at 37°C with constant gentle shaking.
5. Extract the DNA using the phenol:chloroform extraction protocol described below[a]. Add 2 ml of phenol for tissue culture cells and 4 ml of phenol for a chopped tissue sample. Invert the tube 20–30 times to mix.
6. Centrifuge the samples for 10 min at 16000 *g*.
7. Transfer the upper aqueous layer (containing the DNA) to a fresh 15 ml tube. Add an equal volume of phenol:chloroform:isoamyl alcohol (25:24:1, v/v) and invert the tube 20–30 times to mix. Centrifuge the samples for 10 min at 16000 *g*.
8. Transfer the upper aqueous layer (containing the DNA) to a fresh 15 ml tube, add an equal volume of chloroform, and invert tube 20–30 times to mix. Centrifuge the samples for 10 min at 16000 *g*.
9. Transfer the upper aqueous layer (containing the DNA) to a fresh 15 ml tube. To precipitate the DNA, add 0.8 vols of isopropanol and invert tube 20–30 times to mix.
10. If the DNA is visible in the isopropanol solution, remove the DNA by spooling it around a sterile inoculation loop, place the DNA in a new tube containing 70% ethanol, and centrifuge for 10 min at 16000 *g*. If no DNA is visible, incubate the sample at −70°C for 20 min or −20°C overnight, and centrifuge for 10 min at 16000 *g*.
11. Decant and discard the supernatant and wash the DNA pellet with 1 ml of 70% ethanol.
12. Centrifuge the samples at 16000 *g* for 10 min, carefully remove and discard the supernatant, and allow the pellet to air dry[b].
13. Dissolve the DNA pellet in 200 μl of sterile dH_2O and measure the concentration using a spectrophotometer[c].

Notes

[a]Avoid vortexing the solution during the DNA extraction steps as this can cause shearing of the genomic DNA.

[b]It is essential that all of the ethanol has evaporated, as residual ethanol can interfere with dissolving the DNA.

[c]It is advisable to leave the DNA at 4°C overnight to aid dissolving. If the DNA does not dissolve completely, gently pipette the solution using a 1 ml pipette tip with the end cut off to increase the size of the tip bore, which prevents DNA shearing. If the DNA concentration is too high for the 200 μl volume of sterile dH_2O, add an additional 100–200 μl to dissolve the DNA.

Protocol 2

DNA restriction digest

Equipment and Reagents

- PCR tubes (0.2 or 0.5 ml; ABgene)
- *RsaI* and accompanying 10× digest buffer (New England Biolabs)
- Sterile water (DNase/RNase-free; Promega)
- Heating block capable of holding 0.2 and 0.5 ml tubes
- 1% Agarose gel containing 10 ng/ml ethidium bromide (Invitrogen)
- 100 bp DNA ladder (Fermentas)
- 6× Orange loading dye solution (Fermentas)
- Equipment and reagents for agarose gel electrophoresis including 1× TBE agarose gel running buffer (10.8 g/l Tris base; 5.5 g/l boric acid; 4 ml/l 0.5 M EDTA, pH 8.0, diluted from a 10× stock; Sigma)
- UV light source

Method

1. Prepare genomic DNA as outlined in *Protocol 1*[a].
2. For tumor or normal reference DNA, combine 2.5 μg of genomic DNA, 5 μl of restriction enzyme reaction buffer (10×), 2 μl of *RsaI*[b] and sterile water up 50 μl.
3. Incubate the restriction digest reactions overnight in a heating block set at 37°C.
4. Stop the reaction by incubating at 72°C for 10 min.
5. Determine the size of the digest products by resolving a 5 μl aliquot of the restriction digest reaction by agarose gel electrophoresis (1% agarose gel containing 10 ng/ml of ethidium bromide) alongside 1 μl of a molecular mass marker (100 bp DNA ladder).
6. Perform electrophoresis at 100 V for 30 min in 0.5× TBE. Visualize the DNA using a UV light source (see *Fig. 1*).

Notes

[a]It is also possible that DNA obtained using various DNA extraction kits (e.g. Qiagen, Invitrogen, etc.) will be suitable for microarray CGH. However, we have not tested these kits, so we cannot recommend any particular manufacturer and cannot comment on the results that will be obtained.

[b]Any restriction enzyme that generates a uniform DNA smear as assessed by agarose gel electrophoresis will be suitable.

Protocol 3

Purification of digestion products

Equipment and Reagents

- TE buffer (10 mM Tris/HCl, pH 7.5, 1 mM EDTA)
- MinElute 96 UF PCR Purification kit (Qiagen)
- Vacuum pump
- Sterile water (Sigma)
- EB buffer (10 mM Tris/HCl, pH 8.5)
- Microplate shaker
- RediPlate 96 PicoGreen dsDNA Quantitation kit (Invitrogen)
- Fluorescence-based microplate reader or fluorometer

Method

1. For the manual purification of digestion products, bring the digest volume up to 500 µl with TE buffer and mix by gently flicking the tube. Briefly centrifuge the tubes to move the samples to the bottom of the tubes.
2. Purify the DNA using the phenol : chloroform extraction protocol described in *Protocol 1*.
3. For the purification of digestion products using a kit-based approach, add the digested DNA from *Protocol 2* directly onto the microplate membrane of the MinElute kit. Apply a vacuum (~800 millibar) for 15–20 min until all of the wells are completely dry.
4. Add 50 µl of sterile water to each well and apply a vacuum (~800 millibar) for 15-20 min until all of the wells are completely dry[a]. Add a second 50 µl aliquot of sterile water to each well and apply a vacuum (~800 millibar) for 15–20 min until all of the wells are completely dry.
5. Add 100 µl of EB buffer and shake on a microplate shaker for 5–10 min to dissolve the DNA[b].
6. Transfer the solution to a 1.5 ml tube and store at –20°C until required.
7. Determine the DNA concentration using a RediPlate 96 PicoGreen dsDNA Quantitation kit (or similar kit) in conjunction with a fluorescence-based microplate reader, following the instructions of the manufacturer[c].

Notes

[a]We followed the protocol provided by the manufacturer (Qiagen); however, we recommend performing at least two washes in order to guarantee removal of salts, incorporated nucleotides, primers, etc.
[b]The DNA can also be dissolved by pipetting the samples up and down.
[c]As an alternative, the amplified DNA can be quantified by measuring absorbance at 260 nm using a spectrophotometer.

Protocol 4

DNA labeling for cDNA microarray CGH analysis[a,b]

Equipment and Reagents

- PCR tubes (0.2 or 0.5 ml; ABgene)
- Restriction-digested high-molecular-mass DNA (tumor and normal reference)
- Sterile water (DNAse/RNAse-free; Promega)
- BioPrime DNA labeling system (Invitrogen)
- 10× dNTP mix containing 12 µl each of 100 mM dATP, 100 mM dGTP, and 100 mM dTTP, 6 µl of 100 mM dCTP, and 958 µl of TE buffer (10 mM Tris/HCl, 1 mM EDTA, pH 8.0) (dNTPs were purchased individually from Invitrogen)
- 1 mM Cy5–dCTP and 1 mM Cy3–dCTP (Applied Biosystems)
- Heating block capable of holding 0.2 or 0.5 ml tubes
- 0.5 M EDTA (pH 8.0; Sigma)
- 1% Agarose gel containing 10 ng/ml ethidium bromide (Invitrogen)
- 6× Orange loading dye solution (Fermentas)
- 100 bp DNA ladder (Fermentas)
- NanoDrop ND-1000 spectrophotometer
- Equipment and reagents for agarose gel electrophoresis including 1× TBE agarose gel running buffer (10.8 g/l Tris base; 5.5 g/l boric acid; 4 ml/l 0.5 M EDTA, pH 8.0; diluted from a 10× stock; Sigma)
- UV light source

Method

1. Set up four separate 0.2 or 0.5 ml PCR tubes, two containing 2.5 µg of tumor DNA and two containing 2.5 µg of normal DNA.
2. To each tube containing DNA, add sterile water to a volume of 21.5 µl and 20 µl of 2.5× random primers (BioPrime DNA labeling system). Mix the tube contents by vortexing for 5 s. Incubate the tubes at 100°C in a boiling water bath for 5 min and then place them on ice immediately.
3. Add 5 µl of 10× dNTP mix[c] and 2.5 µl of either Cy3–dCTP (1 mM) or Cy5–dCTP[d] (1 mM) to each tube and mix well by gently flicking the tube. Centrifuge the tubes briefly to move the sample to the bottom of the tube. Note that separate labeling reactions with Cy3 and Cy5 are prepared for both the tumor and normal DNA. Thus, each sample (tumor or normal) has two labeling reactions, one for Cy3 and one for Cy5.
4. Add 1 µl of Klenow fragment (BioPrime DNA labeling system) to each reaction and mix by gently flicking the tube. Briefly centrifuge the tubes to move the liquid contents to the bottom of the tube.
5. Incubate for 2 h at 37°C. Add 5 µl of 0.5 M EDTA to stop the reaction.
6. Assess the DNA labeling efficiency using a NanoDrop ND-1000 spectrophotometer[e].
7. Alternatively DNA labeling can be assessed by resolving a 5 µl aliquot of the probe labeling reaction by agarose gel electrophoresis (1% agarose gel containing 10 ng/ml of ethidium bromide) alongside 1 µl of a molecular mass marker (100 bp DNA ladder)[f]. Perform the electrophoresis at 100 V for 30 min in 0.5× TBE and visualize the DNA using a UV light source.

Notes

[a]It is possible that a number of the cDNA clones may be annotated incorrectly. It is therefore important to verify microarray CGH results by fluorescent *in situ* hybridization.

[b]This protocol is optimized for use with cDNA microarrays provided by the University Health Network, Microarray Centre (Toronto, Canada). It is probably compatible with most cDNA microarray platforms, but we recommend performing a series of male versus female hybridizations prior to using precious test samples. In a male versus female hybridization, a change in the copy number of the X and Y chromosomes is easily distinguishable.

[c]Do not use the dNTPs from BioPrime DNA labeling system.

[d]Cy3 and Cy5 dyes are light sensitive so avoid prolonged exposure to light.

[e]For instructions on use of this instrument, please view the user manual at the following web link: http://www.nanodrop.com/pdf/nd-1000-users-manual.pdf.

[f]Gel electrophoresis will reveal a bright smear at approximately 100–200 bp corresponding to newly synthesized nucleic acid, which is not present on the restriction digest gel (see *Fig. 1*).

Protocol 5

Probe purification and hybridization for cDNA microarray CGH analysis

Equipment and Reagents

- Microcon 30 filter (Millipore)
- TE buffer (10 mM Tris/HCl, pH 7.5, 1 mM EDTA)
- Yeast tRNA (10 mg/ml; Invitrogen)
- Poly(dA/dT) (4 mg/ml; Invitrogen)
- Human Cot-1 DNA (1 mg/ml; Invitrogen)
- DIG Easy Hyb (Roche)
- cDNA microarrays (University Health Network, Microarray Centre)
- 24 × 50 mm cover slips
- Hybridization oven

Method

1. Purify each labeling reaction separately using a Microcon 30 filter as follows.
2. Bring the reaction volume up to 500 µl with TE buffer and pipette the solution directly onto a Microcon 30 filter.
3. Centrifuge for 8–10 min at 8000 *g*.
4. Invert the filter and transfer to a fresh tube, then centrifuge for 1 min at 8000 *g* to recover the purified probe. The probe volume should be 20–40 µl.
5. Combine one Cy3-labeled tumor sample with one Cy5-labeled normal sample, and one Cy5-labeled tumor sample with one Cy3-labeled normal sample. The two combined sample volumes should be 40–80 µl.
6. To each combined probe mixture, add 50 µl of human Cot-1 DNA, 20 µl of yeast tRNA, and 4 µl of poly(dA/dT)[a].

7. Concentrate each probe mixture down to 10 μl by vacuum centrifugation (e.g. using a SpeedVac)[b,c]. Add small volumes of dH_2O to bring the volume up to exactly 10 μl.
8. Add 30 μl of DIG buffer.
9. Denature the hybridization mixture by heating to 100°C for 1.5 min and then incubate for 30 min at 37°C[d].
10. Centrifuge the tubes at 16000 *g* for 5 min to pellet the unwanted precipitates. Note that the probe molecules of interest are soluble and remain in solution after this centrifugation step. Take care not to disturb the precipitates at the bottom of the tube during step 11 below.
11. Apply the probe mixture to a 24 × 50 mm cover slip and then carefully overlay the cDNA microarray onto the cover slip, being careful not to trap air bubbles between the cover slip and the microarray substrate.
12. Hybridize for 12–16 h at 37°C in a humid chamber[e].

Notes

[a]Cot-1 DNA blocks hybridization to repetitive sequences, yeast tRNA blocks nonspecific DNA binding, and poly(dA/dT) blocks hybridization to the poly(A) tails of cDNA microarray targets.
[b]As an alternative to using a SpeedVac, the DNA can be precipitated by adding 0.1 vols of 3 M sodium acetate and 2.5 vols of 100% ethanol, and spinning the mixture at full speed in a microfuge (16000 *g*) for 10 min to pellet the DNA.
[c]The pellet or solution should have a purplish coloration indicating that there are equal amounts of Cy3- and Cy5-labeled DNA. If the pellet is too pink or too blue, this suggests that either the tumor or the reference DNA was not labeled effectively.
[d]Incubation at 37°C allows annealing of Cot-1 DNA to repetitive sequences within the probe molecules.
[e]Corning provides microarray hybridization chambers that we have found to be ideally suited for our experiments. Hybridization cassettes from other vendors (e.g. ArrayIt) may work equally well for this step.

Protocol 6

Washing procedure for cDNA microarray CGH analysis

Equipment and Reagents

- Coplin jars
- Wash buffer A (1 ml of 10% SDS, sterile dH_2O up to 100 ml)
- Wash buffer B (2.5 ml of 20× SSC, 0.1 ml of 10% SDS, sterile dH_2O up to 100 ml)
- Wash buffer C (0.3 ml of 20× SSC, sterile dH_2O up to 100 ml)
- Slide centrifuge
- Microarray scanner

Method

1. Perform all washes in glass coplin jars at room temperature as follows.
2. Detach cover slips by gently dipping the slides into and out of wash buffer A.

3. Wash the slides for 2 min in wash buffer A, followed by 2 min in wash buffer B, and 2 min in wash buffer C.
4. Centrifuge the slides immediately for 3–5 min at 85 *g* at room temp to dry them[a].
5. The slides are now ready for scanning. If slides are not be scanned immediately, they should be stored dry and in the dark[b].

Notes

[a]As an alternative, the slides can be dried using compressed gas, and ArrayIt offers a Microarray High-Speed Centrifuge that works well for this application.
[b]The data produced by CGH analysis represents an average of all cells from a sample. If the specimen is a heterogeneous mixture of normal and tumor cells, this will reduce the sensitivity of the assay. In those instances, it is important to dissect out the specific cells of interest.

Protocol 7

DNA labeling for CGH analysis

Equipment and Reagents

- PCR tubes (0.2 or 0.5 ml; ABgene)
- Sterile water (DNase/RNase-free; Promega)
- BioPrime DNA labeling system (Invitrogen)
- 10× dNTP mix containing 12 µl each of 100 mM dATP, 100 mM dGTP, and 100 mM dTTP, 6 µl of 100 mM dCTP, and 958 µl of TE buffer (10 mM Tris/HCl, 1 mM EDTA, pH 8.0) (dNTPs were purchased individually from Invitrogen)
- 1 mM Cy5-dCTP and 1 mM Cy3-dCTP (Applied Biosystems)
- Heating block capable of holding 0.2 or 0.5 ml tubes
- 0.5 M EDTA (pH 8.0; Sigma)
- 1% Agarose gel containing 10 ng/ml ethidium bromide (Invitrogen)
- 6× Orange loading dye solution (Fermentas)
- 100 bp DNA ladder (Fermentas)
- Equipment and reagents for agarose gel electrophoresis including 1× TBE agarose gel running buffer (10.8 g/l Tris base; 5.5 g/l boric acid; 4 ml/l 0.5 M EDTA, pH 8.0; diluted from a 10× stock; Sigma)
- UV light source
- (Optional) Spectral Random Prime kit (Spectral Genomics)

Method

1. Set up four reactions in separate tubes (0.2 or 0.5 ml PCR tubes), with two containing 2.5 µg of tumor DNA and two containing 2.5 µg of normal DNA.
2. To each tube containing DNA, add sterile water up to a volume of 21.5 µl and then add 20 µl of 2.5× random primers (BioPrime DNA labeling system). Mix the tube contents by brief vortexing for 5 s. Place the tubes in a boiling water bath for 5 min to denature the DNA, and then immediately place the tubes on ice.

3. To each tube containing denatured DNA[a], add 5 μl of 10× dNTP mix and 2.5 μl of either 1 mM Cy3–dCTP or 1 mM Cy5–dCTP. Mix well by gently flicking the tube and then spin the tubes briefly by centrifugation to consolidate the samples at the bottom of the tubes. Both the tumor and normal DNA samples should be labeled separately with Cy3 and Cy5, generating a total of four labeling reactions for the two samples.
4. Add 1 μl of Klenow fragment (BioPrime DNA labeling system) to each reaction and mix by gently flicking the tube. Spin the tubes briefly by centrifugation to consolidate the samples at the bottom of the tubes.
5. Incubate for 2 h at 37°C and then stop the reaction by adding 5 μl of 0.5 M EDTA.
6. Assess DNA labeling using the procedure described in *Protocol 4*, step 6.
7. Alternatively, perform electrophoresis at 100 V for 30 min in 0.5× TBE, as described in *Protocol 4*, step 7, and visualize the stained DNA using a UV light source (see *Fig. 1*).

Note

[a]When using Spectral Genomics microarrays, the first part of step 3 should be as follows: to each tube, add 2.5 μl of Spectral labeling buffer (Spectral Random Prime kit) and 1.5 μl of either 1 mM Cy3–dCTP or 1 mM Cy5–dCTP, and mix well by pipetting.

Protocol 8

Probe hybridization for CGH analysis[a]

Equipment and Reagents

- Human Cot-1 DNA (1 mg/ml; Invitrogen)
- Yeast tRNA (10 mg/ml; Invitrogen)
- 3 M Sodium acetate (Sigma)
- 70 and 100% Ethanol (Sigma)
- Sterile water (DNase/RNase free; Promega)
- Microarray Hybridization Solution (Amersham)
- Deionized formamide (Sigma)
- Human BAC microarrays (Cancer Research UK Microarray Centre or Spectral Genomics)
- 24 × 60 mm cover slips
- Hybridization oven
- (Optional) Spectral Hybridization Buffer I (Spectral Random Prime kit; Spectral Genomics)
- (Optional) 5 M NaCl
- (Optional) Isopropanol

Method

1. Combine the Cy3-labeled tumor sample with the Cy5-labeled normal sample and vice versa. This allows the use of so-called 'dye-swap' experiments, wherein the two samples are labeled in a reciprocal manner with the two fluors. Combining the four individually labeled samples produces two dual-color samples for the dye-swap analysis.

2. Add 100 µl of 1 mg/ml human Cot-1 DNA and 25 µl of 10 mg/ml yeast tRNA to each tube and mix well. Add 17.5 µl (0.1 vols) of 3 M sodium acetate and 480 µl (2.5 vols) of 100% ethanol and mix well[b].
3. Centrifuge the samples in a microfuge at full speed (16000 *g*) for 10 min. Decant and discard the supernatant carefully to avoid disrupting the DNA pellet at the bottom of the tube.
4. Wash the pellet with 500 µl of 70% ethanol.
5. Centrifuge the samples in a microfuge at full speed (16000 *g*) for 10 min. Decant and discard the supernatant carefully to avoid disrupting the DNA pellet at the bottom of the tube.
6. Recentrifuge the tubes in a microfuge at full speed (16000 *g*) for 30 s. Remove any residual ethanol using a fine pipette tip and allow the pellet to air dry at room temperature in the dark[c].
7. Add 10 µl of sterile water to the pellet, incubate at room temperature for 5 min in the dark, and then resuspend the pellet by vortexing[d].
8. Once resuspended, add 10 µl of Microarray Hybridization Solution and 20 µl of deionized formamide, and mix well by pipetting.
9. Denature the hybridization mixture at 72°C for 10 min, place on ice for 5 min, and then incubate at 37°C for 30 min.
10. Apply the probe mixture to a 24 × 60 mm cover slip and then carefully overlay the BAC microarray onto the cover slip, being careful not to trap any air bubbles.
11. Hybridize the BAC microarray for 12–16 h at 37°C in a humid chamber.

Notes

[a]When using BAC microarrays, we have not found it necessary to purify the probe prior to hybridization.
[b]When using Spectral Genomics microarrays, step 2 of this protocol should be performed by adding 45 µl of Spectral Hybridization Buffer I (Spectral Random Prime kit) to each tube, and then 11.3 µl of 5 M NaCl and 110 µl of room temperature isopropanol.
[c]This should take approximately 10 min.
[d]When using Spectral Genomics microarrays, step 8 of this protocol should be performed as follows: once resuspended, add 30 µl of Spectral Hybridization Buffer II (Spectral Random Prime kit) and mix well by pipetting.

Protocol 9

Probe washes for CGH analysis

Equipment and Reagents

- Wash buffer I (50 ml of 20× SSC, 5 ml of 10% SDS, and sterile water up to 500 ml, pH 7.5)
- Wash buffer II (50 ml of 20× SSC, 250 ml of deionized formamide, and sterile water up to 500 ml, pH 7.5)

- Wash buffer III (5 ml of 20× SSC and sterile water up to 500 ml, pH 7.5)
- Hybridization oven fitted with a 50 ml conical tube rotator
- Distilled deionized water

Method

1. Warm the three wash buffers (wash buffers I, II, and III) to 42°C in separate 50 ml tubes prior to use. All washes are performed at 42°C in a hybridization oven fitted with a 50 ml conical tube rotator[a].
2. Detach the cover slips by gently dipping the slides in and out of wash buffer I. The cover slips should release from the microarray surface in 5–10 s[b].
3. Transfer the slides to a second tube containing pre-warmed wash buffer I and incubate for 15 min at 42°C in a hybridization oven fitted with a rotator.
4. After step 3, wash once for 15 min in wash buffer II, once for 30 min in wash buffer I, and once for 15 min in wash buffer III.
5. Briefly rinse the slides in distilled deionized water for 5–10 s.
6. Dry the slides by spinning in a centrifuge for 3–5 min (85 *g*) at room temperature[c]. At this point, the slides are ready for fluorescent scanning[d].

Notes

[a]As an alternative, washes can be performed at 42°C on a rocking platform or by using ArrayIt High-Throughput Wash Stations (TeleChem International).
[b]If the cover slip fails to release easily from the surface, this may indicate that drying has occurred during the hybridization reaction, which will lead to high fluorescent background. Increasing the probe volume, reducing the hybridization time, or increasing the humidity in the incubation chamber will reduce dehydration during the hybridization reaction (see *Protocol 8*, steps 7–11).
[c]As an alternative to this step, ArrayIt offers a Microarray High-Speed Centrifuge that works very well for slide drying (TeleChem International).
[d]Microarray slides should be scanned immediately or stored in the dark to avoid photobleaching.

Protocol 10

Data acquisition and analysis for CGH

Equipment

- Microarray scanner
- Data quantification and analysis software

Method

1. CGH analysis involves three steps: (i) image acquisition, (ii) fluorescence intensity quantification, and (iii) data interpretation as follows.
2. Acquire a CGH microarray image using a fluorescence scanner or imager. Commercial instruments are available from a number of different vendors including Axon Instruments, PerkinElmer, ArrayIt, Applied Precision, Agilent, Bio-Rad, and others[a,b].

3. Quantify the fluorescence intensity values in the 16-bit data files using quantification software. Quantification software for microarray analysis is available from a number of different commercial vendors including BioDiscovery, Axon Instruments, PerkinElmer, ArrayIt, Applied Precision, Agilent, Bio-Rad, and others[c].
4. Interpret chromosomal gain or loss in the two samples by assessing the red–green overlay data and the genomic sequence.

Notes

[a]In contrast to gene expression microarrays, which produce a relatively large number of green and red spots in two-color overlays, the majority of spots should be yellow in CGH microarray images because most of the genome has not undergone a gain or loss of genetic material in the two samples of interest. For some cell lines and patient samples, the proportion of red and green spots will be far higher, indicating extensive genomic aberrations. In most instances, however, a large number of red and green spots in CGH microarray experiments indicates a failed experiment such as inefficient labeling due to poor-quality DNA or unequal labeling due to different starting amounts of test and control DNA. This is only obvious once the data has been analyzed and will be characterized by 'noisy' genomic plots.

[b]The steps required for image acquisition and fluorescence intensity quantification can be performed using the software provided with the microarray scanner. For the Axon GenePix 4000A scanner (Axon Instruments) and the Perkin Elmer systems, GENEPIX PRO 3.0 and SCANARRAY EXPRESS are the corresponding quantification packages. A quick reference guide for use of these two machines and software packages is available at http://www.microarrays.ca/support/proto.html. When using these scanners, the Cy3 and Cy5 data are collected separately.

[c]To identify regions of chromosomal gain or loss, the quantified fluorescence intensities require normalization prior to further analysis. Currently, many commercial microarray scanners provide data filtering and normalization as a built-in option. The minimum set of information used to describe the microarray experiment and its features includes: microarray configuration (block, column, row), spot foreground and background intensities for each of the two channels, spot diameter, and feature flag. Once a data file is loaded into such software, user-configurable pre-normalization filtering is usually used and options include removal of spots that are of small diameter, low intensity, saturated, or zero intensity, or those that do not contain a given foreground-to-background ratio. This filtering process operates on each printed feature individually and independently of duplicates that may exist on the microarray. Subsequently, the user specifies the experiment configuration (dye-swap), and then may normalize the output files. Such software tools are used sequentially to load scanned microarray CGH images with the ultimate goal of producing a graphical output of experimental findings analogous to chromosome CGH plots. The detailed technical aspects of these procedures are specific to the various software packages and are beyond the scope of this chapter. However, we would like to direct the reader to a number of recent publications describing computer programs that are suitable for this analysis (8–15).

The application of CGH to genomic analysis has allowed the identification of copy number changes in a wide range of tumor samples and cell lines. The microarray CGH results depicted in *Fig. 2* for the neuroblastoma cell line NUB7 were generated using the methods described in this chapter. The summary image in *Fig. 3* displays the CGH results obtained for a series of uveal melanoma patients, and demonstrates the ability to identify

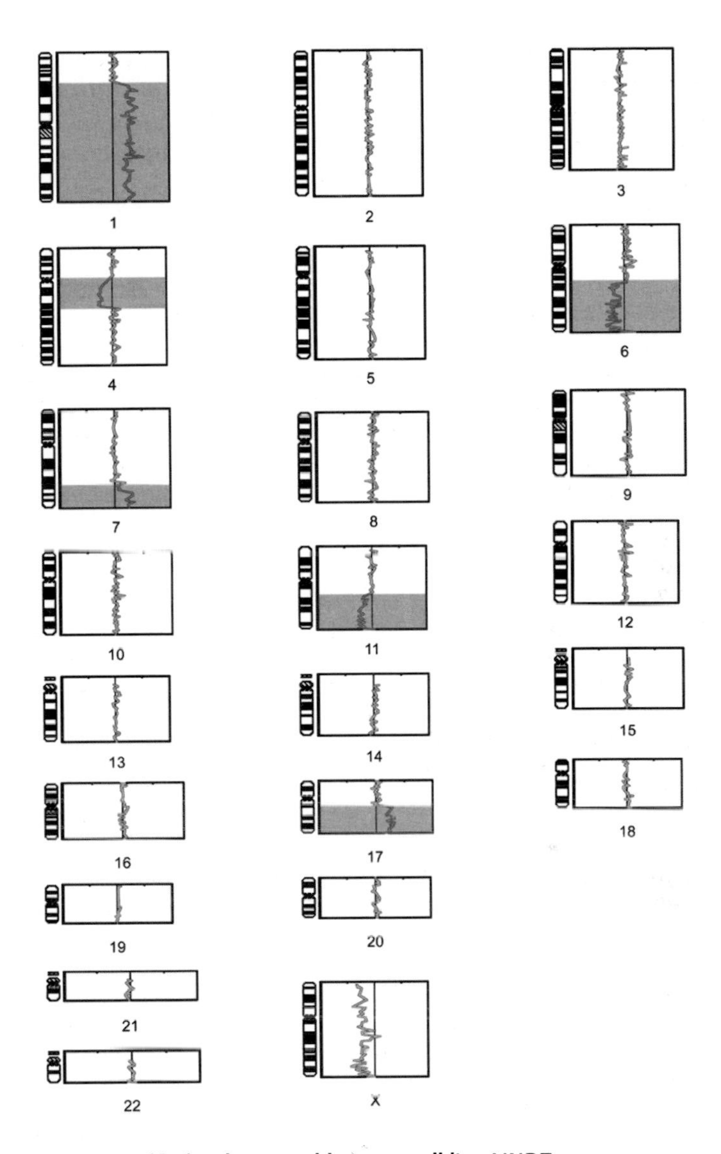

Figure 2. CGH profile for the neuroblastoma cell line NUB7.
Microarray CGH was performed using standard protocols. A gain of genetic material at 1p, 1q, 2p, 7q, and 17q and a loss of genetic material from 4q, 6q, and 11q can be seen. The vertical central axis for each chromosome indicates a 1 : 1 ratio of test DNA to control DNA and therefore no loss or gain. A shift of the line to the left indicates copy number loss, whilst a shift of the line to the right indicates copy number gain. Gray shading indicates regions of copy number loss or gain. This figure was generated using NORMALISE SUITE 2.4 (http://www.utoronto.ca/cancyto) (16).

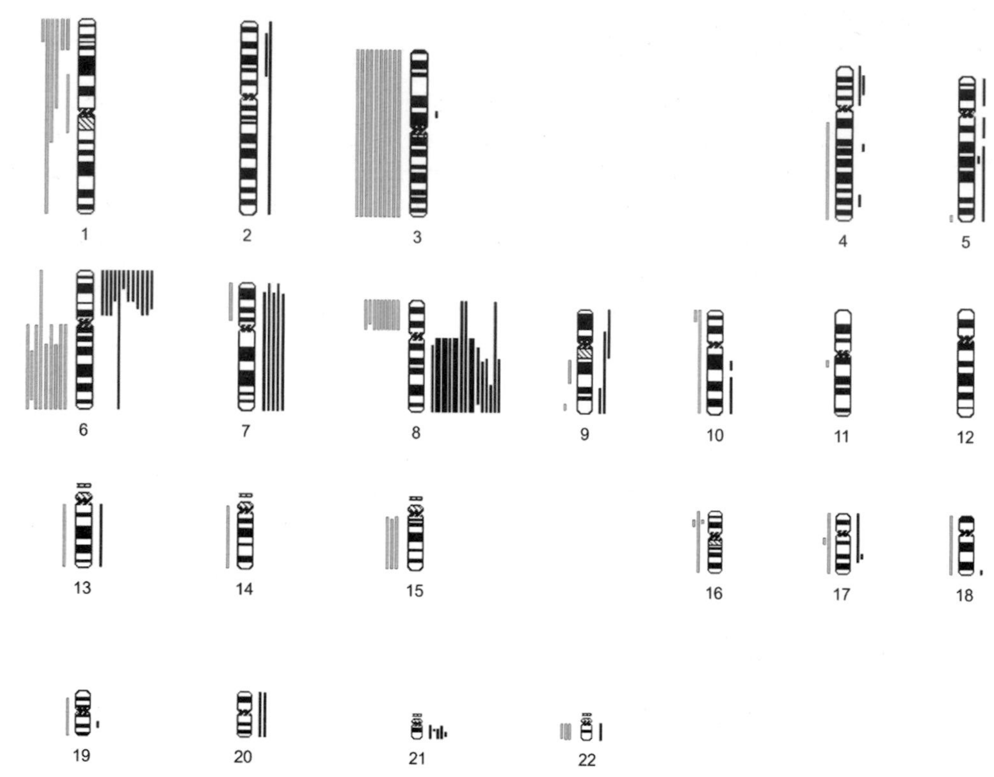

Figure 3. Chromosomal alterations observed for uveal melanoma DNA samples.
A series of 16 uveal melanomas was studied by microarray CGH using BAC microarrays to identify chromosomal abnormalities. Chromosomal loss (gray bars) and chromosomal gain (black bars) are indicated in these samples.

a number of common changes as well as rarely reported copy number alterations.

In this chapter, we have described protocols for cDNA and BAC microarrays. Despite all of the advantages of cDNA and BAC microarrays for CGH, it is important to be aware of the limitations of the technology. The use of cDNA microarrays can lead to a decrease in sensitivity when attempting to detect single-copy changes. With BAC microarrays, limitations of BAC size (greater than 100 kb) and resolution (80–100 kb at best) can make it difficult to detect copy number differences between closely spaced genes. The utilization of oligonucleotide microarrays provides an alternative platform offering the same high resolution of cDNA microarrays and the uniformity of hybridization of the BAC microarrays. None the less, the microarray methods described here, which have permitted a decrease in target size from whole chromosomes to shorter DNA sequences, have provided a wealth of information for molecular biologists.

3. TROUBLESHOOTING

The use of DNA obtained from fixed tissue is problematic for microarray CGH. The fixation process normally causes moderate to severe DNA shearing, as well as DNA and protein cross-linking. As a result, the quality of the CGH results obtained from fixed tissue is generally inferior compared with the data seen with DNA from fresh tissue. When using DNA from fixed tissue, moderately degraded DNA (average size range >2 kb) is usually suitable, whereas highly degraded DNA (average size range <1 kb) is not recommended. With fixed tissue DNA sources, it is important to match the DNA quality of the test and control samples such that if the test is DNA that is mildly degraded, the control DNA should be similarly degraded.

Validation of CGH microarray results is important. The use of tissue microarrays or tissue sections in conjunction with immunohistochemistry can be used as a consistency benchmark. Alternatively, chromosomal gains or losses can be confirmed by the use of quantitative PCR.

4. REFERENCES

★ 1. **Pollack JR, Sorlie T, Perou CM, *et al.*** (2002) *Proc. Natl. Acad. Sci. U.S.A.* **99**, 12963–12968. – *Important publication showing alterations in DNA copy number can lead to changes in gene expression.*

★ 2. **Hui AB, Lo KW, Teo PM, To KF & Huang DP** (2002) *Int. J. Oncol.* **20**, 467–473. – *Important paper providing the first genome-wide survey of multiple oncogene amplifications involved in nasopharyngeal carcinoma.*

3. **Forozan F, Karhu R, Kononen J, Kallioniemi A & Kallioniemi OP** (1997) *Trends Genet.* **13**, 405–409.

4. **James LA** (1999) *J. Pathol.* **187**, 385–395.

★★ 5. **Kallioniemi A, Kallioniemi OP, Sudar D, *et al.*** (1992) *Science*, **258**, 818–821. – *Landmark publication showing the use of two-color fluorescence labeling and detection for CGH.*

★ 6. **Ishkanian AS, Malloff CA, Watson SK, *et al.*** (2004) *Nat. Genet.* **36**, 299–303. – *Impressive publication describing BAC coverage of the entire human genome.*

7. **Matsuzaki H, Dong S, Loi H, *et al.*** (2004) *Nat. Methods*, **1**, 109–111.

8. **Autio R, Hautaniemi S, Kauraniemi P, *et al.*** (2003) *Bioinformatics*, **19**, 1714–1715.

9. **Chen W, Erdogan F, Ropers HH, Lenzner S & Ullmann R** (2005) *BMC Bioinformatics*, **6**, 85.

10. **Chi B, DeLeeuw RJ, Coe BP, MacAulay C & Lam WL** (2004) *BMC Bioinformatics*, **5**, 13.

11. **Kim SY, Nam SW, Lee SH, *et al.*** (2005) *Bioinformatics*, **21**, 2554–2555.

12. **Lingjaerde OC, Baumbusch LO, Liestol K, Glad IK & Borresen-Dale AL** (2005) *Bioinformatics*, **21**, 821–822.

13. **Margolin AA, Greshock J, Naylor TL, *et al.*** (2005) *Bioinformatics*, **21**, 3308–3311.

14. **Price TS, Regan R, Mott R, *et al.*** (2005) *Nucleic Acids Res.* **33**, 3455–3464.

15. **Wang J, Meza-Zepeda LA, Kresse SH, Myklebost O** (2004) *BMC Bioinformatics*, **5**, 74.

★★ 16. **Beheshti B, Braude I, Marrano P, Thorner P, Zielenska M & Squire JA** (2003) *Neoplasia*, **5**, 53–62. – *Key publication showing the use of high-density cDNA microarrays for CGH profiling of neuroblastomas.*

CHAPTER 10

RNA sample preparation and small-quantity RNA profiling for microarray biomarker discovery

Jianyong Shou and Lawrence M. Gelbert

1. INTRODUCTION

Genomics is the comprehensive analysis of the entire genetic content of an organism, and forms part of the newly defined discipline called systems biology. The recent completion of the human genome sequence and the development of associated technologies, such as microarrays, have allowed the first global views of human biological systems (1). Genomic approaches are now being applied to those biological systems where the genome has been sequenced, and have been embraced by all fields of biology including translational biomedical research and the biopharmaceutical industry.

The initial focus of genomics in drug discovery was on the identification of new drug targets; with current drug therapies focused on fewer than 500 targets, the human genome sequence promised to provide thousands of new, higher-quality targets (2). This initial focus has expanded to include more traditional areas of drug discovery and development including compound screening (chemogenomics) (3–5), toxicology (toxicogenomics) (6, 7), and clinical drug development (pharmacogenetics, pharmacogenomics) (8, 9). Genomic applications in drug discovery and development have been driven in great part by the desire to develop therapies for the complex diseases where there is still an unmet medical need (10). The acceptance of genomic approaches by the biopharmaceutical industry has also been driven by the need to address the increasing costs, high attrition rates, and increased regulatory and safety concerns facing the development of new drugs. It is now estimated that the cost of developing a new drug exceeds US $800 million (11, 12). This high cost is due in great part to approximately 90% of

DNA Microarrays: *Methods Express* (M. Schena, ed.)

new drugs failing in late clinical trials (13). Toxicity, adverse drug events, and lack of efficacy in man are the primary reasons for these failures (14).

2. METHODS AND APPROACHES

The development and use of biomarkers for safety and efficacy is now being pursued as one solution to the current challenges. A biomarker is defined as a characteristic that is measured objectively and evaluated as an indicator of normal and/or pathological biological processes, or pharmacologic responses to a therapeutic intervention. There are several subclasses of biomarkers (see *Table 1*), and the use of biomarkers in clinical practice and in drug discovery and development has been recognized recently as a critical success factor, by both industry and the US Food and Drug Administration (15, 16, 19–21).

Biomarker strategies have now been incorporated into the contemporary drug discovery and development process. The relationship of different types of biomarker in drug discovery and how they translate to drug development is summarized in *Fig. 1(a)*. Microarray analysis of human tissue and model systems to determine patterns of normal/pathological gene expression not only allows for target identification and validation, but also provides the foundation for disease and target-specific biomarkers. Microarray analysis of compound mechanisms of action, both

Table 1. Biomarker definitions (15–18)

Type of marker or endpoint	Description
Biomarker	A characteristic that is objectively measured and evaluated as an indicator of normal and/or pathological biological processes, or pharmacologic responses to a therapeutic intervention.
Disease biomarker	A biomarker that indicates the presence or likelihood of a disease in a patient. Also called type 0 biomarkers.
Drug activity biomarker	A biomarker that measures the modulation of the target by a therapeutic intervention (on-target). Also called type I biomarkers. Can also measure other activities (off-target, novel mechanism of action, toxicity biomarker).
Clinical endpoint	A characteristic of a variable that reflects the feelings, function, or survival of a patient.
Surrogate endpoint	A biomarker that is intended to substitute for a clinical endpoint. Also called a type II biomarker. Must predict a clinical endpoint.

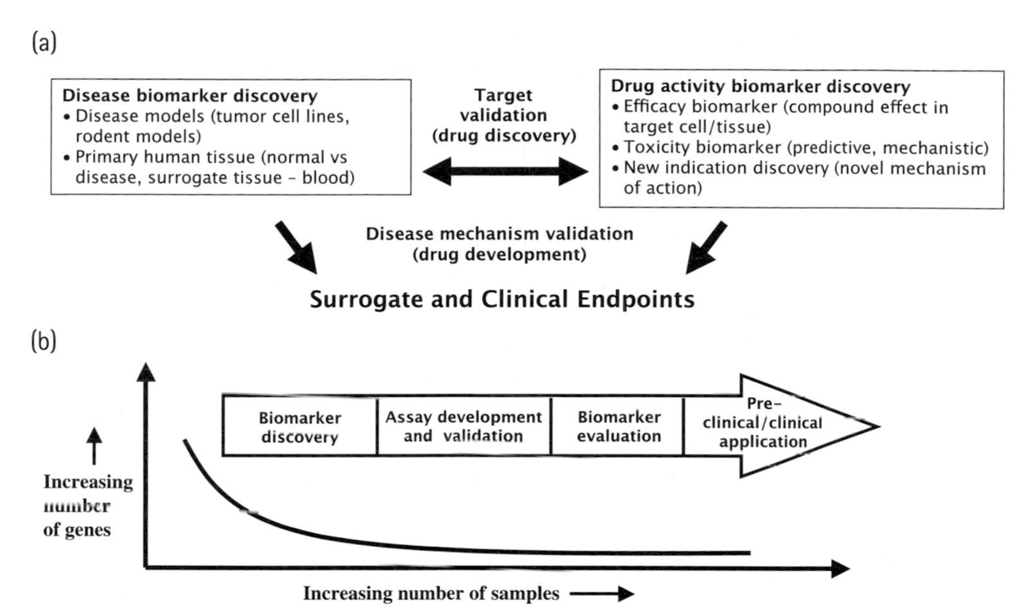

Figure 1. Graphical representation depicting biomarkers in drug discovery and development. (*a*) Role of biomarkers in drug discovery and development and their translation to surrogate endpoints. (*b*) Microarray biomarker discovery process and the relationship between the number of genes studied in each step and the number of samples. Whilst microarray use is extensive to the discovery step, they can also be used in subsequent steps when the number of genes to be analyzed is large (e.g. hundreds to thousands).

on-target (efficacy) and off-target (toxicity), allows the development of drug activity biomarkers. A recent example of biomarker discovery using microarrays is highlighted by the work of Sawada *et al.* (22). Drug-induced phospholipidosis is a lipid storage disorder affecting multiple organs and is associated with several major classes of drug including antibiotics, antihistamines, and tamoxifen (23, 24). Whilst the adverse effects of phospholipidosis are not completely clear, it is still an issue in the development of new drugs. Using microarrays to analyze compounds known to induce phospholipidosis, Sawada *et al.* both established the mechanism of drug-induced phospholipidosis and developed an *in vitro* reverse transcriptase polymerase chain reaction (RT-PCR) assay based on the microarray data for the rapid screening of new compounds.

Development of biomarkers during drug discovery also allows the timely development of surrogate endpoints, which enable 'quick win/quick kill' decision-making in early clinical trials of drug candidates. The co-development of surrogate endpoints for drug activity will not only allow for the early identification of drug candidates that will fail, but is also critical for the new generation of targeted cancer therapies where dose selection requires the determination of the optimal efficacious dose rather than the maximum tolerated dose (25).

It has been a decade since the first publications describing microarrays (26, 27), and microarray technology was first used for disease biomarker discovery in the molecular classification of tumors to predict clinical outcome (28, 29), including response to drug therapy (30). Soon after, transcript profiling studies for drug mechanisms of action were described to discover drug activity biomarkers (31–34). *Fig. 1(b)* summarizes the biomarker discovery and development process using microarray data. The process consists of several steps:

1. Microarray experiments drive the biomarker discovery phase when it is necessary to analyze a large number of genes in clinical samples or in pre-clinical model systems.
2. Assay development and validation involves the development of specific assays to measure a subset of gene expression changes that are predictive of a biological process or outcome. This can include continued measurement of mRNA changes using microarrays or by quantitative RT-PCR (TaqMan), but also includes approaches that measure proteins (Western blotting, enzyme-linked immunosorbent assays, immunohistochemistry). Assay validation should be performed in both the original sample set and an independent set of samples.
3. Biomarker evaluation tests to determine whether the assay is robust enough to work in the appropriate biological sample (fluid, tissue) under a variety of conditions seen in clinical applications (sample shipping, temperature variation, etc.).

The results of this process are candidate biomarkers that can be used to assess disease status and drug activity in pre-clinical and clinical drug development. Several recent reports highlight the value of microarrays to discover biomarkers and the process to convert them to assays. Using a combination of comparative genomic hybridization and expression profiling, Platzer *et al.* (35) identified genes whose expression is significantly overexpressed in colon cancers. In a subsequent report, these researchers showed that one of the genes, colon cancer secreted protein-2 (CCSP-2), encodes a novel secreted protein that is not normally expressed, but has high expression in colorectal carcinomas. In these studies, Northern blotting, quantitative RT-PCR, and protein expression were used to confirm that CCSP-2 expression is specific to human colon cancers and encodes a secreted protein that can be detected in the blood of xenografted mice (36). These reports highlighted the first two steps of biomarker development using microarrays (discovery and assay development, and validation).

Another recent example that highlights the microarray biomarker process is the development of a multi-gene expression assay that predicts breast cancer recurrence after tamoxifen adjuvant therapy. Being able to stratify breast cancer patients is important, as less than 20% of node-negative patients receiving tamoxifen adjuvant therapy have a recurrence, but there is currently no way to identify these patients to spare them the toxicity and other side effects associated with subsequent chemotherapy (37). Paik and co-workers (38) analyzed several sets of breast cancer microarray data and identified 250 genes whose expression predicted cancer progression (biomarker discovery), then tested them in an independent set of samples for validation. The results of the validation study identified 21 genes whose expression pattern predicts recurrence, and a high-throughput RT-PCR assay was developed and tested that allowed analysis of these surrogate markers in clinically accessible samples (assay development evaluation of paraffin-embedded tissue blocks).

One challenge to using microarray expression profiling as a biomarker discovery tool is accessibility to a homogenous and sufficient quantity of tissue (39). The standard sample labeling protocol for Affymetrix microarrays requires 5–10 μg of total RNA, and protocols such as laser-capture microdissection are sometimes required to isolate the desired sample from heterogeneous samples (40). The use of readily accessible surrogate tissue analysis, such as peripheral blood cells, is one solution (41–43). Another alternative is RNA amplification and labeling protocols that allow the use of nanogram amounts of starting RNA (44, 45).

Herein we present a review of and detail protocols for RNA preparation and labeling from whole blood for small-quantity RNA profiling. These protocols have been optimized and validated in our laboratory and used successfully in biomarker discovery projects. Having a high turnover, blood is a dynamic tissue that is in contact with virtually every tissue in the body and has been recognized as a 'sentinel tissue' that systematically reflects disease progression in the body (46, 47). Gene expression profiling in peripheral blood cells has emerged as an attractive biomarker discovery strategy for various disease indications (15, 43, 47–52). In contrast to target tissue biopsy-based approaches, which are limited by restricted access to many tissues, blood is readily accessible and allows repeated biomarker measurements, thus enabling pharmacodynamic studies with drug activity biomarkers. However, given the heterogeneous blood cell population, artifacts from *ex vivo* activation of the lymphocytes during cell fractionation, and technical difficulties in isolating and labeling whole-blood RNA, it remains technically challenging to profile peripheral blood cells for biomarker discovery (50).

There are currently several blood RNA isolation protocols used to isolate RNA from different subsets of cells in blood, including from peripheral blood mononuclear cells (PBMCs) or from whole blood. It is important to select the appropriate protocol to take into account the advantages and disadvantages of a given method. The PBMC method is a blood cell profiling strategy with the focus on peripheral lymphocytes (53–55). However, isolation of PBMCs is time-consuming and labor-intensive (a consideration for experiments with a large number of samples), and may involve undesired artifacts from *ex vivo* activation of lymphocytes during cell fractionation. The BD Vacutainer CPT tube (Becton Dickinson) provides a simplified approach to isolate RNA from white blood cells. The phase separation in the CPT tube makes the isolation of the white blood cell population more efficient. However, the CPT tube does not contain RNA stabilizing reagent, and in our experience RNA yield is not consistent.

For applications in which the erythrocytes or granulocytes represent the cell population of interest, a whole blood isolation approach should be used. However, hemoglobin interference is a common problem that limits the use of whole blood gene expression profiling (50). We have found the Tempus blood collection and RNA extraction system from Applied BioSystems is best suited for biomarker identification needs, providing high-quality, reproducible RNA yields and satisfactory microarray performance with minimal hemoglobin interference.

For blood RNA sample processing for Affymetrix microarray use, the RNA isolated using the Tempus protocol (see *Protocol 1*) is compatible with the standard Affymetrix sample labeling protocol without subsequent purification. As RNA is usually eluted in elution buffer containing EDTA, concentrating RNA by drying is not recommended as the EDTA in the RNA elution buffer will interfere with the subsequent labeling reaction. For samples that need to be concentrated, we precipitated 2 μg of total RNA along with 5 μg of glycogen carrier (Ambion) using 2 volumes of cold absolute ethanol. The resulting RNA pellet is then resuspended in 10 μl of distilled water treated with 0.1% diethylpyrocarbonate (DEPC) and processed for Affymetrix GeneChip use following the standard RNA labeling procedure. When sample concentration is used, we suggest using the GeneChip Sample Cleanup Module (P/N 900371; Affymetrix) for cDNA and RNA purification to ensure sufficient *in vitro* transcription (IVT) yield and subsequent microarray performance.

Using 2 μg of input total RNA, we routinely obtain more than 30 μg of IVT from human samples, and more than 55 μg of IVT from rat samples. In our experience, the quality control data (including percentage of present calls, background, and scaling factor) using blood RNA is similar to various nonblood tissue sources. Significantly, we achieved approximately 40%

of present calls, indicating minimal hemoglobin interference using this approach. Such detection sensitivity is sufficient to detect subtle changes in statistically designed experiments (56).

The effects of freezing and thawing samples on blood RNA isolation and microarray performance have been explored. Immediate RNA isolation is not always possible in clinical biomarker studies, and sample storage and stability are critical factors to be considered. For such biomarker studies, we have found that the Tempus tube RNA collection and isolation protocol (see *Protocol 1*) is successful when RNA isolation is delayed due to shipping or other factors. The RNA is stabilized in the blood cell lysate and can be processed immediately after blood collection when stored at 4°C for a short period of time (<12 h), or stored at −80°C for longer periods of time. To assess sample stability, we performed RNA extraction 3 weeks after blood collection with the sample stored at −80°C. These studies showed that one round of sample freeze–thaw had no effect on RNA yield or quality, and did not affect the mRNA representation in the samples as assessed by microarray hybridization. We compared the overall microarray data by correlation analysis and found the correlation coefficient between the fresh and frozen samples was nearly identical to that of the technical replicates within a group. Shown in *Fig. 2* is the scatter plot of the microarray signal intensities obtained from the fresh or frozen samples from one representative blood sample. Taken together, we believe that the quality, stability, throughput, and ease of sample handling make this approach an attractive method for pre-clinical or clinical blood biomarker discovery (56).

Small-quantity RNA amplification is also essential for applications such as biomarker discovery, where limited sample quantity is a common challenge. The standard Affymetrix GeneChip labeling protocol requires a significant amount of input RNA (5–10 μg), which is not possible when using approaches such as laser-capture microdissection or when analyzing small biopsies. A few RNA amplification and labeling protocols to overcome this limitation have been described. PCR-based amplification is technically robust (44, 45, 57), but generally is not suitable for global RNA amplification due to the exponential amplification bias introduced by GC content within certain sequences (58, 59). T7 RNA polymerase-based linear amplification has been described for RNA amplification in microarray studies (58, 60–64). However, discrepancies in microarray performance such as dropped present calls that affect the sensitivity of the microarray study have been reported (45), and optimization and standardization of the process to improve the accuracy and reproducibility of microarray data are necessary for biomarker discovery. As described below (see *Protocol 2*), we have adapted a commercially available double-round amplification system, the Arcturus RiboAmp technology, for genomic biomarker discovery using small-quantity RNA.

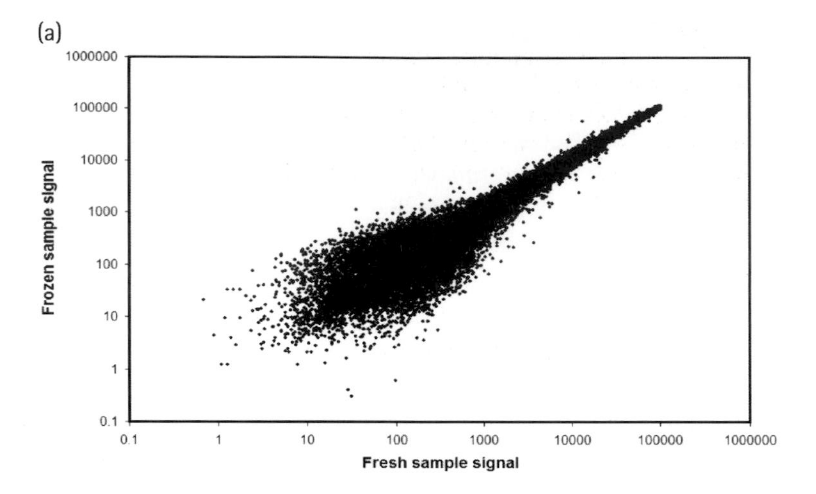

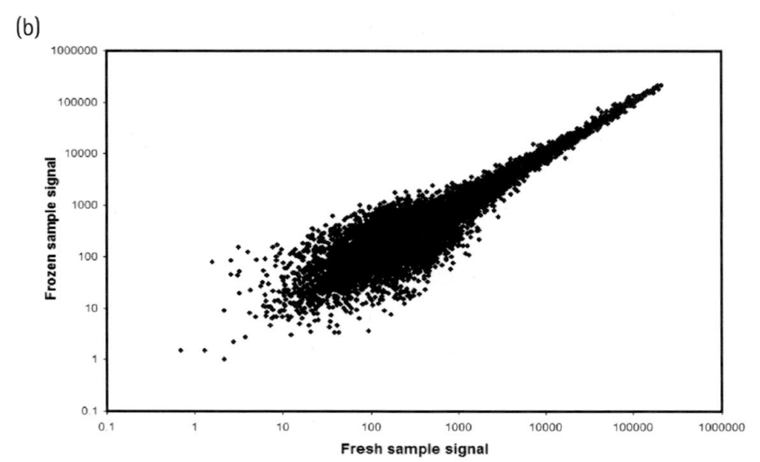

Figure 2. Scatter plot data of fluorescent signals from representative microarrays hybridized with rat (*a*) and human (*b*) samples, respectively.
The *x*-axis represents signals from microarrays processed using fresh samples and the *y*-axis represents signals from microarrays processed using frozen samples. Good correlation is observed between the fresh and frozen samples.

2.1 Recommended protocols

Protocol 1

Tempus tube blood collection and RNA isolation[a]

Equipment and Reagents

- 6100 Nucleic Acid PrepStation (ABI)
- Tempus tube (ABI)

- Lysis buffer
- PBS (calcium- and magnesium-free)
- RNA purification tray (ABI)
- RNA purification wash solution 1 (ABI)
- RNA purification wash solution 2 (ABI)
- Nucleic acid purification elution solution (ABI)
- RNA 6000 Nano Gel System (Agilent)
- 2100 Bioanalyzer (Agilent)

Method

1. Draw 3 ml of blood directly into the Tempus tube containing 6 ml of lysis buffer[b].
2. Shake vigorously for 30 s. Store the lysate at –80°C or proceed directly with RNA isolation.
3. Dilute the cell lysate using 3 ml (1 vol.) of calcium/magnesium-free PBS for a blood : lysis buffer : PBS volume-to-volume ratio of 1 : 2 : 1.
4. Load the sample into the RNA purification tray of an ABI 6100 Nucleic Acid PrepStation, programming the procedure according to the instructions of the manufacturer.
5. Pre-wet all of the wells with 40 µl per well of RNA purification wash solution 1.
6. Load the samples at a volume of 20–3000 µl of lysate per single well. Multiple rounds of sample loading are needed for lysate volumes in excess of 650 µl. Apply the vacuum at 80% strength for 180 s.
7. Rinse the wells with 650 µl of RNA purification wash solution 1. Apply the vacuum at 80% strength for 180 s.
8. Rinse the wells twice with 650 µl of RNA purification wash solution 2. Apply the vacuum at 80% strength for 180 s.
9. Rinse the wells with 400 µl of RNA purification wash solution 2. Apply the vacuum at 80% strength for 180 s.
10. As a pre-elution step to remove residual wash buffer from the wells, apply the vacuum to the wells at 90% strength for 300 s.
11. Move the carriage from the waste position into the collection position. Elute the RNA samples from the wells with 150 µl of nucleic acid purification elution solution.
12. Quantify the RNA and run the samples on an RNA 6000 Nano Gel System using the Agilent 2100 Bioanalyzer for RNA quality determination. Store the RNA at –80°C until the labeling step used for Affymetrix GeneChip microarray study.

Notes

[a]The following protocol using the 6100 Nucleic Acid PrepStation to isolate total RNA from whole blood is adapted from the ABI user instructions.

[b]A 3 ml volume of blood was collected into the tube containing 6 ml of Applied BioSystem's RNA stabilization reagent. When 3 ml of blood is not available, we lysed a smaller blood volume with the RNA stabilization buffer at a ratio of 1 : 2. The smallest volume of blood we have tested is 500 µl. To ensure that the blood cells are thoroughly lysed and the RNA is stabilized, we mix the blood with lysis buffer vigorously for 30 s by inverting the tube up and down immediately after the blood is drawn. The lysate can be processed for RNA extraction immediately or stored at –80°C for later use. In our experience, samples can be frozen at –80°C for at least 3 weeks

with no effect on RNA yield, quality, or microarray hybridization. This RNA extraction procedure can also be fully automated using the ABI 6700 Nuclear Acid Workstation or semi-automated using the ABI PRISM 6100 Nucleic Acid PrepStation in 96-well format (ABI), which provides sufficient throughput for core laboratories or large studies. We find that the RNA yield is sufficient from a single well of the RNA isolation tray for labeling and microarray hybridization. Although RNA isolation can be performed using the large filter and reservoir, we found it not to be reliable and the RNA yield was less than optimal. We recommend extracting the RNA using the 96-well format. Using this protocol, an on-column DNase treatment is available to reduce the genomic DNA contamination for RT-PCR analysis. RNA was usually eluted in the ABI RNA elution buffer, or in RNAse-free H_2O. From 1.0 ml of blood, we can normally obtain an average of 23.7 µg from rat whole blood or 2.7 µg from human. *Fig. 3* shows the RNA yield from representative human and rat samples.

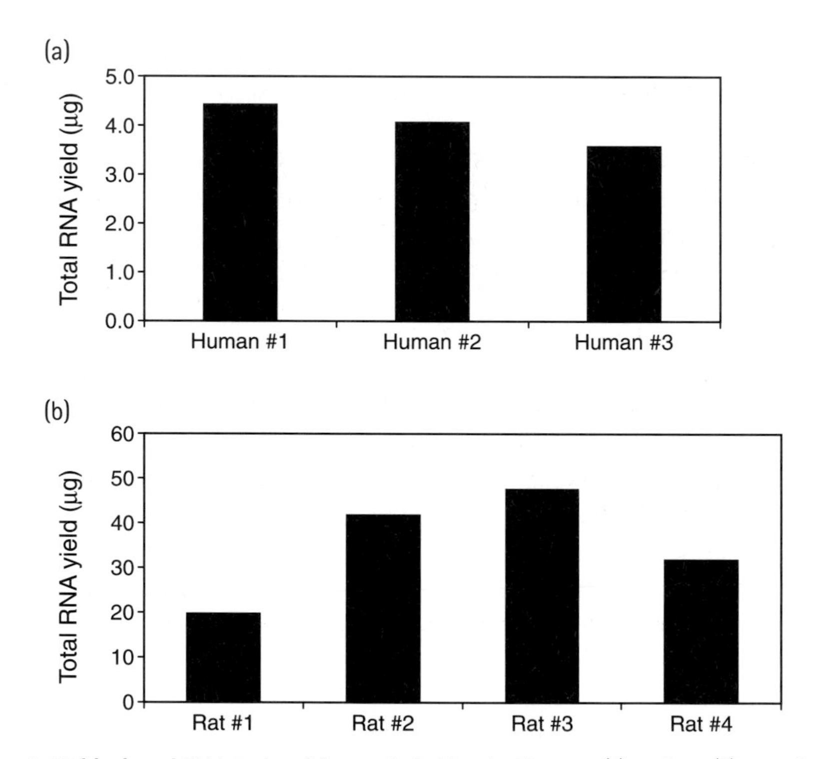

Figure 3. Yield of total RNA isolated from whole blood of human (*a*) and rat (*b*) samples. A 3 ml volume of blood from normal male rat or healthy human donor was drawn into Tempus tubes containing 6 ml of RNA lysis buffer and lysis was performed immediately. Half (4.5 ml) of the cell lysate, equivalent to 1.5 ml of blood, was mixed with 1.5 ml of calcium- and magnesium-free PBS and applied to the ABI PRISM 6100 Nucleic Acid PrepStation for RNA isolation. Total RNA was eluted in elution buffer and quantified by measuring the absorbance at 260 nm.

Protocol 2

First-strand cDNA synthesis for first-round amplification of small-quantity RNA[a]

Equipment and Reagents

- RiboAmp kit (Arcturus)
- 0.5 ml RNase-free microfuge tubes
- Poly(dI-C) carrier RNA (Roche Diagnostics)
- DEPC-treated dH_2O

Method

1. To a 0.5 ml RNase-free microfuge tube, add 40–1000 ng of RNA and 200 ng of poly(dI-C) carrier RNA to a total reaction volume of 10 μl. Volume adjustments up to 10 μl should be made using DEPC-treated dH_2O.
2. Add 1.0 μl of primer A, mix well, and spin the tube down briefly in a microfuge.
3. Incubate the mixture of RNA and primer at 65°C for 5 min, chill the samples on ice for 1 min, and spin the down briefly in a microfuge.
4. Thaw all of the first-strand kit components except the enzymes. Place the thawed components on ice and prepare a first-strand cocktail containing 7 μl of first-strand buffer and 2 μl of first-strand enzyme mix. Mix well.
5. To the chilled RNA/primer samples, add 9 μl of first-strand cocktail per sample. Mix the samples gently, spin down briefly, and incubate at 42°C for 45 min.
6. Chill the samples on ice for 1 min and spin down briefly.
7. Add 2.0 μl of first-strand nuclease mix to the sample. Mix gently, spin down briefly, and incubate at 37°C for 10 min.
8. Incubate at 95°C for 5 min.
9. Chill the samples on ice for 1 min and spin down briefly. Proceed to the second-strand synthesis protocol (*Protocol 3*) or store at –20°C overnight.

Note

[a]The following protocols (*Protocols 2–6*) were adapted from the manufacturer-recommended Arcturus RiboAmp protocol. All reagents are included in the RiboAmp kit (Arcturus) unless otherwise stated.

Protocol 3

Second-strand cDNA synthesis for first-round amplification of small-quantity RNA

Reagents

- RiboAmp kit (Arcturus)

Method

1. Obtain the tube from the previous protocol (*Protocol 2*, step 9) and place on ice. Add 1.0 µl of primer B to the reaction tube at 4°C, mix well, and spin down briefly.
2. Incubate the tube at 95°C for 2 min to denature the cDNA and then chill the sample on ice for at least 2 min.
3. Thaw all of the second-strand components on ice except for the enzyme, and prepare a second-strand cocktail by mixing 29 µl of second-strand master mix and 1 µl of second-strand enzyme mix.
4. Add 30 µl of second-strand cocktail to the sample, mix well, and spin down briefly.
5. Incubate the reaction for 5 min at 25°C, for 10 min at 37°C, for 5 min at 70°C, and then hold at 4°C until ready to proceed to the next step.
6. Add 250 µl of DNA binding buffer to a DNA/RNA purification column, incubate at room temperature for 5 min to equilibrate the column, and centrifuge the column tube assembly at full speed in a microfuge (16 000 *g*) for 1 min.
7. Add 200 µl of DNA binding buffer to the second-strand sample tubes, mix well, and apply the entire sample to the column.
8. Centrifuge the column tube assembly at 100 ***g*** for 2 min, followed immediately by a second spin at 10 000 ***g*** for 30 s.
9. Add 250 µl of DNA washing buffer, spin the column tube assembly at 16 000 ***g*** for 2 min. Recentrifuge at 16 000 ***g*** for 1 min (optional).
10. Discard the flow-through and the collection tube, and place the column onto a new collection tube.
11. Add 16 µl of DNA elution buffer to the column and incubate for 1 min at room temperature.
12. Centrifuge the column tube assembly at 1000 ***g*** for 1 min, followed by 16 000 ***g*** for 1 min. Proceed to the IVT step or store the cDNA at –20°C.

Protocol 4

First-round IVT and purification

Reagents

- RiboAmp kit (Arcturus)

Method

1. Thaw all of the IVT reaction components except the enzyme. Prepare the IVT cocktail by mixing 8 μl of IVT buffer, 12 μl of IVT master mix, and 4 μl of IVT enzyme mix.
2. Add 24 μl of the IVT cocktail to each sample tube, mix well, and spin down briefly.
3. Incubate the reaction at 42°C for 3–4 h and then cool the samples to 4°C.
4. Add 2 μl of DNase mixture, mix gently, and spin down briefly. Incubate at 37°C for 15 min and then cool to 4°C.
5. Add 250 μl of RNA binding buffer to a DNA/RNA purification column, incubate for 5 min at room temperature, and centrifuge the tube at 16000 ***g*** for 1 min.
6. Add 200 μl of RNA binding buffer to the IVT sample tubes, mix well, and apply the entire volume of sample to the equilibrated column.
7. Centrifuge the column tube assembly at 100 ***g*** for 2 min, followed immediately by a second spin at 10000 ***g*** for 30 s.
8. Add 200 μl of fresh RNA washing buffer and spin at 10000 ***g*** for 1 min.
9. Add 200 μl of RNA washing buffer and spin at 16000 ***g*** for 2 min.
10. Recentrifuge the column tube assembly at 16000 ***g*** for 1 min (optional).
11. Discard the flow-through and the collection tube, and place the column into a new collection tube.
12. Add 11 μl of RNA elution buffer and incubate the column tube assembly at room temperature for 1 min.
13. Centrifuge the column tube assembly at 1000 ***g*** for 1 min and then at 16000 ***g*** for 1 min to elute the RNA.
14. Measure the amplified RNA (aRNA) yield and store the aRNA at –80°C until ready to use.

Protocol 5

First-strand cDNA synthesis for second-round amplification

Reagents

- RiboAmp kit (Arcturus)

Method

1. Obtain the aRNA sample stored at −80°C from the previous step (*Protocol 4*, step 14). Thaw the sample and place 10 µl of aRNA into an RNase-free tube.
2. Add 1.0 µl of primer B, mix well, and spin down briefly.
3. Incubate the sample at 65°C for 5 min, chill the samples on ice for 1 min, and spin down briefly.
4. Thaw all of the first-strand components except the enzymes. Place the thawed components on ice and prepare the first-strand cocktail by mixing 7 µl of first-strand buffer and 2 µl of first-strand enzyme mix.
5. To the chilled sample, add 9.0 µl of first-strand cocktail. Mix gently by tapping the tube and spin down. Incubate the reaction at 25°C for 10 min and then at 37°C for 45 min.
6. Chill the sample on ice for 1 min (up to 30 min) and spin down briefly. The sample can also be stored overnight at −20°C.

Protocol 6

Second-strand cDNA synthesis for second-round amplification

Reagents

- RiboAmp kit (Arcturus)

Method

1. Obtain the sample from the previous step (*Protocol 5*, step 6) and place it on ice. Add 1.0 µl of primer A to the tube, mix well, and spin down briefly.
2. Incubate the reaction at 95°C for 2 min, then place the sample on ice and chill for ≥2 min.
3. Thaw all of the second-strand components except the enzymes and place them on ice. Prepare the second-strand cocktail by mixing 29 µl of second-strand master mix and 1 µl of second-strand enzyme mix.
4. Add 30 µl of second-strand cocktail to the sample on ice, mix gently by tapping the tube, and spin down briefly.
5. Incubate the tube at 37°C for 15 min, for 70°C for 5 min, and then hold at 4°C until ready to proceed to the next step (<30 min).

6. Add 250 µl of DNA binding buffer to a DNA/RNA purification column, equilibrate the column for 5 min at room temperature, and spin at full speed (16 000 *g*) in a microfuge for 1 min.
7. Add 200 µl of DNA binding buffer to the second-strand sample tubes, mix well, and apply the entire volume to the equilibrated column.
8. Centrifuge the column tube assembly in a microfuge at 100 *g* for 2 min, followed immediately by a second spin at 10 000 *g* for 30 s.
9. Add 250 µl of DNA washing buffer and spin in a microfuge at 16 000 *g* for 2 min. Recentrifuge the column tube assembly at 16 000 *g* for 1 min (optional).
10. Discard the flow-through and the collection tube, and place the column containing the sample into a new collection tube.
11. Add 16 µl of DNA elution buffer and incubate the assembly at room temperature for 1 min.
12. Centrifuge the column tube assembly at 1000 *g* for 1 min, followed by 16 000 *g* for 1 min. Store the eluted cDNA sample at –20°C until ready for use.

Protocol 7

Second-round IVT[a]

Reagents

- Enzo BioArray High Yield Transcription kit (Affymetrix)

Method

1. Prepare the following IVT master mix at room temperature: 4 µl of 10× HY buffer, 4 µl of 10× biotinylated nucleotides, 4 µl of 10× DTT, 4 µl of 10× RNase inhibitor, 2 µl of 10× T7 RNA polymerase, and 12 µl of DEPC-treated dH_2O.
2. Add 30 µl of IVT master mix to 10 µl of double-strand cDNA sample from the previous protocol (*Protocol 6*, step 12). Mix gently by tapping the tube and incubate the reaction at 37°C for 5 h. Mix the sample by gently vortexing at 450 r.p.m. every 30 min.
3. Place the sample on ice until ready to proceed to the clean-up steps (*Protocol 8*).

Note

[a]This method uses the standard Affymetrix protocol and the Enzo BioArray High Yield Transcription kit (Affymetrix).

Protocol 8

Second-round IVT sample purification[a]

Reagents

- RNeasy Mini kit (Qiagen)

Method

1. Obtain the sample from the previous protocol (*Protocol 7*, step 3). Add 310 μl of buffer RLT. Vortex to mix.
2. Add 350 μl of 70% ethanol diluted from 100% ethanol with DEPC-treated dH_2O. Mix by pipetting up and down.
3. Apply the entire volume to the column. Centrifuge in a microfuge at 8000 ***g*** for 15 s.
4. Transfer the RNeasy column to a new collection tube.
5. Rinse the column with 700 μl of buffer RW1. Centrifuge at 8000 ***g*** for 15 s.
6. Transfer the column to new 2 ml collection tube and add 500 μl of buffer RPE. Centrifuge the column tube assembly at 8000 ***g*** for 15 s. Discard the flow-through.
7. Add 500 μl of buffer RPE to the column. Centrifuge the column tube assembly at 16 000 ***g*** for 2 min.
8. Add 50 μl of RNase-free dH_2O to the column and incubate at room temperature for 1 min. Centrifuge the column tube assembly for 1 min at 8000 ***g*** to elute the purified RNA.
9. Repeat the RNA elution step (step 8) using 20 μl of RNase-free dH_2O. Combine the eluted samples to obtain approximately 70 μl of purified RNA.
10. Determine the RNA concentration and yield, and store the sample at –80°C.

Note

[a]For samples in which RNA quantity is not limiting, RNA isolation is performed using Trizol reagent (Life Technologies) according to the instructions of the manufacturer (Invitrogen), followed by purification using RNeasy spin columns (Qiagen).

3. TROUBLESHOOTING

For small-quantity RNA isolation, an alternative RNA isolation method might be necessary depending on the amount and the type of tissue. The smallest amount of RNA we have tried to label using the double-round RNA amplification is 40 ng of total RNA per reaction, an amount that meets most biomarker needs. We have not optimized this protocol for smaller RNA quantities, although it should work with less than 40 ng of total RNA. The IVT fragmentation, microarray hybridization, and data processing were performed using the standard Affymetrix GeneChip protocols. Double-round amplification using 40 ng of input RNA yielded an average of 36 μg of labeled IVT material. The quality assessment of the microarray data showed satisfactory microarray performance. For example, we obtained low background, within normal range scaling factor, and close to 45% present call. The only exception was the higher than normal 3′:5′ ratios, which was in the 20s for both the actin and glyceraldehyde 3-phosphate dehydrogenase control probe sets. This was not due to RNA degradation, as we tested RNA quality in a parallel experiment in which the same samples were processed using the standard Affymetrix protocol. The lower 3′:5′ ratio observed with our protocol was the result of truncated IVT products. *Fig. 4* shows the size distribution of the IVT products separated by agarose gel electrophoresis. Shortened enzyme processivity or biased enrichment of short IVT products pertaining to purification methods could both lead to this observation.

Assessment of the reproducibility of microarray data generated from double-round RNA amplification is a key consideration when using these protocols. Although the double-round RNA amplification produces shortened IVT products, which subsequently results in a higher 3′:5′ ratio, this seems to have little effect on the resulting microarray data. To assess the impact of this on microarray data, we compared data when the starting material was reduced from 5–10 μg to 40 ng. We have used a robust, previously reported tumor necrosis factor α response in HUVEC cells to address this issue systematically (32). As expected, the double-round amplification generated some variability, as any perturbation in the enzymatic reaction is expected to increase the noise. The coefficient

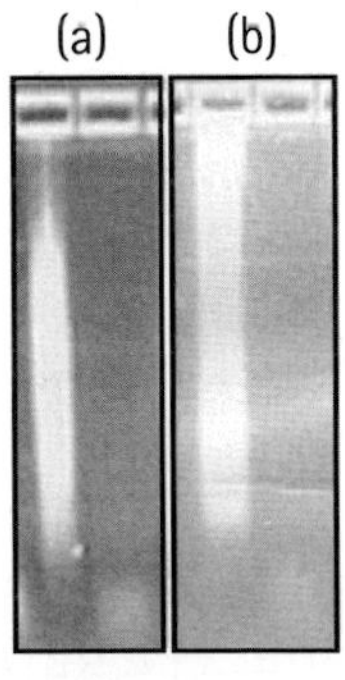

Figure 4. IVT products separated by agarose gel electrophoresis.
Samples were prepared using the Arcturus technology (*a*) and the standard Affymetrix protocol (*b*). The high-molecular-mass RNA products are under-represented in the Arcturus sample and shorter cRNA products will produce a higher 3′ : 5′ ratio when hybridized to the microarray.

of variance increases as the amount of starting material is reduced. However, the coefficient of variance between the control and tumor necrosis factor-α-treated samples was comparable among different RNA labeling platforms. The error appeared to be systematic and can probably be managed through data normalization and the appropriate data analysis matrix. We were able to observe an 80% agreement between a 40 ng double-round amplification study and a previous study using 10 μg of input RNA (65).

Taken together, cross-microarray, cross-experiment or cross-platform (e.g. double-round amplification versus single-round amplification) absolute gene expression analysis is generally not recommended when analyzing microarray data. However, comparison analyses generated from small sample amplification will be comparable to those generated from standard approaches.

Automation of sample handling is another important issue and a potential source of variability in these studies. In our experience, approximately 50% of the laboratory effort for microarray biomarker identification is spent on sample labeling, and the need to run a large number of samples for biomarker studies has led to the need to automate microarray sample labeling. Several commercial and custom automation solutions are now available, and we have experience with the TheOnyx system from MWG Biotech and Aviso (www.aviso-gmbh.de/). This system is shown in *Fig. 5*(*a*). The entire labeling procedure, from total RNA through to IVT quantitation and normalization, has been automated on the TheOnyx platform. The system uses magnetic bead separation to purify cDNA and cRNA, temperature-controlled racks for reagent storage, a thermocycler block for cDNA and cRNA synthesis, and an integrated plate reader for cRNA quantitation. Integrated software controls the entire process and completes the labeling of 95 samples in approximately 36 h. Comparison of manual and automated sample labeling has shown that the results obtained with automated labeling are similar to those from manual labeling approaches (*Fig. 5b*). We recommend a complete validation of all automated systems prior to routine use.

As with all RNA-based protocols, extreme attention and care must be made to avoid and eliminate ribonuclease contaminants from reagents and equipment including buffers, columns, pipette tips, tubes, and centrifuges. Gloves should be worn at all times when working with RNA. Autoclaving and treatment with DEPC are among the ways of inactivating ribonucleases for these studies.

(a)

(b)

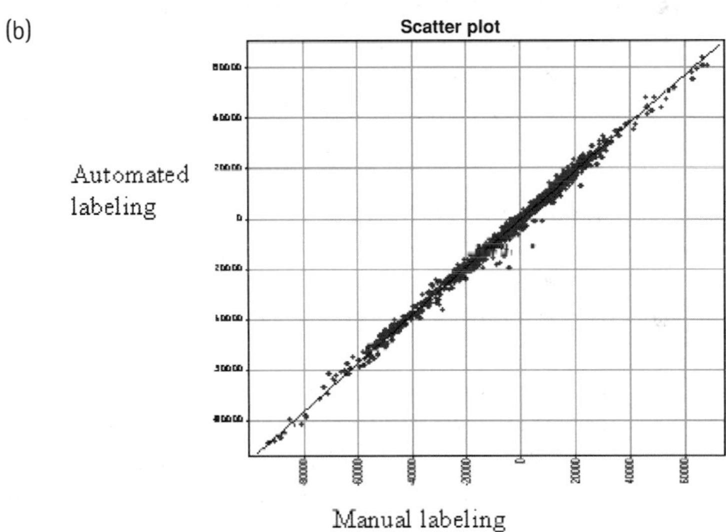

Figure 5. Automation system for microarray sample labeling.
(*a*) The TheOnyx liquid handling robot configured for automated Affymetrix sample labeling: (1) magnetic bead reservoir and 96-well microplate magnet; (2) thermocycler block; (3) gripper to move reaction plates to different positions on the robot deck; (4) plate reader for concentration measurements; (5) disposable tip storage; (6) temperature-controlled reagent storage and vacuum manifold; (7) eight-channel pipetting head with liquid level sensing; and (8) instrument controller. (*b*) Comparison of manual and automated sample labeling. Rat brain and spleen RNA were labeled and the signal change values were plotted. The R^2 value for the data was 0.990.

4. CONCLUSIONS

Microarray expression profiling has evolved into a robust, sensitive, and highly quantitative tool that is providing novel and comprehensive views of biological systems. Biomarker discovery is an area where the use of microarray technology is having a positive impact. Due to its genetic basis, this impact has occurred first in cancer where new biomarkers have been discovered for cancer diagnosis and prognosis, and to predict response to therapeutic intervention. From these initial discoveries, microarrays are now being used to find biomarkers for other diseases and for biological assessment of the efficacy and toxicity of new drugs. The use of biomarkers is now considered an essential part of the modern drug discovery and development process, and biomarkers are being used to improve the success and reduce the cost of developing new drugs. Along with the methods for sample labeling that we have described, improvements in gene annotation and statistical and bioinformatic analysis of microarray data will continue to expand the use of microarray biomarker discovery in translational research.

Acknowledgements

We would like to thank the members of the Functional Genomics Group at Eli Lilly and Company for their technical assistance and helpful suggestions.

5. REFERENCES

★★ 1. **Lander ES** (1996) *Science*, **274**, 536–539. – *Excellent review of the promise of genomic approaches to biomedical research.*
2. **Drews J** (1996) *Nat. Biotechnol.* **14**, 1516–1518.
3. **Bredel M & Jacoby E** (2004) *Nat. Rev. Genet.* **5**, 262–275.
4. **Giaever G, Flaherty P & Kumm J, *et al.*** (2004) *Proc. Natl. Acad. Sci. U.S.A.* **101**, 793–798.
5. **Vieth M, Higgs RE , Robertson DH, Shapiro M, Gragg EA & Hemmerle H** (2004) *Biochim. Biophys. Acta*, **1697**, 243–257.
6. **Aardema MJ & MacGregor JT** (2002) *Mutat. Res.* **499**, 13–25.
7. **Waters MD & Fostel JM** (2004) *Nat. Rev. Genet.* **5**, 936–948.
8. **Norton RM** (2001) *Drug Discov. Today*, **6**, 180–185.
9. **Roses AD** (2001) *Drug Discov. Today*, **6**, 59–60.
10. **Bumol TF & Watanabe AM** (2001) *JAMA*, **285**, 551–555.
★ 11. **Service RF** (2004) *Science*, **303**, 1796–1799. – *Recent review of current drug discovery and development challenges.*
12. **DiMasi JA, Hansen RW & Grabowski HG** (2003) *J. Health Econ.* **22**, 151–185.
13. **Kola I & Landis J** (2004) *Nat. Rev. Drug Discov.* **3**, 711–715.
14. **Kubinyi H** (2003) *Nat. Rev. Drug Discov.* **2**, 665–668.
★★ 15. **Frank R & Hargreaves R** (2003) *Nat. Rev. Drug Discov.* **2**, 566–580. – *Excellent review of biomarker applications in drug discovery and development.*

16. **Baker M** (2005) *Nat. Biotechnol.* **23**, 297–304.
17. **De Gruttola VG, Clax P, DeMets DL, *et al.*** (2001) *Control Clin. Trials*, **22**, 485–502.
18. **Biomarkers Definitions Working Group** (2001) *Clin. Pharmacol. Ther.* **69**, 89–95.
★★ 19. **FDA** (2004) *Challenge and Opportunity on the Critical Path to New Medical Products.* http://www.fda.gov/oc/initiatives/criticalpath/whitepaper.html. – *Regulatory whitepaper describing the role of biomarkers in drug development.*
20. **Rolan P, Atkinson AJ Jr & Lesko LJ** (2003) *Clin. Pharmacol. Ther.* **73**, 284–291.
21. **Littman BH & Williams SA** (2005) *Nat. Rev. Drug Discov.* **4**, 631–638.
22. **Sawada H, Takami K & Asahi S** (2005) *Toxicol. Sci.* **83**, 282–292.
23. **Cunningham ML & Lehman-McKeeman L** (2005) *Toxicol. Sci.* **83**, 205–206.
24. **Reasor MJ & Kacew S** (2001) *Exp. Biol. Med. (Maywood)*, **226**, 825–830.
25. **Yingling JM, Blanchard KL & Sawyer JS** (2004) *Nat. Rev. Drug Discov.* **3**, 1011–1022.
26. **Schena M, Shalon D, Davis RW & Brown PO** (1995) *Science*, **270**, 467–470.
27. **Lockhart DJ, Dong H, Byrne MC, *et al.*** (1996) *Nat. Biotechnol.* **14**, 1675–1680.
28. **Golub TR, Slonim DK, Tamayo P, *et al.*** (1999) *Science*, **286**, 531–537.
29. **Rhodes DR & Chinnaiyan AM** (2005) *Nat. Genet.* **37** (Suppl.), S31–S37.
30. **Cheok MH, Yang W, Pui CH, *et al.*** (2003) *Nat. Genet.* **34**, 85–90.
31. **Grinnell BW & Joyce D** (2001) *Crit. Care Med.* **29** (Suppl.), S53–S60; discussion S60–S61.
32. **Joyce DE, Gelbert L, Ciaccia A, DeHoff B & Grinnell BW** (2001) *J. Biol. Chem.* **276**, 11199–11203.
33. **Onyia JE, Helvering LM, Gelbert L, *et al.*** (2005) *J. Cell. Biochem.* **95**, 403–418.
34. **Kulkarni NH, Halladay DL, Miles RR, *et al.*** (2005) *J. Cell. Biochem.* **95**, 1178–1190.
35. **Platzer P, Upender MB, Wilson K, *et al.*** (2002) *Cancer Res.* **62**, 1134–1138.
36. **Xin B, Platzer P, Fink SP, *et al.*** (2005) *Oncogene*, **24**, 724–731.
37. **Bast RC Jr & Hortobagyi GN** (2004) *N. Engl. J. Med.* **351**, 2865–2867.
38. **Paik S, Shak S, Tang G, *et al.*** (2004) *N. Engl. J. Med.* **351**, 2817–2826.
39. **Ilyin SE, Belkowski SM & Plata-Salaman CR** (2004) *Trends Biotechnol.* **22**, 411–416.
40. **Emmert-Buck MR, Bonner RF, Smith PD, *et al.*** (1996) *Science*, **274**, 998–1001.
41. **Burczynski ME, Twine MC, Dujart G, *et al.*** (2005) *Clin. Cancer Res.* **11**, 1181–1189.
42. **Rockett JC, Burczynski ME, Fornace AJ Jr, Herrmann PC, Krawetz SA & Dix DJ** (2004) *Toxicol. Appl. Pharmacol.* **194**, 189.
43. **Twine NC, Stover JA, Marshall B, *et al.*** (2003) *Cancer Res.* **63**, 6069–6075.
44. **Iscove NN, Barbara M, Gu M, Gibson M, Modi C & Winegarden M** (2002) *Nat. Biotechnol.* **20**, 940–943.
45. **Klur S, Toy K, Williams MP & Certa U** (2004) *Genomics*, **83**, 508–517.
46. **Ma J & Liew CC** (2003) *J. Mol. Cell Cardiol.* **35**, 993–998.
47. **Ogawa M** (1993) *Blood*, **81**, 2844–2853.
48. **Alcorta D, Preston G, Munger W, *et al.*** (2002) *Exp. Nephrol.* **10**, 139–149.
49. **Amundson SA, Grace MB, MeLeland CB, *et al.*** (2004) *Cancer Res.* **64**, 6368–6371.
★★ 50. **Fan H & Hegde PS** (2005) *Curr. Mol. Med.* **5**, 3–10. – *Excellent review on gene expression profiling in blood.*
51. **Tsuang MT, Nossova N, Yager T, *et al.*** (2005) *Am. J. Med. Genet. B Neuropsychiatr. Genet.* **133**, 1–5.
52. **Xu T, Shu CT, Purdom E, *et al.*** (2004) *Cancer Res.* **64**, 3661–3667.
53. **DePrimo SE, Wong LM, Khatry DB, *et al.*** (2003) *BMC Cancer*, **3**, 3.
54. **Jison ML, Munson PJ, Barb JJ, *et al.*** (2004) *Blood*, **104**, 270–280.
55. **Rus V, Chen H, Zernetkina V, *et al.*** (2004) *Clin. Immunol.* **112**, 231–234.
56. **Shou J, Dotson C, Qian HR, *et al.*** (2005) *Biomarkers*, **10**, 310–320.
57. **Makrigiorgos GM, Chakrabarti S, Zhang Y, Kaur M & Price BD** (2002) *Nat. Biotechnol.* **20**, 936–939.

★★ **58. Glanzer JG & Eberwine JH** (2004) *Br. J. Cancer*, **90**, 1111–1114. – *Great overview of small-quantity sample amplification for microarray studies.*

59. Polz MF & Cavanaugh CM (1998) *Appl. Environ. Microbiol.* **64**, 3724–3730.

60. Kenzelmann M, Klaren R, Hergenhalm M, *et al.* (2004) *Genomics*, **83**, 550–558.

61. Marciano PG, Brettschneider J, Manduchi E, *et al.* (2004) *J. Neurosci.* **24**, 2866–2876.

62. McClintick JN, Jerome RE, Nicholson CR, Crabb DW & Edenberg HJ (2003) *BMC Genomics*, **4**, 4.

★★ **63. Wilson CL, Pepper SD, Hey Y & Miller CJ** (2004) *Biotechniques*, **36**, 498–506. – *An excellent study on errors introduced by RNA amplification in microarray studies.*

64. Zhu G, Bertrand JR & Malvy C (2003) *Oncogene*, **22**, 3742–3748.

65. Shou J, Qian HR, Lin X, Stewart T, Onyia JE & Gelbert LM (2006) *J. Pharmacol. Toxicol. Methods*, **53**, 152–159.

CHAPTER 11

DNA microarrays to study nonhuman primate gene expression

Stephen J. Walker

1. INTRODUCTION

Nonhuman primates (NHPs), used widely in biomedical research, often serve as ideal experimental models for human diseases that cannot be modeled adequately in other species. For example, NHPs provide a relevant and robust model system for studying alcoholism, a complex behavioral disorder that produces long-term changes in cellular function through alterations in gene expression (1–3). Long-term alcohol use can lead to tolerance, physical dependence, and addiction – outcomes that may well reflect alterations in various signal transduction pathways, leading to neuroadaptive and neurotoxic effects (4, 5). Understanding the addiction process in general and alcoholism in particular requires examination of changes in gene expression (6). Studying gene expression profiles in macaque monkeys that chronically self-administer alcohol is being used as a means of gaining a deeper understanding of the biology, both metabolic and behavioral, that underlies alcoholism. The identification of significant changes in gene expression induced by alcohol, both peripherally (in blood) and within the central nervous system, has the potential to identify important molecular and cellular targets for medications that could be used to treat and/or prevent alcohol-induced neurotoxicity.

Macaque monkeys represent an important model for studying the biological basis of complex diseases such as alcoholism for several reasons. First, their genome is estimated to share >95% DNA sequence identity with humans (7). Therefore, the ability to interpret much of the data derived from NHP gene expression analysis, performed with human sequence-based methods, is often uncomplicated. Secondly, macaque neuroanatomy, neurochemistry, and physiology are similar to humans and underlie similar behaviors. Particularly notable are the response of the hypothalamic–

DNA Microarrays: *Methods Express* (M. Schena, ed.)

pituitary–adrenal axis to stress and the hypothalamic–pituitary–gonadal axis control of sexual physiology (8). Both of these endocrine axes produce steroid derivatives that mimic the subjective effects of alcohol in animal models and may represent the bases for stress-induced and gender-related risk factors for alcoholism, respectively. Thirdly, the average life span of the macaque is long (>25 years), as are critical periods and stages of development. Experimental designs, therefore, can incorporate long-term exposures and sophisticated behavioral studies to capture the effects of developmental stage on risk for alcohol effects. Fourthly, many macaques used in biomedical research are not, strictly speaking, inbred and therefore display a wide range of phenotypic responses to alcohol that are useful in experimental designs of genotypic risk for alcoholism. Finally, these monkeys display kinetics for the absorption and metabolism of alcohol that are very similar to human beings (9). They can be trained to drink alcohol, they have complex social relationships, and they are capable of performing complex cognitive tasks.

The ability to study functional genomic sequelae that underlie the risk for, and the biology of, complex disease is often hindered either by the limited availability of relevant tissues and/or, in the case of NHP research, limited research instruments (e.g. NHP sequence-specific microarrays and primer sets for *post hoc* validation of interesting targets by reverse transcriptase polymerase chain reaction (RT-PCR)). Species-specific tools will become more widely available as more NHP sequence information is generated and shared. In the meantime, accessing relevant tissue still remains a problem, both in animal research and in human disease research. It is becoming increasing popular, therefore, to examine gene expression changes in peripheral tissues, searching for biochemical or biological markers (biomarkers) as a means of understanding disease states without having to query, for example, brain tissue directly. Characterization of the variation in gene expression in healthy tissues is an essential foundation for the recognition and interpretation of the changes in these patterns associated with stress and disease. One tissue in particular, peripheral blood, is ideally suited because it is a uniquely accessible tissue in which to examine the variations in both patients and 'normal' controls (10). Especially powerful is the ability to measure gene expression sequentially in the same individual, under various physiological conditions, and then to examine correlations between disease and peripheral gene expression patterns. Future investigations of human gene expression programs associated with disease, and their potential application to detection and diagnosis, will depend on an understanding of their normal variation within and between individuals, over time, and with age, gender, and other aspects of the human condition (10).

The high degree of sequence similarity between human and NHP genomic DNA suggests that human genome sequence-based DNA microarrays may be used effectively to study gene expression in NHP disease models. This chapter details the preparation of RNA from macaque tissue (whole blood) and the methods associated with the use of two distinct commercially available human genome microarray platforms, the Affymetrix HG U133A GeneChip System utilizing Human Genome U133A microarrays and the Applied Biosystems Expression Array System utilizing the Human Genome Survey Microarray for gene expression studies (see *Table 1*).

Table 1. Comparative specifications of two commercial microarray platforms

	U133A GeneChip	Human Whole Genome Survey Arrays v1.0
Company name and location	Affymetrix, Inc., Santa Clara, CA, USA	Applied Biosystems, Foster City, CA, USA
Type of substrate	Activated glass wafer	Nylon-coated glass substrate
Method of manufacture	Oligonucleotides synthesized *in situ*	Oligonucleotides synthesized, subjected to quality control, and printed
Oligonucleotide length	25mers	60mers
Total number of oligonucleotides	22 283 sets containing a total of 490 226 oligonucleotides	33 096 oligonucleotides
Number of oligonucleotides per mRNA	22 (11 pairs of oligonucleotides)	1
Oligonucleotide variants	Perfect match and single-base mismatch for each mRNA	Perfectly matching sequences only
Type of detection	Single-color fluorescence detection	Single-color chemiluminescence detection
Quantification and analysis software	MAS v5.0	Applied Biosystems 1700 Chemiluminescence Microarray Analyzer Software v1.1

2. METHODS AND APPROACHES

2.1 Human sequence-based microarrays

Increasingly, functional genomic screening tools are being applied to animal model systems to interrogate changes in gene expression that can provide insight into the disease processes. The DNA microarray is uniquely suited to this task. Since its introduction into the scientific literature more than 10 years ago (11), the DNA microarray has become ubiquitous as a means of evaluating changes in gene expression, owing in part to its ability to interrogate the entire transcriptome in a single experiment. Because they are such useful experimental tools, microarrays have been constructed for use on myriad animal, plant, and microbial systems. However, there are many species being studied for which whole genome sequence information, and therefore microarrays based on that information, are not yet available. In particular, most NHP species cannot currently be examined with whole genome species-specific commercially available microarrays because they do not yet exist. As an alternative to species-specific microarrays, NHP studies typically have employed microarrays constructed from human sequence information, but with mixed success.

Microarrays consisting of human target sequences have been used successfully to study NHP gene expression in a number of tissues (12–15). Human sequence-based microarrays have also been used for expression profiling of NHP brain tissue. For example, Affymetrix Human U95A chips were used to survey the transcriptome of pre-frontal cortex samples from human, chimpanzee, orangutan, macaque, and marmoset in order to determine whether NHP models of neurodegenerative diseases were valid, based on the presence of commonly expressed genes (16). Whilst the majority of transcripts were shared among the NHP species, many, including those of several key genes (such as *apoE*) already implicated in those diseases, were not detected in some species. It is conceivable that these genes were not expressed in the animal models, but a more likely explanation is that interspecies sequence differences were sufficient to prevent efficient binding of those RNA probes to the human DNA oligonucleotide targets. This issue can potentially be resolved once more complete NHP genomic information is available.

Investigators have been using human sequence-based microarrays to examine gene expression in NHP tissues with variable success. It is generally accepted that, due to species dissimilarity at the level of DNA and RNA, the use of cross-species microarrays can result in an under-representation of the number of detectably expressed genes. However, as the NHP genome sequence is incomplete, it is not yet possible to know how much information is being missed when NHP RNA is analyzed using human microarrays. By evaluating results generated using the same (NHP) RNA on different microarray platforms or, alternatively, many different RNAs (NHP species and human) measured on the same platform, it should be possible to resolve this question.

2.2 Chapter aims

This chapter will provide detailed methods for extracting high-quality RNA from whole blood. The purified RNA can be used for any downstream application, including RT-PCR and microarray analysis. This chapter will also present detailed methods for preparing RNA for use in two distinct human sequence-based microarray platforms – one that uses short oligonucleotide targets (25mers) and fluorescence detection, and one that uses long oligonucleotide targets (60mers) and chemiluminescence detection (see *Table 1*). Although the RNA isolation, quality control, labeling, and so forth can be performed in any laboratory equipped with standard molecular biology reagents and instrumentation, the scanners needed to image the microarrays are distinct for many of the platforms (including the two discussed here) and are quite expensive. Therefore, it is not uncommon for an investigator to prepare RNA 'in house' and then contract a core services laboratory for the labeling, hybridization, and image capture. In one scenario, the core laboratory sends the data files back to the investigator for analysis with any of a number of available bioinformatics approaches.

Although data analysis is a key component of any successful microarray experiment, this chapter will not go beyond a cursory description of bioinformatics. An in-depth treatment of microarray data analysis, the subject of much debate and discussed in great detail in numerous excellent books and journal articles, is beyond the scope of this chapter.

2.3 Recommended protocols

Protocols 1–4 describe methods for the isolation and characterization of RNA from whole blood. Due to the increasing interest in the identification of peripheral markers for disease, there are numerous commercial products available for RNA isolation from blood. The descriptions that follow are not meant to be comprehensive, nor are they meant to be an endorsement for any specific product(s) other than to point out that what we have described works 'in our hands'.

Protocol 1

Preparing and storing whole blood

Equipment and Reagents

- Whole blood – freshly collected[a]
- 1.5 ml Nuclease-free microfuge tubes
- TRI Reagent BD[b] ('blood derivatives'; Molecular Research Inc.)
- 5 N Acetic acid

Method

Whole blood (usually 3–5 ml) is delivered soon after the draw and is immediately aliquotted and frozen until further use as follows:

1. Pipette 750 μl of TRI Reagent BD into each of ten 1.5 ml microfuge tubes.
2. Add 20 μl of 5 N acetic acid to each tube.
3. Add 500 μl of whole blood to each tube and mix vigorously.
4. Store the tubes at −80°C until needed.

Notes

[a] The blood should be processed as soon as possible after it is drawn to minimize gene expression variation due to changes that occur following the draw and before the inactivating TRI Reagent BD is added.
[b] As an alternative to this, other commercial reagents (there are many available) used to isolate RNA from blood can be substituted.

Protocol 2

RNA extraction from whole blood[a]

Equipment and Reagents

- 1-Bromo-3-chloropropane[b]
- Microcentrifuge capable of cooling to 4°C
- Nuclease-free dH_2O
- Isopropanol
- 75% Ethanol

Method

1. Obtain the tubes from −80°C (*Protocol 1*, step 4) and thaw the contents at room temperature. A 2 ml blood sample typically yields four tubes.
2. To each tube, add 100 μl of 1-bromo-3-chloropropane and mix the contents vigorously by vortexing.
3. Incubate the tubes at room temperature for 5 min. Spin for 15 min at 12 000 ***g*** in a microcentrifuge set at 4°C.
4. Transfer the aqueous (top) layer to a fresh nuclease-free tube, being careful not to disturb the interface.
5. Add 500 μl of isopropanol and mix well.
6. Incubate the samples at room temperature for 20 min. Spin for 15 min at 12 000 ***g*** in a microcentrifuge set at 4°C to pellet the RNA.
7. Remove and discard the supernatant, being careful not to disturb the RNA pellet at the bottom of the tube.

8. Add 1 ml of 75% ethanol and vortex gently, being careful not to disturb or dislodge the RNA pellet. Spin for 15 min at 12 000 ***g*** in a microcentrifuge set at 4°C.
9. Decant and discard the supernatant by inverting the tube, making certain to pour away from the pellet. Place the inverted tube on a clean paper towel for 10–15 min to dry the pellet.
10. Resuspend each pellet in 25 µl of DEPC-treated dH_2O and heat in a water bath for 5 min at 60°C to aid resuspension.
11. Pool the contents of the four tubes (~100 µl total volume) and proceed to *Protocol 3*.

Notes

[a]The RNA extraction protocol is based on a modification of Chomczynski and Sacchi (17).
[b]This is a less-toxic alternative to using chloroform for phase separation.

Protocol 3

RNA purification[a]

Equipment and Reagents

- RNeasy kit (Qiagen)
- NanoDrop spectrophotometer
- 100% Ethanol
- DEPC-treated dH_2O

Method

1. Obtain the 100 µl RNA samples from the previous protocol (*Protocol 2*, step 11).
2. Add 350 µl of buffer RLT to each 100 µl RNA sample and mix well.
3. Add 250 µl of 100% ethanol to each sample and mix by pipetting.
4. Apply the sample to an RNeasy mini column and centrifuge for 15 s at ≥12 000 ***g***. The RNA binds to the column matrix under these conditions.
5. Transfer the column to a fresh microfuge tube, pipette 500 µl of buffer RPE (RNeasy kit) onto the column, and centrifuge for 15 s at ≥12 000 ***g***. Discard the eluent. The RNA remains bound to the column at this step in the procedure.
6. Add an additional 500 µl of buffer RPE to the column, centrifuge for 2 min at ≥12 000 ***g***, and discard the eluent.
7. Transfer the column to a fresh 1.5 ml tube, add 20 µl of DEPC-treated dH_2O directly to the column, and centrifuge for 15 s at ≥12 000 ***g*** to elute the RNA sample.
8. Measure concentration using NanoDrop spectrophotometer.
9. Store RNA at –80°C until use.

Note

[a]This step is absolutely critical if RNA derived from whole blood is to be used for downstream applications such as microarray.

Protocol 4

RNA denaturing gel electrophoresis[a]

Equipment and Reagents

- Bromophenol blue
- Agarose
- Agarose gel electrophoresis apparatus
- Formamide
- 10× MOPS buffer
- Formaldehyde
- 10× Agarose gel running buffer (200 mM 3-(*N*-morpholino)propanesulfonic acid (MOPS) free acid, 50 mM sodium acetate, 10 mM EDTA, adjusted to pH 7.0 with NaOH)
- 1× Formaldehyde agarose gel running buffer (100 ml of 10× agarose gel running buffer, 20 ml of 37% (12.3 M) formaldehyde, 880 ml of RNase-free dH_2O per litre)
- Ethidium bromide (1 mg/ml)
- Microwave or hotplate
- Stir plate
- Fume hood
- Denaturing solution (200 μl of formamide, 80 μl of formaldehyde, 30 μl of 10× MOPS buffer, 15 μl of ethidium bromide). Mix well and make fresh before each use

Method

1. Combine 15 ml of 10× agarose gel running buffer and 109.5 ml DEPC-treated dH_2O in a 250 ml wide-mouth conical flask. Set mixing on a stir plate[b].
2. With the buffer stirring, add 1.5 g of agarose powder and mix well.
3. Heat the agarose mixture to boiling point in a microwave oven and swirl the flask until the agarose melts completely. Several boiling and swirling cycles may be needed to ensure complete melting of the agarose.
4. Cool the molten agarose mixture to 65°C by swirling the flask for several minutes and transfer the flask to a fume hood. Add 25.5 ml of formaldehyde, mix well by swirling the flask, and pour the mixture into the gel tray. The final agarose concentration of the gel will be 1.0%.
5. Insert the gel comb into the correct position in the molten agarose and allow the agarose to harden for 30 min in the hood. Gently remove the comb from the hardened agarose, being careful not to tear the newly formed wells. Fill the electrophoresis tank with approximately 300 ml of 1× formaldehyde agarose gel running buffer until the buffer level exceeds the top surface of the gel by approximately 0.5 cm.
6. To prcpare the RNA samples for gel electrophoresis, transfer 0.5–1.0 μg of each RNA sample to a microfuge tube and add DEPC-treated dH_2O to a final volume of 7 μl. Mix well.
7. Add 7 μl of fresh denaturing solution and mix well.
8. Add 2 μl of 0.015% bromophenol blue and mix well.

9. Incubate the samples in a 65°C water bath for 5 min to denature the RNA.
10. Using a fine pipette tip, load the entire 16 µl volume of each RNA sample into a separate well in the gel. Run the gel for 40 min at 50 V to separate the RNA species. Data from a typical RNA gel electrophoresis run are shown in *Fig. 1(a)* (also available in the color section).

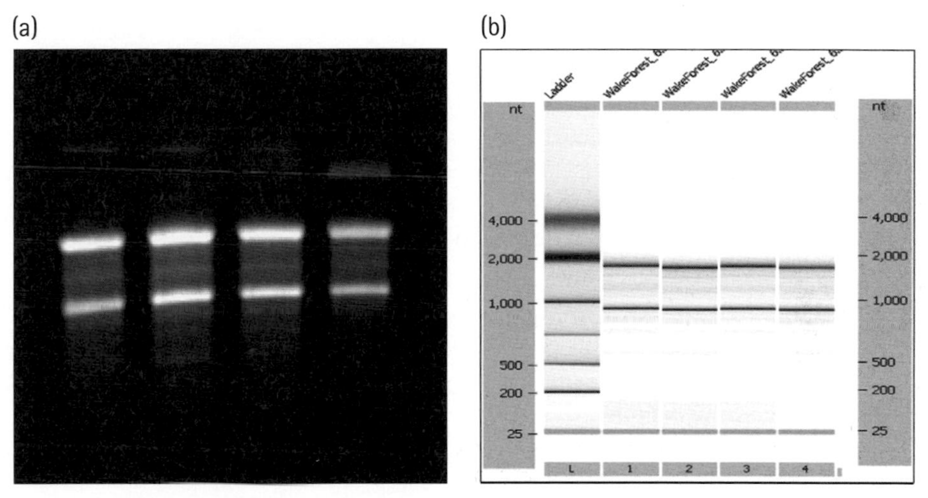

Figure 1. Analysis of RNA preparations from NHP blood (see page xxx for color version). (*a*) Total RNA was isolated from 2 ml of whole blood drawn from four individual animals and purified using TRI Reagent BD (Molecular Research Inc.). Following purification clean-up with an RNeasy kit (Qiagen), the RNA was eluted with 25 µl of nuclease-free dH_2O. A volume of 1 µl of RNA was used to determine concentration. Approximately 1 µg of RNA, based on NanoDrop analysis, was separated on a 1% formamide gel and the stained gel was photographed as shown. (*b*) A Bioanalyzer instrument (Agilent) was used to analyze four NHP RNA samples prior to labeling for microarray analysis: lane 1, sample 89, 1.2 ratio and 9.2 RNA integrity number (RIN); lane 2, sample 92, 1.2 ratio and 9.3 RIN; lane 3, sample 94, 1.2 ratio and 9.3 RIN; and lane 4, sample 96, 1.0 ratio and 9.4 RIN.

Notes

[a]We routinely assess RNA quality and quantity in this way. However, many laboratories and nearly all microarray facilities assess RNA using the Agilent Bioanalyzer (see *Fig. 1b*).
[b]This protocol is used for preparing RunOne apparatus gels and is sufficient for at least one full tray consisting of two large gels and four small gels.

Protocol 5

First- and second-strand cDNA synthesis for microarray analysis[a]

Equipment and Reagents

- 0.6 ml Microfuge tubes
- Affymetrix GeneChip System including fluidics workstation and scanner
- SuperScript Choice System for cDNA Synthesis (Invitrogen)
- 100 pmole/μl T7-oligo(dT) primer with the following primary sequence: 5′-GGCCAGTGAATTGTAATACGACTCACTATAGGGAGGCGG-$(T)_{24}$-3′

Method

1. Transfer 10 μl of total RNA (5–40 μg) to a 0.6 ml microfuge tube and add 1.0 μl of 100 pmole/μl T7-oligo(dT) primer.
2. Heat to 70°C for 10 min to denature the RNA.
3. Place the tubes on ice and add 4 μl of 5× first-strand buffer, 2 μl of 0.1 M dithiothreitol, and 1 μl of 10 mM dNTP mix (SuperScript Choice System for DNA Synthesis). Mix well by gently tapping the tubes.
4. Heat to 37°C for 2 min and add 2 μl of SuperScript II reverse transcriptase (200 units/μl) (SuperScript Choice System for DNA Synthesis).
5. Incubate at 42°C for 60 min to allow first-strand synthesis. Transfer the tubes to ice.
6. To each first-strand synthesis tube, add 91 μl of DEPC-treated dH_2O, 30 μl of 5× second-strand buffer, 4 μl of *E. coli* DNA polymerase I (40 units), 3 μl of 10 mM dNTP mix, 1 μl of *E. coli* DNA ligase (10 units), and 1 μl of RNase H (2 units) (SuperScript Choice System for DNA Synthesis).
7. Incubate the samples at 16°C for 2 h to allow second-strand synthesis.

Note

[a]*Protocols 5–10* describe cDNA synthesis methods for microarray analysis using the Affymetrix GeneChip platform. Most of the steps can be accomplished in a standard molecular biology laboratory. It is not uncommon for these steps to be subcontracted to a core laboratory that has the necessary instrumentation for image capture. Once the microarrays have been processed, the data can be analyzed using any of a large number of commercial and noncommercial software packages. As stated before, these protocols do not imply endorsement of any particular commercial platform or reagent set, but are presented simply to illustrate methods that have been used successfully in our laboratory.

Protocol 6

Purification of double-stranded cDNA synthesis products for microarray analysis

Equipment and Reagents

- 1.7 ml Microfuge tubes
- Tris-saturated phenol : chloroform : isoamyl alcohol (25 : 24 : 1, v/v)
- 7.5 M Ammonium acetate
- 100% Ethanol
- 5 mg/ml Glycogen (Ambion)
- DEPC-treated dH_2O
- Microcentrifuge

Method

1. Transfer the entire 150 µl volume of each second-strand synthesis reaction (*Protocol 5*, step 7) to a 1.7 ml microfuge tube.
2. Add 162 µl of Tris-saturated phenol : chloroform : isoamyl alcohol (25 : 24 : 1) and mix well by vortexing.
3. Separate the phases by centrifugation at ≥14 000 ***g*** for 2 min in a microcentrifuge.
4. Transfer the upper (aqueous) phase containing the cDNA product to a fresh 1.7 ml tube and add 75 µl of 7.5 M ammonium acetate, 375 µl of 100% ethanol, and 4 µl of 5 mg/ml glycogen. Mix well by vortexing.
5. Incubate the samples on ice for 30 min to allow precipitation.
6. Pellet the cDNA by room temperature centrifugation at ≥14 000 ***g*** for 10 min in a microcentrifuge.
7. Wash the pellet twice with 80% ethanol, dry the pellet by vacuum centrifugation, and resuspend in 22 µl of DEPC-treated dH_2O. Proceed immediately to the *in vitro* transcription step or store at −80°C until ready to use.

Protocol 7

In vitro transcription (IVT) and RNA labeling for microarray analysis[a]

Equipment and Reagents

- 0.6 ml RNase-free microfuge tubes
- RNA transcript labeling kit (Enzo; Affymetrix)
- Chroma Spin-100 (Clontech) or Sephadex G-25 gel filtration columns
- Microcentrifuge

Method

1. Obtain the 22 μl cDNA samples from the previous step (*Protocol 6*, step 7) and transfer to an RNase-free 0.6 ml microfuge tube.
2. To each 22 μl sample, add 4 μl of 10× HY reaction buffer, 4 μl of 10× biotin-labeled ribonucleotides, 4 μl of 10× dithiothreitol, 4 μl of 10× RNase inhibitor mix, and 2 μl of 20× T7 RNA polymerase (RNA transcript labeling kit).
3. Mix the reagents by gently flicking the tube and spin for 5 s in a microcentrifuge to consolidate the reagents to the tube bottom.
4. Incubate at 37°C for 4–5 h to allow IVT to proceed, gently mixing every 30-45 min to prevent condensation from forming on the sides of the tubes.
5. After the 4–5 h incubation, transfer the IVT reactions to pre-spun Chroma Spin-100 or Sephadex G-25 gel filtration columns. Centrifuge the spin columns according to the instructions of the manufacturer to collect the labeled cRNA in the 1.5 ml sample collection tubes.
6. Determine cRNA concentration and purity by measuring the optical density at 260 and 280 nm.

Note

[a]This IVT protocol requires the cDNA material obtained from reverse transcription of 5–15 μg of total RNA or 0.2–0.5 μg of poly(A)$^+$ RNA.

Protocol 8

cRNA fragmentation and probe preparation

Equipment and Reagents

- 1.7 ml Microfuge tubes
- DEPC-treated dH_2O
- 5× Fragmentation buffer (4.0 ml of 1 M Tris/acetate adjusted to pH 8.1 with glacial acetic acid, 0.64 g of magnesium acetate, 0.98 g of potassium acetate, DEPC-treated dH_2O to a final volume of 20 ml; filter sterilize using 0.2 μM filtration)
- 3 nM Control oligo B2 (Affymetrix)
- 20× Eukaryotic hybridization control mix containing 30 pM *BioB*, 100 pM *BioC*, 500 pM *BioD*, and 2 nM *CRE* (Affymetrix)
- 10 mg/ml Herring sperm DNA (Promega)
- 50 mg/ml Acetylated bovine serum albumin (BSA; Invitrogen)
- MES free acid monohydrate (Sigma)
- MES sodium salt (Sigma)
- 12× MES stock buffer (70.4 g of MES free acid monohydrate (1.22 M final concentration), 193.3 g of MES sodium salt (0.89 M final), and 800 ml DEPC-treated dH_2O. Stir with heating at ~50°C and adjust to 1000 ml after cooling to room temperature. The pH should be 6.5–6.7. Sterilize using a 0.2 μM filter unit and store at 4°C)

- 2× Hybridization buffer (8.3 ml of 12× MES stock buffer (100 mM MES final), 17.7 ml of 5 M NaCl (1 M Na^+ final), 4 ml of 0.5 M EDTA (20 mM final), 0.1 ml of 10% Tween 20 (0.01% final), and 19.9 ml of DEPC-treated dH_2O)

Method

1. Obtain the cRNA sample from the previous protocol (*Protocol 7*, step 5). To a 1.7 ml microfuge tube, add 15 μg of cRNA in 15–30 μl of DEPC-treated dH_2O, 8 μl of 5× fragmentation buffer, and DEPC-treated dH_2O to a final volume of 40 μl.
2. Incubate the cRNA sample at 95°C for 35 min to fragment the cRNA. Proceed directly to step 3 or store at −20°C.
3. Prepare 300 μl of probe solution by mixing 10–15 μg of fragmented cRNA (0.05 μg/μl final), 5 μl of 3 nM control oligo B2 (50 pM final), 15 μl of 20× eukaryotic hybridization control mix (1.5 pM *BioB*, 5 pM *BioC*, 25 pM *BioD* and 100 pM *CRE* final), 3 μl of 10 mg/ml herring sperm DNA (0.1 mg/ml final), 3 μl of 50 mg/ml acetylated BSA (0.5 mg/ml final), 150 μl of 2× hybridization buffer, and 84 μl of DEPC-treated dH_2O. Mix well and use immediately or store at −20°C until ready to use.

Protocol 9

Microarray hybridization

Equipment and Reagents

- 2× Hybridization buffer (8.3 ml of 12× MES stock buffer (100 mM MES final concentration), 17.7 ml of 5 M NaCl (1 M Na^+ final), 4 ml of 0.5 M EDTA (20 mM final), 0.1 ml of 10% Tween 20 (0.01% final), and 19.9 ml of DEPC-treated dH_2O)
- 500 ml 0.2 μM Bottle-top sterile filters (Fisher Scientific)
- Antifoam 0-30 (Sigma)
- 1× Non-stringent wash buffer (300 ml of 20× SSPE (6× final concentration), 1.0 ml of 10% Tween 20 (0.01% final), and 698 ml of DEPC-treated dH_2O. Filter through a 500 ml 0.2 μM bottle-top filter to sterilize, and add 1.0 ml of 5% antifoam 0-30 to prevent foaming of the wash buffer during use)
- Microcentrifuge
- Hybridization oven

Method

1. Allow the microarray to equilibrate to room temperature before use. It takes ~15 min to warm the microarray from 4°C to 20–25°C.
2. Pre-wet the microarray with 200 μl of 1× hybridization buffer for 10–20 min at 45°C with rotational agitation (60 r.p.m.).
3. Denature the 300 μl of labeled and fragmented cRNA probe mixture by heating the sample to 99°C in a water bath for 5 min.
4. Centrifuge the probe mixture for 5 min at 16 800 *g*[a] to remove any particulates present in the sample. When transferring the sample from the microfuge tube to the microarray (see step 6), be careful not to disturb the debris at the bottom of the tube.

5. Remove and discard the 200 μl of 1× hybridization buffer used to pre-wet the microarrays. The microarrays are ready for hybridization at this point.
6. Add the 300 μl of probe mixture to the microarray. Use one microarray for each 300 μl sample.
7. Incubate the microarrays at 45°C for 16 h in a hybridization oven with a 60 r.p.m. mixing speed.
8. Following the 16 h hybridization step, remove the 300 μl probe mixture from the cassette and store it at –20°C for possible reuse.
9. Once the probe mixture is removed, immediately add 200 μl of 1× non-stringent wash buffer to the microarray and process on the fluidics station.

Note

[a]Do not allow the probe mixture to cool below 20–25°C at any point during the centrifugation step.

Protocol 10

Microarray washing, staining, scanning, and data analysis[a]

Equipment and Reagents

- 12× MES stock buffer (70.4 g of MES free acid monohydrate (1.22 M final concentration), 193.3 g of MES sodium salt (0.89 M final), and 800 ml of DEPC-treated dH_2O. Stir with heating at ~50°C and adjust to 1000 ml after cooling to room temperature. The pH should be 6.5–6.7. Sterilize using a 0.2 μM filter unit and store at 4°C)
- 5 M NaCl
- 10% Tween 20
- DEPC-treated dH_2O
- 500 ml 0.2 μM Bottle-top sterile filters (Fisher Scientific)
- Antifoam O-30 (Sigma)
- 50 mg/ml Acetylated BSA (Invitrogen)
- 1 mg/ml Streptavidin–phycoerythrin (Molecular Probes)
- 10 mg/ml Goat IgG (Sigma)
- 0.5 mg/ml Biotinylated anti-streptavidin antibody (Vector Laboratories)
- 2× Stain buffer (41.7 ml of 12× MES stock buffer (100 mM MES final concentration, 92.5 ml of 5 M NaCl (1 M Na^+ final), 2.5 ml of 10% Tween 20 (0.05% final), and 112.8 ml of DEPC-treated dH_2O. Sterilize using a 500 ml 0.2 μM bottle-top sterile filter, and add 0.5 ml of 5% Antifoam O-30)
- Streptavidin phycoerythrin stain used for the first and third staining steps (600 μl of 2× stain buffer, 48 μl of acetylated BSA (50 mg/ml; Invitrogen), 12 μl of streptavidin–phycoerythrin (1 mg/ml; Molecular Probes), and 540 μl of DEPC-treated dH_2O
- Second antibody stain (300 μl of 2× stain buffer, 24 μl of acetylated BSA (50 mg/ml; Invitrogen), 6 μl of goat IgG (10 mg/ml; Sigma), 3.6 μl of biotinylated anti-streptavidin antibody (0.5 mg/ml; Vector Laboratories), and 266.4 μl of DEPC-treated dH_2O.
- Stringent wash buffer (83.3 ml of 12× MES stock buffer (100 mM MES final concentration), 5.2 ml of 5 M NaCl (0.1M Na^+ final), 1.0 of ml 10% Tween 20 (0.01%

final), and 910.5 ml of DEPC-treated dH_2O. Sterilize by filtering through a 500 ml 0.2 μM bottle-top filter)

Method

1. Register each microarray sample in the Affymetrix GeneChip Operating Software (GCOS).
2. Run the recommended Affymetrix GeneChip Fluidics station 400 or 450 wash protocol (EukGE-WS4 or -5).
3. Scan and analyze the microarrays using GCOS.
4. Examine expression profiles across the genome. For transcriptome analysis, the detection thresholds (present/marginal/absent calls) used are those specified by the manufacturer of each platform. For Affymetrix microarrays, the detection threshold is set as the 'present' call output from Affymetrix MAS software v5.0 ($P < 0.05$).
5. Normalize the expression data. A global median normalization, which normalizes signal intensities across all microarrays to achieve the same median signal intensities for each microarray, is performed on the data set. The normalization can take into account all of the data (from perfect match *and* mismatch probes; MAS v5.0) or only the perfect match data (e.g. DCHIP or RMA). Scatter plots from both types of normalization analyses are shown (see *Fig. 2a* and *b*, also available in the color section).

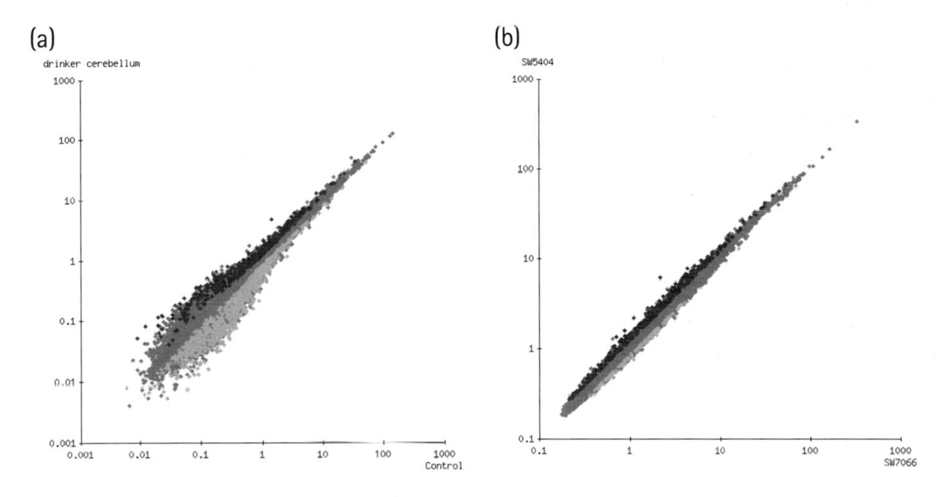

Figure 2. Scatter plots showing NHP RNA-generated microarray datasets normalized using MAS v5.0 software (*a*) or RMA (*b*) (see page xxx for color version).

6. Generate expression data text files for further analysis. Data from an Affymetrix GeneChip experiment are supplied as five individual text files, including a 'Report' file that lists how many transcripts in a given RNA sample were called present, absent, or marginal, based on hybridization to the microarray. Some of these files (e.g. the CHP or CEL files) can then be uploaded to appropriate software packages such as GENESPRING (Agilent Technologies), SPOTFIRE (Spotfire, Inc.), and GENESIFTER (VizX Labs) for differential gene expression analysis, and DAVID/EASE (Scripps Research Institute) and PANTHER (Panther Software Publishing) for pathway integration.

Note

[a]The washing, staining, scanning, and data analysis steps were performed exactly as specified in the Affymetrix Expression Manual. Please consult the Expression Manual for additional protocol details. *Table 2* tabulates some of the data obtained for microarray analysis of blood mRNA from cynomolgus monkeys.

Table 2. Microarray analysis of blood mRNA isolated from adult male cynomolgus monkeys using the Affymetrix U133A Human GeneChip technology

The control animals (44, 66, 67, 68 and 6107) were alcohol naïve, whereas the 'alcohol' animals (34, 35, 36, 38 and 5498) had been chronically self-administering alcohol for 18 months. Pres., present; % Pres., percentage present; Marg., marginal; % Marg., percentage marginal; Abs, absent; and % Abs., percentage absent.

Sample	Scaling factor	Pres.	% Pres.	Marg.	% Marg.	Abs.	% Abs.
Control							
44	49	2098	9.4	264	1.2	19921	89.4
66	23	3690	16.6	341	1.5	18252	81.9
67	23	3089	13.9	332	1.5	18862	84.6
68	22	3072	13.8	318	1.4	18893	84.6
6107	16	3557	16	385	1.7	18341	82.3
Alcohol							
34	59	2679	12	232	1.0	19372	86.9
35	52	2119	9.5	253	1.1	19911	89.4
36	31	1911	8.6	195	0.5	20177	90.5
38	40	2965	13.3	274	1.2	19044	85.5
5498	26	3082	13.8	350	1.6	18851	84.6

Protocol 11

Reverse transcription of RNA for microarray analysis[a]

Equipment and Reagents

- Applied Biosystems 1700 Chemiluminescent Microarray Analyzer
- Applied Biosystems Chemiluminescent RT-IVT labeling kit[b]
- Reverse transcription reagents (blue-capped tubes) including 10× first-strand buffer mix, control RNA, reverse transcriptase enzyme mix, and T7-oligo(dT) primer
- Nuclease-free dH_2O
- 0.2 ml MicroAmp reaction tubes
- Manual pipette with a 1–20 μl range
- Microcentrifuge
- Thermal cycler

Method

1. Thaw the RNA samples and reagents including 1–10 µg of total RNA or 0.05–2 µg of mRNA, T7-oligo(dT) primer, control RNA, nuclease-free dH_2O, and 10× first-strand buffer mix, and place the tubes on ice.
2. Mix the RNA samples and reagents by gentle vortexing, and then spin the tubes by brief centrifugation to consolidate the liquid to the bottom of the tubes.
3. To a 0.2 ml MicroAmp reaction tube on ice, add 2.0 µl of T7-oligo(dT) primer, 4.0 of control RNA, 9.0 µl of RNA (1–10 µg total RNA or 0.05–2 µg poly$(A)^+$ mRNA), and nuclease-free dH_2O as required to a total volume of 15.0 µl.
4. Transfer the 0.2 ml reaction tubes to a thermal cycler and heat the samples to 70°C for 5 min and then cool to 4°C indefinitely (or until ready to use).
5. After the thermal cycler run, place the tubes on ice.
6. Check the 10× first-strand buffer mix for precipitates[c].
7. To each 0.2 ml reaction tube, add 2.0 µl of 10× first-strand buffer mix and 3.0 µl of reverse transcriptase enzyme mix, and mix thoroughly by pipetting.
8. Place the 0.2 ml reaction tubes in a thermal cycler and incubate the 20 µl samples as follows: stage 1, 25°C for 10 min; stage 2, 42°C for 2 h; stage 3, 70°C for 15 min; and stage 4, hold at 4°C indefinitely.
9. After the thermal cycler run, place the reaction tubes on ice.

Notes

[a]*Protocols 11–17* are used in conjunction with the Human Genome Survey Microarray Platform from Applied Biosystems.
[b]For safety and biohazard guidelines, refer to the 'Safety' section in the Applied Biosystems Chemiluminescent RT-IVT labeling kit protocol (PN 4339629). For all chemicals, read the material data safety sheet (MSDS) information and follow the handling instructions. Wear appropriate protective eyewear and gloves at all times when handling hazardous chemicals.
[c]If precipitates are present, warm the buffer to 37°C for 2–3 min and mix briefly by vortexing before use.

Protocol 12

Second-strand synthesis of cDNA for microarray analysis

Equipment and Reagents

- Chemiluminescent RT-IVT Labeling kit (Applied Biosystems)
- Second-strand synthesis reagents (yellow-capped tubes)[a]
- Second-strand enzyme mix
- 5× Second-strand buffer mix
- Nuclease-free dH_2O
- Microcentrifuge
- Thermal cycler

Method

1. Obtain the 20 μl first-strand cDNA samples from the previous step (*Protocol 11*, step 9).
2. Thaw the 5× second-strand buffer mix on ice, mix by vortexing, centrifuge the tube briefly, and check for precipitates[a].
3. To the 20 μl cDNA samples on ice, add 95 μl of nuclease-free dH_2O, 30 μl of 5× second-strand buffer mix, and 5 μl of second-strand enzyme mix.
4. Place the reaction tubes in a thermal cycler and perform second-strand cDNA synthesis using the following regime: stage 1, 2 h at 16°C; stage 2, 15 min at 70°C; and stage 3, indefinitely at 4°C.
5. After the second-strand synthesis run, place the tubes on ice.

Note

[a]If precipitates are present, warm the buffer to 37°C for 2–3 min and mix briefly by vortexing before use.

Protocol 13

Purification of double-stranded cDNA for microarray analysis

Equipment and Reagents

- Chemiluminescent RT-IVT Labeling kit (Applied Biosystems) including the kit components listed below:
  - DNA binding buffer
  - DNA wash buffer
  - DNA elution buffer
  - 1.5 ml Nuclease-free microfuge tubes
  - DNA purification columns
- 2.0 ml Receptacle tubes
- 1.5 ml Elution tubes
- Microcentrifuge

Method

1. To a 1.5 ml nuclease-free microfuge tube, add the 150 μl double-stranded cDNA synthesis product from the previous step (*Protocol 12*, step 5) and 150 μl of DNA binding buffer.
2. Insert a DNA purification column into a 2.0 ml receptacle tube and transfer the entire 300 μl contents from step 1 onto the column. Close the tube and spin the column and tube at 13 000 ***g*** for 1 min in a microfuge, making certain that the entire volume of liquid passes through the column. If sample remains in the column, spin the column and tube again at 13 000 ***g*** for 1 min. Remove the column from the tube, discard the eluent, then re-insert the column containing the cDNA into the tube.

3. Wash the cDNA bound to the column by adding 700 μl of DNA wash buffer to the column, closing the tube, and spinning at 13 000 *g* for 1 min in a microfuge. Remove the column from the tube, discard the eluent, and re-insert the column into the tube.
4. Repeat step 3 for a second wash step.
5. To elute the purified cDNA, transfer the column to a new 1.5 ml elution tube, pipette 30 μl of DNA elution buffer onto the fiber matrix at the bottom of the column[a], and close the tube. Incubate the column at room temperature for 1 min to render the cDNA soluble, then spin the column and tube at 13 000 **g** for 1 min in a microfuge.
6. Add another 30 μl of DNA elution buffer to the column, incubate for 1 min, spin for 13 000 *g* for 1 min, and pool the eluted fractions to obtain 60 μl of purified cDNA.
7. Repeat step 6 to obtain 90 μl of purified cDNA.
8. Discard the column and close the tube. Store at –20°C for short-term storage (<2 months) or at –80°C for long-term storage (>2 months).

Note

[a]When pipetting onto the spin column fiber membrane, be sure to avoid contacting the membrane directly with the pipette tip as this may damage the membrane and contaminate the sample.

Protocol 14

IVT labeling for microarray analysis

Equipment and Reagents

- Chemiluminescent RT-IVT Labeling kit (Applied Biosystems) including the kit components listed below:
  - IVT labeling reagents (green-capped tubes)
  - 5× IVT buffer mix
  - IVT enzyme mix
  - DIG–UTP
- Vacuum concentrator
- 0.2 ml MicroAmp reaction tubes
- Thermal cycler

Method

1. Obtain the 90 μl purified cDNA sample from the previous step (*Protocol 13*, step 8). Estimate the volume of cDNA required for IVT labeling based on the amount of RNA used in step 3 of the reverse transcription reaction (*Protocol 11*, step 3). If ≤2.0 μg of total RNA or ≤0.05 μg mRNA was used for reverse transcription, it may be necessary to concentrate the 90 μl purified cDNA sample down to 24 μl using a vacuum-based centrifuge. If >2.0 μg of total RNA or >0.05 μg of mRNA was used for reverse transcription, it may be necessary to use only a portion of the 90 μl purified cDNA sample, increasing the volume to 24 μl using dH_2O.

2. Check the 5× IVT buffer mix for precipitates and, if precipitates are present, warm the buffer to 37°C for 2–3 min and vortex briefly to mix before using.
3. Set up the IVT reaction by mixing in a 0.2 ml MicroAmp reaction tube the following components: 24.0 µl of purified cDNA, 8.0 µl of 5× IVT buffer mix, 4.0 µl of DIG-UTP (14 nmoles), and 4.0 µl of IVT enzyme mix for a total reaction volume of 40 µl.
4. Perform the IVT reaction in the thermal cycler by incubating for 9 h at 37°C and then at 4°C for an indefinite time until ready to proceed to the next step.
5. After the IVT reaction, remove the tube from the thermal cycler and proceed to the next protocol or store at –80°C.

Protocol 15

Purifying cRNA transcription (IVT) labeling for microarray analysis

Equipment and Reagents

- 1.5 ml Nuclease-free microfuge tube
- Nuclease-free dH_2O
- Chemiluminescent RT-IVT Labeling kit (Applied Biosystems) including the kit components listed below:
  - RNA binding buffer
  - 100% Ethanol
  - RNA wash buffer
  - RNA elution buffer
  - RNA purification columns
- 2.0 ml Receptacle tubes
- 1.5 ml Elution tubes
- Microcentrifuge

Method

1. To a 1.5 ml nuclease-free microfuge tube, add the entire 40 µl IVT reaction from the previous step (*Protocol 14*, step 5) and 20 µl of nuclease-free dH_2O, and mix briefly by vortexing.
2. Add 200 µl of RNA binding buffer and 140 µl of 100% ethanol, and mix well by vortexing to provide a 400 µl sample.
3. Insert an RNA purification column into a 2.0 ml receptacle tube and transfer the 400 µl sample from step 2 onto the column. Close the tube and centrifuge the column and tube at 13 000 ***g*** for 1 min in a microfuge. If the entire 400 µl of sample did not pass through the column, centrifuge the column and tube at 13 000 ***g*** for an additional 1 min. Remove the column from the tube, decant and discard the eluent, and re-insert the column into the tube.
4. Add 500 µl of RNA wash buffer to the column and close the tube. Centrifuge the column and tube at 13 000 ***g*** in a microfuge for 1 min. Remove the column containing the cRNA from the tube, discard the eluent, and re-insert the column into the tube.

5. Repeat step 4 to achieve a second wash step.
6. Elute the washed cRNA as follows. Transfer the column containing the cRNA to a new 1.5 ml elution tube. Pipette 50 µl of RNA elution buffer onto the fiber matrix at the bottom of the column and close the tube. Incubate the column at room temperature for 2 min to render the cRNA soluble. Centrifuge the column and tube at 13 000 ***g*** for 1 min in a microfuge to obtain a 50 µl sample of eluted cDNA.
7. Pipette a second 50 µl of elution buffer onto the fiber matrix at the bottom of the column and close the tube. Centrifuge the column and tube at 13 000 ***g*** for 1 min in a microfuge to obtain a 100 µl sample of eluted cDNA.
8. Discard the column and close the tube. Keep the cRNA product on ice during quantity and quality assessment. The cRNA should be stored at –20°C for short-term storage and at –80°C for long-term storage.

Protocol 16

Assessing cRNA quantity and quality for microarray analysis

Equipment and Reagents

- Microfuge tubes
- UV spectrophotometer
- Quartz cuvette
- 1–2% Agarose gel
- TE buffer
- Ethidium bromide

Method

1. Obtain the cRNA sample from the previous step (*Protocol 15*, step 8). Transfer 3 µl of cRNA to a microfuge tube containing 97 µl of TE buffer for a 30-fold dilution of the RNA and measure the UV absorbance at 260 and 320 nm using a UV spectrophotometer. Calculate the cRNA concentration and yield as follows: cRNA concentration (µg/µl) = $(A_{260} - A_{320}) \times 1.2$, and cRNA yield (µg) = cRNA concentration (µg/µl) × 100 µl.
2. Prepare a 1–2% agarose gel, load 0.5–1.0 µl of cRNA per well, run the gel to fractionate the cRNA, and then stain with ethidium bromide to visualize the cRNA product.

Protocol 17

Final steps of microarray use and data analysis

Reagents

- Human Genome Survey Microarray v2.0 (Applied Biosystems)[a]

Method

1. Pre-hybridize the microarrays to prevent nonspecific binding.
2. Fragment the cRNA to facilitate more efficient hybridization.
3. Hybridize the fragmented cRNA samples to the microarrays.
4. Wash the microarrays to remove unbound probe.
5. Perform antibody binding and wash steps to detect specific cRNA probe sequences bound to the microarray.
6. Perform the chemiluminescence reaction and detection steps to detect antibody binding and hybridization signals.
7. For transcriptome analysis, set the thresholds as specified by Applied Biosystems, in which the detection threshold is set to a signal-to-noise ratio of greater than 3.0 (S/N > 3) and a 'quality flag' of <5000 is used. An experimental data summary using these settings is shown in *Fig. 3*.
8. Apply global median normalization to the transcriptome data set to normalize signal intensities across all of the microarrays. This data transformation is used to achieve the same median signal intensities for each microarray.
9. Output the gene expression data as tab-delimited text files. These files can then be uploaded to appropriate software packages such as GENESPRING (Agilent Technologies), SPOTFIRE (Spotfire, Inc.), and GENESIFTER (VizX Labs) for differential gene expression analysis, and DAVID/EASE (Scripps Research Institute) and PANTHER (Panther Software Publishing) for pathway integration. A sample output in the form of a molecular function diagram using PANTHER software is shown (see *Fig. 4*, also available in the color section).

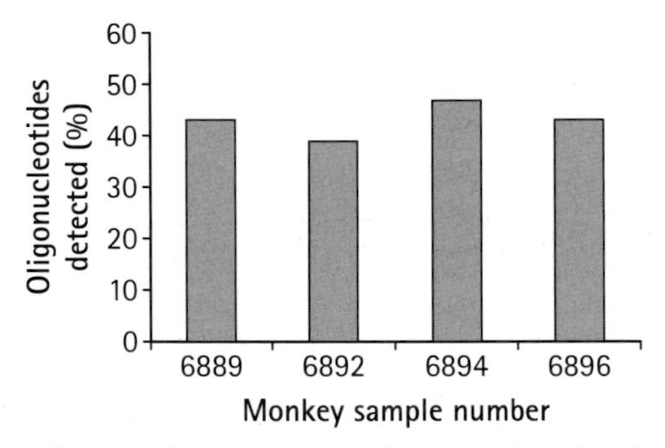

Figure 3. Data summary showing the percentage of oligonucleotides detected on human genome survey microarrays using RNA extracted from whole blood from four individual cynomolgus macaque monkeys who had been self-administering alcohol for 18 months. The average percentage of oligonucleotides detected was approximately 45% or 12 680 out of 27 868 total genes. The percentage of oligonucleotides detected was calculated using S/N >3 and flags <5000.

Note

[a]A complete set of detailed protocols for the Human Genome Survey Microarray v2.0 are available on line from Applied Biosystems.

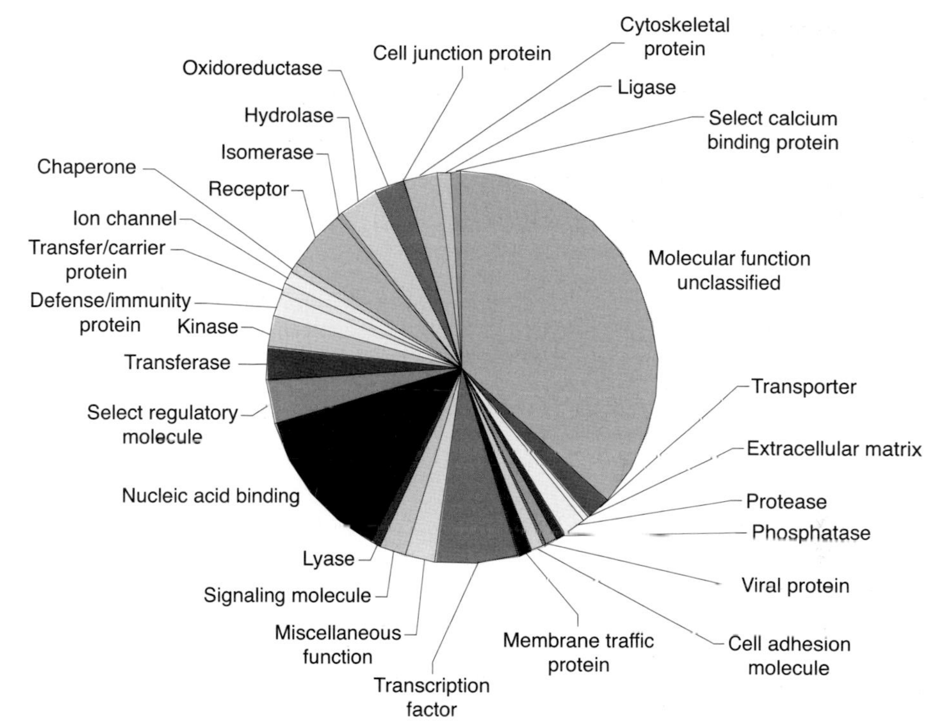

Figure 4. Molecular function diagram from analysis using PANTHER software and microarray results from four cynomolgus monkey whole blood RNA samples (see page xxxi for color version).

3. TROUBLESHOOTING

The most important aspect of any experiment designed to measure gene expression, assuming that the tissue has been collected appropriately, is the quality of the RNA. Exacting attention and proficiency in isolating high-quality RNA will provide increased confidence in downstream measurements from microarray and PCR. Any protocol that involves as many individual steps as a microarray experiment has great potential for the introduction of experimental errors. It is therefore important to follow the protocols carefully and to validate the endpoints (e.g. RNA quantity and quality) whenever possible.

It is absolutely necessary to give considerable thought to experimental design considerations before engaging in this (very expensive) endeavor. There are numerous resources available to assist with this. A search of the term 'MIAME' (Minimum Information About a Microarray Experiment) provides a good place to start in terms of the importance of quality standards for various aspects of the microarray process.

NHP microarrays are currently not available for most monkey species. However, human microarrays work quite well in our experience. As with any microarray experiment, it is useful and necessary to validate at least some of the expression results using a secondary method (e.g. real-time PCR) to ensure that the human microarrays are yielding valid data for the monkey samples.

There are numerous programs and algorithms available for microarray data analysis. To maximize the information obtained from microarray experimentation, it is necessary to spend a great deal of time on the computational aspects of microarray analysis.

4. REFERENCES

★★ 1. **Vivian JA, Green HL, Young JE, *et al.*** (2001) *Alcohol Clin. Exp. Res.* **25**, 1087–1097. *– This paper describes the NHP model system we are interrogating (blood, brain, and other tissues) with microarray experiments.*

2. **O'Brien CP** (1996) In *The Pharmacological Basis of Therapeutics*, 9th edn, pp 557–577. Edited by JG Hardman, LE Limbird, RB Molinoff, RW Ruddon & AG Gilman. McGraw-Hill, New York.

3. **Grant KA & Bennett AJ** (2003) *Pharmacol. Ther.* **100**, 235–255.

4. **Lewohl JM, Wang L, Miles MF, Zhang L, Dodd PR & Harris RA** (2000) *Alcohol Clin, Exp. Res.* **12**, 1873–1882.

5. **Mayfield RD, Lewohl JM, Dodd PR, Herlihy A, Liu J & Harris RA** (2002) *J. Neurochem.* **81**, 802–813.

6. **Nestler EJ** (2001) *J. Neurosci.* **21**, 8324–8327.

★ 7. **Hacia JG, Makalowski W, Edgemon K, *et al.*** (1998) *Nat. Genet.* **18**, 155–158. *– This paper provides some background to explain why cross-species microarray experiments should work.*

8. **Dukelow WR** (1997) *J. Med. Primatol.* **6**, 33–42.

9. **Green KL, Szeliga KT, Bowen CA, Kautz MA, Azarov AV & Grant KA** (1999) *Alcohol Clin. Exp. Res.* **23**, 611–616.

10. **Whitney AR, Diehn M, Popper SJ, *et al.*** (2003) *Proc. Natl. Acad. Sci. U.S.A.* **100**, 1896–1901.

11. **Schena M, Shalon D, Davis RW & Brown PO** (1995) *Science,* **270**, 467–470.

12. **Bigger CB, Brasky KM & Lanford RE** (2001) *J. Virol.* **75**, 7059–7066.

13. **Iizuka N, Oka M, Yamada-Okabe H, *et al.*** (2003) *Oncogene,* **22**, 3007–3014.

★★ 14. **Chismar JD, Mondala T, Fox HS, *et al.*** (2002) *BioTechniques,* **33**, 516–522. *– This paper explains some of the important issues one needs to be aware of when doing cross-species microarray experiments.*

15. **Huff JL, Hansen LM & Solnick JV** (2004) *Infect. Immun.* **72**, 5216–5226.

★ 16. **Marvanová M, Ménager J, Bezard E, Bontrop RE, Pradier L & Wong G** (2003) *FASEB J.* **17**, 929–931. *– This paper explains the significance of using NHPs to examine important clinical conditions that affect humans.*

17. **Chomczynski P & Sacchi N** (1987) *Anal. Biochem.* **162**, 156–159.

CHAPTER 12

Enhanced microarray hybridization using surface acoustic wave mixing

Natalie Stickle, Kelly Jackson, Andreas Toegl, Frank Feist, and Neil Winegarden

1. INTRODUCTION

Microarray technology relies on efficient and specific hybridization to thousands of different target elements that comprise the microarray. Microarray hybridization presents several challenges. First, it is important to find conditions that promote specific hybridization across potentially highly variant targets. Differences in target length and GC content (particularly in the case of cDNA microarrays) can lead to the use of 'compromise' conditions that maximize the number of specific hybridization events and minimize the number of cross-hybridization events without ever achieving completely optimal conditions. Oligonucleotide microarrays allow rational target design, which can partially overcome this challenge by allowing a choice of targets with a very narrow range of lengths and GC content.

Another challenge common to virtually all microarray platforms is that very little mixing occurs during the hybridization reaction, particularly when a sample is placed under a cover slip and hybridized to a glass substrate. The use of a cover slip has the advantage of maintaining a low hybridization volume and therefore a high probe concentration, but any mixing of the solution is limited to the rate of diffusion (1). It has been suggested that, using a passive hybridization process, any given element or spot on an 18 mm × 54 mm microarray is exposed to less than 0.3% of the available probe molecules in solution (1). Furthermore, a typical DNA probe molecule diffuses only 1–3 mm during a static 24 h hybridization (2) and, as such, equilibrium is not reached for weeks after the hybridization has been initiated (3, 4). This problem is generally best dealt with via modifications to the hybridization apparatus. Cover slips that have raised edges (e.g. LifterSlip; Erie Scientific) provide a thin 20-100 μm lumen and

DNA Microarrays: *Methods Express* (M. Schena, ed.)

therefore allow increased volumes and potentially greater mixing; however, this approach alone does not provide full mixing.

The issue of limited mixing of the sample is often not apparent when measuring highly expressed messages, as the local concentration of probe molecules may be sufficient to obtain strong signals. The situation is much more critical with respect to low-abundance messages, where the lack of diffusion and mixing may cause a local depletion of probe molecules near the cognate microarray target elements (1). This may, in part, explain why signals from low-intensity spots tend to be more variable than high-intensity spots. Oligonucleotide microarrays present an additional level of concern with respect to different binding efficiencies and kinetics for sequences that are perfectly complementary to probe molecules versus sequences that have one or more mismatches. Dai and colleagues (5) demonstrated that, under standard hybridization conditions with 60mer oligonucleotide microarrays and a complex sample, it takes longer for perfectly matched targets to reach equilibrium than targets bearing a sequence mismatch.

Another common hardware design used to mitigate the effects of limited sample mixing involves injecting the hybridization solution into a closed cassette containing the microarray and mixing the solution inside the cassette lumen using a rotary mixing device. To facilitate mixing of the hybridization solution on the rotary device, an air bubble is introduced into the lumen of the cassette. Whilst this simple method is relatively effective and proven, it does have some limitations: (i) generally large sample volumes (~200 μl) are required (6); (ii) if the air bubble becomes trapped during the incubation, portions of the microarray can dry out leading to very high background; and (iii) the air bubble causes oxidation of dye molecules at the sample–bubble interface leading to slightly elevated background and weaker signals, particularly when using extended hybridization times.

Mechanical mixing systems have also been developed and adopted in many laboratories. Mechanical mixing systems generally involve the pumping of sample and wash buffers in and out of the hybridization chamber. Again, one of the limitations of this type of system is that hybridization volumes need to be increased to ~200 μl, which reduces hybridization signals relative to the use of smaller volumes. Although mixing is expedited with mechanical mixing systems, the mixing can be uneven across the slide surface, with a greater effect at the periphery of the slide (proximal to the inlet/outlet) and a reduced effect towards the middle of the slide.

One novel method of enhancing microarray sample mixing involves the use of acoustic waves to pulse the hybridization solution across the microarray surface. This technology has been incorporated into the SlideBooster and ArrayBooster instruments provided by Advalytix. Here, we present methods that take advantage of the mixing provided by the SlideBooster instrument, with slightly different methods recommended for cDNA and oligonucleotide microarrays.

2. METHODS AND APPROACHES

2.1 Labeling of cDNA

Although the labeling method in and of itself may have little impact on the overall hybridization results, we have presented our current preferred protocol (based on aminoallyl incorporation), as this protocol has proved to be effective when combined with the Advalytix SlideBooster

Certain labeling methods can affect the ability of the labeled cDNA probe molecules to hybridize to the target molecules on the microarray. Certainly the most common methods of direct or indirect (aminoallyl) labeling have proved effective; however, alternative methods have begun to be presented that can complicate the experimental process. The use of large labeling molecules such as phycoerythrin or QuantumDots can create steric problems. When these detection molecules are used, it is generally necessary to add these molecules post-hybridization, such as is the case when using the Affymetrix platform. If methods using large labeling molecules were to be employed, the protocols presented here would need to be modified accordingly.

2.2 Surface acoustic wave mixing

Surface acoustic waves (SAWs) can be created by incorporating small interdigital transducers into piezoelectric substrates (7). These SAW elements can be used to transfer momentum into a liquid placed in juxtaposition to the piezoelectric transducers. For example, the SAW elements can be used to transfer momentum into a microarray substrate, allowing efficient mixing of a probe sample placed on the substrate. In the case of microarray hybridization, the SAWs function to mix the hybridization solution under a specialized cover slip (e.g. LifterSlip; Erie Scientific) by causing streaming of the hybridization solution (7). In the SlideBooster instrument from Advalytix, three mixing chips are located under each microarray slide, held in fluid contact to the piezoelectric transducers via a coupling liquid termed AdvaSon. This mechanical configuration has no macroscopic moving parts and does not require liquid tubing connections, which reduces clogging and maintenance. Evaporation of the hybridization solution is prevented by sealing the reaction chambers with round elastomer gaskets (i.e. o-rings) and through the use of humidifying solutions (e.g. dilute saline sodium citrate buffer) to create a fully saturated atmosphere (100% humidity) inside the reaction chamber. It has been reported previously that this system can enhance the kinetics of hybridization as well as the overall signal-to-noise ratio (8). Here, we present an integrated set of protocols for using the SlideBooster instrument to enhance hybridization of both cDNA and oligonucleotide microarrays. We also discuss briefly the applicability of this technology to other microarray types including those provided by Agilent and ArrayIt.

2.3 Recommended protocols

Protocol 1[a]

Reverse transcription and aminoallyl incorporation

Equipment and Reagents

- Reverse transcriptase (SuperScript II; Invitrogen)
- 5× First-strand buffer (SuperScript II; Invitrogen)
- AncT primer (5′-T_{20}VN) (100 pmol/µl)
- dNTP mix (6.67 mM each of dATP, dGTP, and dCTP; Invitrogen)
- 2 mM dTTP (Invitrogen)
- 20 mM Aminoallyl-dUTP (Sigma)
- 0.1 M dithiothreitol (DTT)
- 2–10 ng/µl Control RNA (artificial *Arabidopsis* transcripts)
- Nuclease-free dH_2O (Sigma)
- 1 M NaOH
- 1 M HCl
- 1 M Tris/HCl (pH 7.5)
- Block heater
- Water bath
- Thermal cycler

Method

1. In a 0.5 ml microfuge tube, combine the following reagents: 8 µl of 5× first-strand buffer, 1.5 µl of AncT primer, 3 µl of 6.67 mM dNTP mix, 3 µl of 2 mM dTTP, 3 µl of 2 mM aminoallyl–dUTP, 4 µl of 0.1 M DTT, 1 µl of control RNA (optional), 10 µl of mRNA or total RNA (0.1–0.5 µg of mRNA or 5–10 µg of total RNA), and 6.5 µl of nuclease-free dH_2O.
2. Incubate the reaction mixture at 65°C for 5 min to denature the RNA and then at 42°C for 2 min to cool the mixture before adding the enzyme.
3. Add 2 µl of reverse transcriptase and incubate at 42°C for 2 h to allow synthesis of aminoallyl-containing cDNA molecules from the RNA templates.
4. Add 8 µl of 1 M NaOH and heat to 65°C for 15 min to hydrolyze the RNA. This step leaves the aminoallyl-containing cDNA intact.

5. Add 4 μl of 1 M Tris/HCl (pH 7.5) and 8 μl of 1 M HCl to neutralize the solution.
6. Purify the 62 μl of newly synthesized cDNA mixture according to *Protocol 2*.

Note

[a]We have successfully used the Advalytix SlideBooster and SAW mixing technology to hybridize cDNA and oligonucleotide microarrays. Upstream preparation of labeled cDNA can be performed using a number of methods including many different commercial kits. We present a series of protocols for aminoallyl labeling, purification, and fluorophore conjugation that we have shown to be highly successful for both cDNA and oligonucleotide microarrays. Both types of microarray are printed using microarrayers equipped with ArrayIt brand microarray printing pins (TeleChem International, Inc.).

Protocol 2

Purification of aminoallyl-containing cDNA

Equipment and Reagents

- CyScribe GFX Purification kit (GE Amersham) including:
  - Capture buffer
  - GFX purification columns
- 80% Ethanol
- 0.017 M Sodium bicarbonate (pH 9)
- Nuclease-free dH_2O
- Microcentrifuge
- Vacuum evaporator (SpeedVac; ThermoElectron)

Method[a]

1. Add 500 μl of capture buffer to each GFX purification column.
2. Add 62 μl of unpurified cDNA sample from the previous step (*Protocol 1*, step 6) and pipette up and down in the column to mix thoroughly[a].
3. Spin the column in a microcentrifuge at 13 800 ***g*** for 30 s and discard the flow through At this point in the procedure, the cDNA molecules remain bound to the GFX column.
4. Add 600 μl of 80% ethanol to the column, place the column in a microcentrifuge, spin at 13 800 ***g*** for 30 s, and discard the flow-through. This washing step removes contaminants and leaves the cDNA molecules bound to the column.
5. Repeat step 4 twice for a total of three column washes.
6. Spin the column in a microcentrifuge for an additional 30 s at full speed to remove the residual ethanol.
7. Transfer the GFX column to a fresh microfuge tube and add 60 μl of 0.017 M sodium bicarbonate (pH 9) elution buffer directly onto the column membrane[b].
8. Incubate the GFX column at room temperature for 1 min to allow the cDNA to become soluble prior to elution.

9. Spin the column at 13 800 *g* for 1 min to elute the labeled and purified cDNA. The sample volume at this step is approximately 60 µl[c].
10. Dry the sample to completion using a vacuum centrifuge (e.g. SpeedVac) with the heat control set on high. Be careful not to dry the sample excessively. Resuspend the purified cDNA sample in 7 µl of nuclease-free dH_2O.

Notes

[a]For this purification, each labeling reaction should be purified using one GFX column.
[b]It is crucial that the elution buffer covers the column membrane to allow efficient elution. The use of 0.017 M sodium bicarbonate represents a 1:6 dilution of a 0.1 M stock solution. Resuspension of the dry sample in 7 µl of dH_2O, followed by the addition of 3 µl of dye/dimethyl sulfoxide (DMSO) solution (see *Protocols 3* and *4*) provides a 0.1 M final concentration of sodium bicarbonate in the 10 µl dye conjugation reaction.
[c]Following purification, the aminoallyl-labeled cDNA sample is stable for several days at –20°C, allowing the researcher some flexibility and convenience in the protocol.

Protocol 3

Preparation of monofunctional reactive cyanine and Alexa dyes

Reagents

- DMSO
- Cy5 and Cy3 monofunctional reactive dyes (GE Healthcare)
- Alexa 647 and Alexa 555 monofunctional reactive dyes (Invitrogen)

Method

1. Obtain monofunctional dyes from the appropriate vendor. Each pack of Cy5 and Cy3 contains five vials of dye, whereas the Alexa dyes are packaged individually.
2. To resuspend the cyanine dyes, add 45 µl of DMSO to one vial of dye. To resuspend the Alexa dyes, add 3 µl of DMSO per tube. Both resuspended dyes can be stored at –20°C for up to 3 months.
3. Use 3 µl of either cyanine or Alexa dye per labeling reaction (see *Protocol* 4).

Protocol 4

Dye coupling to aminoallyl-containing cDNA

Equipment and Reagents

- Cy5 and Cy3 monofunctional reactive dyes (GE Healthcare)
- Alexa 647 and Alexa 555 monofunctional reactive dyes (Invitrogen)
- 4 M Hydroxylamine

Method

1. Obtain the 7 µl of purified aminoallyl-containing cDNA samples from *Protocol 2* (step 10).
2. To each 7 µl cDNA sample, add 3 µl of cyanine or Alexa dye (from *Protocol 3*), mix by pipetting up and down, and incubate in the dark for 60 min at room temperature to allow chemical coupling between dye molecules and aminoallyl groups on the cDNA.
3. After the 60 min coupling reaction, quench the reaction by adding 4.5 µl of 4 M hydroxylamine and incubating for 15 min at room temperature in the dark.

Protocol 5

Purification of fluorescently labeled cDNA

Equipment and Reagents

- illustra CyScribe GFX Purification kit (GE Amersham)
- 80% Ethanol
- Nuclease-free dH_2O
- Microcentrifuge
- Vacuum evaporator (SpeedVac; ThermoElectron)

Method

1. Obtain the 14.5 µl samples of labeled cDNA from the previous step (*Protocol 4*, step 3).
2. Add 35.5 µl of dH_2O water to each cDNA sample for a total sample volume of 50 µl.
3. For two-color experiments, combine the pairs of Cy3- and Cy5-labeled samples or Alexa 647- and Alexa 555-labeled samples in preparation for co-hybridization. Two-color samples have a total volume of 100 µl.
4. Add 500 µl of capture buffer from the GFX Purification kit to each GFX column.
5. Transfer the labeled cDNA product (50 or 100 µl) onto the column, pipetting up and down several times to mix the cDNA sample with the capture buffer.
6. Spin the column in a microcentrifuge at 13 800 ***g*** for 30 s and discard the flow-through. At this point, the cDNA molecules remain bound to the column matrix.
7. To each column, add 600 µl of 80% ethanol, spin at 13 800 ***g*** for 30 s, and discard the flow-through.
8. Repeat step 7 twice more for a total of three washes.
9. Spin the column for an additional 30 s to remove any residual ethanol. This step is important because residual ethanol can complicate downstream steps.
10. Transfer the GFX column to a fresh tube and add 60 µl of the elution buffer provided with the kit.
11. Incubate the GFX column at room temperature for 1 min to render the cDNA soluble.
12. Spin the column at 13800 ***g*** for 1 min to elute the purified labeled cDNA.

13. Evaporate the samples to dryness using vacuum centrifugation, taking care not to dry the samples excessively[a].

Note

[a]Purified samples are stable for several days at −20°C, providing a convenient stopping place for users wishing to pause and resume subsequent steps at a later time.

Protocol 6

Hybridization to oligonucleotide microarrays[a]

Equipment and Reagents

- Nuclease-free dH_2O (Sigma)
- Yeast tRNA (10 mg/ml; Invitrogen)
- Calf thymus DNA (10 mg/ml; Sigma)
- DIG Easy Hyb solution (Roche)
- BSA Fraction V solution (12.5 mg/ml; Sigma)
- 20× Sodium chloride/sodium citrate buffer (SSC)
- 10% Sodium dodecyl sulfate (SDS)
- 24 × 60 mm M-Series LifterSlips (Erie Scientific)
- Printed oligonucleotide microarrays (60–70mers)
- Powder-free nitrile gloves
- AdvaSon (Advalytix)
- AdvaHum 101 for nonformamide-based hybridization buffers (Advalytix)
- SlideBooster (Advalytix)

Method

1. Obtain the dry cDNA samples from the previous protocol (*Protocol 5*, step 13).
2. Resuspend each cDNA sample in 5 µl of nuclease-free dH_2O.
3. Prepare the hybridization solution by mixing 100 µl of DIG Easy Hyb solution, 5 µl of 10 mg/ml yeast tRNA, 5 µl of 10 mg/ml calf thymus DNA, and 4 µl of 12.5 mg/ml BSA. Incubate the hybridization solution at 65°C for 3 min and cool to room temperature.
4. Add 90 µl of hybridization solution (step 3) to each 5 µl cDNA sample (step 2).
5. Incubate the probe mixture at 65°C for 3 min and cool to room temperature.
6. Pipette 15 µl of AdvaSon coupling fluid onto each agitation position of the SlideBooster instrument.
7. Position the microarray slide such that the agitation chips are directly beneath the hybridization area.
8. While wearing powder-free nitrile gloves, push down on the edges of the slide and move gently back and forth to ensure complete contact between the slide and the agitation chips.
9. Load 500 µl of AdvaHum 101 into the humidifying reservoirs located at each end of the chamber (1000 µl total).

10. Close each lid on the SlideBooster instrument, making certain that the lid latch has snapped closed firmly.
11. Load the program for each chamber and press 'Start' to pre-heat the hybridization chambers on the instrument. Follow the on-screen instructions and use the following parameters: Mix Power = 27, Pulse/Pause Ratio = 3:7, Temperature = 37°C, Processing Time = 16 h.
12. Place the LifterSlip on to the array.
13. Slowly pipette the sample at the edge of the LifterSlip allowing the solution to be drawn underneath the LifterSlip by capillary action.

Note

[a]Once the labeled cDNA has been prepared, the hybridization protocol required depends on the type of microarray being used. For oligonucleotide microarrays, we have had excellent success with the DIG Easy Hyb solution from Roche. Hybridization is carried out overnight with continual mixing at 37°C. Microarrays of cDNAs require a slightly different protocol. We have found that AdvaHyb is a more appropriate hybridization buffer when using the SlideBooster instrument. To enhance specificity, the cDNA microarray hybridization protocol uses a 10 min incubation at 75°C with mixing and then a gradual temperature reduction to 42°C for the overnight hybridization.

Protocol 7

Hybridization to cDNA microarrays

Equipment and Reagents

- Nuclease-free dH_2O (Sigma)
- Yeast tRNA (10 mg/ml; Invitrogen)
- Calf thymus DNA (10 mg/ml; Sigma)
- DIG Easy Hyb solution (Roche)
- BSA Fraction V solution (12.5 mg/ml; Sigma)
- 20× SSC
- 10% SDS
- 24 × 60 mm M-Series LifterSlips (Erie Scientific)
- Printed cDNA microarrays
- Powder-free nitrile gloves
- Heating block set at 42°C
- AdvaHyb (Advalytix)
- AdvaSon (Advalytix)
- AdvaHum 102 for formamide-based hybridization buffers (Advalytix)
- SlideBooster (Advalytix)

Method

1. Pre-heat the AdvaHyb solution at 42°C until the precipitate has dissolved.

2. Obtain the purified dry cDNA samples from a previous protocol (*Protocol 5*, step 13). Resuspend the samples in 4.5 μl of 10 mg/ml yeast tRNA and 4.5 μl of 10 mg/ml calf thymus DNA.
3. To each resuspended cDNA sample, add 81 μl of pre-heated AdvaHyb solution from step 1 above. Mix by vortexing.
4. Denature the sample at 95°C for 3 min.
5. Cool the sample to 42°C (5 min).
6. Load 15 μl of AdvaSon coupling fluid onto each SlideBooster agitation chip, one location per microarray hybridized.
7. Position the microarray slide such that the agitation chip is located directly beneath the hybridization area.
8. While wearing powder-free nitrile gloves, push down on the edges of the slide while moving the slide back and forth to ensure complete contact between the slide and the agitation chips.
9. Load 500 μl of AdvaHum 102 into the humidifying reservoirs located at either end of each chamber, for a total volume of 1000 μl per chamber.
10. Close the lid and make sure the catch has snapped closed firmly. This step is very important. Failure to snap the catch closed firmly will lead to a failed experiment.
11. Load the program written for each chamber and push 'Start' to pre-heat the chambers. The on-screen instructions should read as follows: Mix Power = 27; Pulse/Pause Ratio = 3:7; Temperature = Initial Step – 1 min at 42°C with mixing; Step 1: 10 min at 75°C with mixing; passively cools to 42°C; and Final Step: 900 min at 42°C with mixing.
12. Place the LifterSlip on to the array.
13. Slowly pipette the sample at the edge of the LifterSlip allowing the solution to be drawn underneath the LifterSlip by capillary action.

Protocol 8

Washing cDNA and oligonucleotide microarrays

Equipment and Reagents

- Powder-free nitrile gloves
- Staining dishes (Evergreen Scientific) containing (i) 1× SSC, (ii) 1× SSC, (iii) 1× SSC + 0.1% SDS at 50°C, (iv) 1× SSC + 0.1% SDS at 50°C, (v) 1× SSC + 0.1% SDS at 50°C, (vi) 1× SSC, (vii) 1× SSC, and (viii) 0.1× SSC[a]
- Staining racks (Evergreen Scientific)
- Water bath set at 50°C
- Slide box lined with Whatman filter paper
- Swinging bucket centrifuge with microplate holders

Method

1. While wearing powder-free nitrile gloves, open the SlideBooster chambers and remove the microarray slides.
2. Remove the cover slip by quickly but gently dipping the microarray slide into staining dish #1 containing 1× SSC. Hold the slide at the bar-code end with forceps and allow the cover slip to fall off gently.
3. Place the slides into a staining rack and place the rack into staining dish #2 containing 1× SSC.
4. When all of the microarray slides have been removed from the hybridization chambers, transfer the slide rack to staining dish #3 containing 1× SSC + 0.1% SDS at 50°C and wash for 15 min with gentle occasional agitation.
5. Repeat this wash twice more using staining dishes #4 and #5.
6. Rinse the slides twice by dunking four to six times each in staining dishes #6 and #7 containing 1× SSC.
7. Rinse one final time in staining dish #8 containing 0.1× SSC.
8. Spin the slides dry at 70 *g* for 5 min in a slide box lined with Whatman paper.
9. Store the microarrays in the dark, scanning as soon as possible (within 2 days) after the washing steps[b].

Notes

[a]High-throughput wash stations (ArrayIt or Advalytix) can also be used for this step. After hybridization and washing are complete, the arrays can be scanned using any conventional microarray scanner.

[b]Application to other microarray types: whilst we present here the protocols for hybridization using acoustic mixing provided by the SlideBooster instrument, this technology is applicable to a host of other microarray types. Nearly all microarray experiments (DNA, protein, or tissue) suffer from the lack of mixing that occurs under a static cover slip. With appropriate modifications of the protocols presented here, acoustic wave mixing can benefit any of these microarray applications.

Protocol 9

Preparation of hybridization solutions for Agilent microarrays

Equipment and Reagents

- *In situ* hybridization kit (Agilent)
- Nuclease-free dH_2O
- Heating block/water bath
- Vortex

Method

1. Combine 1 μg of each sample with nuclease-free dH_2O to a total volume of 38 μl.
2. Add 10 μl of 10× control target to each sample for a final volume of 48 μl[a].
3. Add 2 μl of fragmentation buffer to each 48 μl sample for a final sample volume of 50 μl. Mix well and incubate at 60°C in the dark for 30 min to fragment the RNA.
4. Add 50 μl of 2× hybridization buffer to each 50 μl sample and prepare to hybridize the samples immediately[b].

Note

[a]Add 0.5 ml of dH_2O to the lyophilized control target, vortex, and store at −20°C to give a 2× target solution with a total volume of 48 μl.
[b]Agilent microarrays have become increasingly popular recently and offer a robust and sensitive solution for microarray analysis. Typically, Agilent microarrays are hybridized using a rotisserie-type device. Whilst this is an effective and proven solution, the Advalytix SlideBooster can also be used to hybridize these microarrays. Labeled cRNA should be prepared following Agilent's instructions, as should downstream washing and scanning steps. The hybridization step is carried out differently, however, if the SlideBooster is used. The traditional method of hybridization to an Agilent microarray generally involves a significantly larger volume of probe solution (often four to ten times more), which dilutes the labeled probe molecules and potentially reduces signals. We have found that the use of the SAW technology can also enhance the signal-to-noise ratio of Agilent microarray hybridization reactions.

Protocol 10

Hybridization to Agilent microarrays

Equipment and Reagents

- Whole human genome oligonucleotide microarrays (Agilent)
- SlideBooster (Advalytix)
- AdvaSon (Advalytix)
- AdvaHum 101 (Advalytix)
- 24 × 60 mm LifterSlips (Erie Scientific)

Method

1. Load 20 μl of AdvaSon coupling buffer onto each agitation chip.
2. Fill each humidifying chamber with 500 μl of AdvaHum 101.
3. Place a microarray on top of the agitation chip and heat the chambers to 65°C using the following program: Mix Power = 27; Pulse/Pause Ratio = 3:7; Temperature = 65°C; and Processing time = 17 h.
4. Place a 24 × 60 mm LifterSlip onto the microarray and load 100 μl of hybridization solution from the previous protocol (*Protocol 9*, step 4) under the LifterSlip.
5. Initiate the SlideBooster program and allow the microarrays to hybridize for 17 h[a].

Note

[a]CpG island microarrays have been used for chromatin immunoprecipitation (ChIP) analysis. ChIP assays helps to identify binding sites for transcription factors and other DNA-binding proteins. Whilst the techniques involved in preparing labeled cDNA for hybridization to CpG island microarrays are very different from traditional gene expression assays, CpG island microarrays are, in effect, simply cDNA microarrays. As such, once the labeled probe material has been prepared, CpG island microarrays can be treated in much the same manner as any other cDNA for hybridization on the SlideBooster instrument.

3. TROUBLESHOOTING

We recommend storing all reaction components in small aliquots as repeated freeze–thaw cycles can affect their quality and produce less-than-optimal results. High-quality RNA is essential for efficient cDNA generation and optimal results. We recommend testing RNA integrity on the Agilent BioAnalyzer 2100 (Agilent Technologies) and RNA concentration using a spectrophotometer or NanoDrop instrument.

4. REFERENCES

1. **Adey NB, Lei M, Howard MT, *et al.*** (2002) *Anal. Chem.* **74**, 6413–6417.
2. **McQuain MK, Seale K, Peek J, *et al.*** (2004) *Anal. Biochem.* **325**, 215–226.
3. **Allison SA, Northrup SH & McCammon JA** (1986) *Biophys. J.* **49**, 167–175.
4. **Chan V, Graves DJ & McKenzie SE** (1995) *Biophys. J.* **69**, 2243–2255.
5. **Dai H, Meyer M, Stepaniants S, Ziman M & Stoughton R** (2002) *Nucleic Acids Res.* **30**, e86.
6. **Liu RH, Lenigk R, Druyor-Sanchez RL, Yang J & Grodzinski P** (2003) *Anal. Chem.* **75**, 1911–1917.
7. **Wixforth A, Gauer C, Scriba J, Wassermeier M & Kirchner R** (2003) In *Proceedings of the SPIE*, vol. 4982, *Microfluidics, BioMEMs, and Medical Microsystems*, pp. 235–242.
8. **Toegl A, Kirchner R, Gauer C & Wixforth A** (2003) *J. Biomol. Tech.* **14**, 197–204.

APPENDIX 1
List of suppliers

ABgene – www.abgene.com
Advalytix – www.advalytix.de
Affymetrix – www.affymetrix.com
Alexis Corporation – www.alexis-corp.com
Amersham Biosciences – www.amershambiosciences.com
Anachem Ltd - www.anachem.co.uk
Appleton Woods Ltd – www.appletonwoods.co.uk
Applied Biosystems (ABI) – www.appliedbiosystems.com
ArrayIt – www.arrayit.com
AutoGen, Inc. – www.autogen.com
Axon Instruments – www.axon.com

Beckman Coulter, Inc. - www.beckman.com
Becton, Dickinson and Company - www.bd.com
Bio-Rad Laboratories, Inc. - www.bio-rad.com
BOC Group – www.boc.com
Brosch direct Ltd – www.broschdirect.com

Calbiochem – www.calbiochemicom
Cambridge Scientific Products – www.cambridgescientific.com
Carl Zeiss – www.zeiss.com
Chemicon International, Inc. – www.chemicon.com
Corning, Inc. – www.corning.com

DakoCytomation – www.dakocytomation.com
DAVID/EASF – http://david.abcc.ncifcrf.gov
Difco Laboratories - www.difco.com
Dionex Corporation – www.dionex.com
DuPont – www.dupont.com

Elliot Scientific Ltd – www.elliotscientific.com
European Collection of Animal Cell Culture - www.ecacc.org.uk

Findel Education Ltd – www.fipd.co.uk
Fisher Scientific International - www.fishersci.com
Fluka – www.sigma-aldrich.com
Fluorochem – www.fluorochem.co.uk

GENESIFTER – www.genesifter.net/web/
GENESPRING – www.silicongenetics.com/cgi/SiG.cgi/Products/GeneSpring/download
Gibco BRL – www.invitrogen.com
Goodfellow Cambridge Ltd – www.goodfellow.com
Greiner Bio-One – www.gbo.com

Harlan – www.harlan.com
Hybaid – www.hybaid.com
HyClone Laboratories – www.hyclone.com

ICN Biomedicals, Inc. – www.icnbiomed.com
Insight Biotechnology – www.insightbio.com
Invitrogen Corporation – www.invitrogen.com

Jencons-PLS – www.jencons.co.uk

Kendro Laboratory Products – www.kendro.com
Kodak: Eastman Fine Chemicals – www.eastman.com

Lab Plant Ltd – www.labplant.com
Lancaster - www.lancastersynthesis.com
Leica – www.leica.com
Life Technologies Inc. – www.lifetech.com
LOT-Oriel – www.lot-oriel.com

Merck, Sharp and Dohme – www.msd.com
MetaChem – www.metachem.com
Millipore Corporation – www.millipore.com
Miltenyi Biotec – www.miltenyibiotec.com
Molecular Research Center, Inc. – www.mrcgene.com
MWG Biotech – www.mwg-biotech.com

National Diagnostics – www.nationaldiagnostics.com
New England BioLabs, Inc. – www.neb.com
Nikon Corporation – www.nikon.com

Olympus Corporation – www.olympus-global.com
Optivision Ltd – optivision.co.uk

PANTHER – http://curation.pantherdb.org/about.jsp
Perbio Science – www.perbio.com
PerkinElmer, Inc. – www.perkinelmer.com
Pharmacia Biotech Europe – www.biochrom.co.uk
Photonic Solutions plc – www.psplc.com
Promega Corporation – www.promega.com

Qiagen N.V. – www.qiagen.com

R&D Systems – www.rndsystems.com
Roche Diagnostics Corporation - www.roche-applied-science.com

Sanyo Gallenkamp – www.sanyogallenkamp.com
Sarstedt – www.sarstedt.com
Schleicher and Schuell Bioscience, Inc. – www.schleicher-schuell.com
Scientifica – www.scientifica.uk.com
Serotec – www.serotec.com
Shandon Scientific Ltd – www.shandon.com
Sigma-Aldrich Company Ltd – www.sigma-aldrich.com
Sorvall – www.sorvall.com
SPOTFIRE – www.spotfire.com
Stratagene Corporation – www.stratagene.com

TeleChem International – www.arrayit.com
Thames Restek – www.thamesrestek.co.uk
Thermo Electron Corporation – www.thermo.com
Thistle Scientific – www.thistlescientific.co.uk

Vector Laboratories – www.vectorlabs.com
VWR International Ltd - www.bdh.com

Wolf Laboratories - www.wolflabs.co.uk
York Glassware Services Ltd – www.ygs.net

Index

Note: Roman numbers indicate the color section.

44K, 6, 7, 29
60mer, 6, 88, 203, 205, 226

ABgene, 165, 167, 170, 238
ABI, 75, 76, 186–8, 238
absolute expression values, 108–10
  calculation, 108–10, 113
  confirming, 110
  troubleshooting and, 113
accuracy, 5, 73, 103, 104, 105, 106, 110, 120, 126, 157, 185
AceView, 7
acoustic, 225, 226, 227, 235
acoustic waves enhance hybridizations, *see* surface acoustic wave mixing
actin, xxiv, 37, 59, 81, 82, 123, 140, 151, 188, 195, 204, 219, 223
adapters (for amplified gene expression microarrays), 52–5
  ligation to *Taq*I-digested cDNA templates, 52, 53, 55, 58–9
ADGE microarrays, *see* differential gene expression microarrays, amplified
adipose, 31
Advalytix, 226, 227, 229, 232, 233, 236, 238
Affymetrix, 124, 144, 151, 162, 183–7, 193, 210–6, 227, 238
  GeneChip, 6, 185
  U95A chips, 204
  U133A 2.0 Human Genome Microarray, 6, 203, 216
    transcriptional start site expression profiling, 88–91
agarose gel electrophoresis xxii, 71, 165, 167, 170, 195, 208
  of macaque RNA, denaturing, xxx, 208–9
Agilent xxiv, xxx, 6–10, 12, 14–7, 20–4, 26, 29, 40–1, 84–5, 99, 173–4, 187, 209, 215, 222, 228
  BAC microarray resources, 117
  Human Genome Microarrays (incl. 44K), 6, 7
    surface acoustic wave mixing studies in hybridization to, 236–7
    transcriptional start site expression profiling, 88, 89
  hybridization to microarray products, solutions, 235–6
Ahr, xx, 32, 33, 34, 36, 48
air filtration, BAC microarrays, 121
alcoholism, primate models, 201–2
aldehyde, 70, 120, 124, 129, 135, 136, 148, 195, 208
  -treated slides for BAC microarrays, 120, 136–7
Alexa dyes for surface acoustic wave mixing studies, 230–1
algorithm, 2, 112–3, 132, 224
Ambion, 17, 23, 24, 26, 39, 40, 41, 44, 184, 211
Amersham, xix, 26, 42, 61, 77, 117, 137, 171, 229, 231, 238
amine linkers, N-terminal modifications, 67
aminoallyl, xix, xxvii, 23–6, 42–4, 55, 61, 227–31
aminoallyl-modified cDNA
  for dioxin toxicogenomics
    dye coupling and purification, 43–4
    preparation, 42–3
  for human genome microarrays
    fluorescent dye labelling, 26
    purification and concentration, 24–5
    synthesis from sense-strand aRNA, 23–4
  for surface acoustic wave mixing studies, 228–9
    purification, 229–30
aminoglycan, 151
amino-modified, 120, 135
ammonium acetate, 95, 97, 211
Amplicon, xxii, 3, 71, 76
amplification, xix, xxiv, xxv, xxvii, 6–12, 14, 17, 19–22, 29, 35, 44, 52, 55, 59, 66, 71–2, 82, 85–8, 99, 116, 119, 125–7, 142, 149, 155, 159, 161, 177, 183, 185, 189–92, 194–6, 200
  of RNA, *see* DNA; RNA; transcription
amplified differential gene expression microarrays, *see* differential gene expression microarrays

analysis of variance (ANOVA), differential gene expression detection, 37–8
animal, xxx, 202–4, 209, 216, 238
anneal, xxv, xxvii, 20, 21, 42, 52, 53, 67, 69, 75, 86–8, 97, 98, 128, 135, 138, 156, 169
annotation, 31, 101, 113, 150, 151, 198
ANOVA analysis, differential gene expression detection, 37–8
antibiotic, 181
antibodies, 3, 214, 222
antifoam, 213, 214
antihistamine, 181
antisense, xxv, 54, 67, 69, 86, 88
apoptosis, 33
Applied Biosystems
  Human Genome Survey Microarray 2.0, 6, 203
  RNA extraction system, 184
Arabidopsis, 1, 228
Arcturus, 189, 190, 191, 192, 195
  RiboAmp, 185
aromatic hydrocarbon receptor (ARH)
  in homeostasis and development, 33–4
  signalling pathway, xx, 32–3
aromatic hydrocarbon receptor nuclear translocator, 33
ArrayBooster, 226
ArrayIt, xix, 17, 19–29, 44, 62, 117, 122, 129, 131, 137–9, 169–70, 173–4, 228–9, 235, 238, 240
  H25K Human Genome Microarray, xix, 6, 7
  artificial chromosome, xxvi, 3, 116, 162
  aryl hydrocarbon receptor, xx, 32, 33
  assay, xxiii, 2, 3, 39, 41, 56, 65–7, 70–5, 79–80, 83, 122, 150, 162, 170, 181–3, 237
  Astec, 75, 77
  atlas, 89–91
automation in microarray sample handling, 3, 80, 196, 197
Axon, xix, 45–6, 78, 173–4, 238
5-aza-dC (and other DNA methyltransferase inhibitors), 147–9, 150, 152–3

BAC, xxvi, xxvii, xxviii, 116–20, 122, 124–8, 130–42, 144, 162, 171–2, 176–7
  see also bacterial artificial chromosome microarrays
  background, high 37, 45, 106, 120, 124, 142, 173–4, 184, 195, 224, 226
  amplified differential gene expression microarrays, 63–4
  HPV DNA microarrays, 80
  sources, 116, 117
bacteria, xxvi, 3, 75, 115–6, 119, 156, 162
bacterial artificial chromosome (BAC) microarrays, xxvi–xxviii, 115–45
  disease-specific, 116
  genome-wide, 116
  methods and approaches, 118–39
  for microarray CGH, 116, 118, 125, 126, 127, 131, 133, 162, 176
    probe hybridization, 171–2
  platform choice, 118
  troubleshooting, 127, 139–40
Bayesian approaches, 37, 106, 107–8
BCP, 39–41
BD Vacutainer CPT tube, 184
bead, xxv, 86–7, 91, 95–6, 196–7
  chip, 6
  bias, xxv, 37, 86–7, 116, 119, 125–7, 132–3, 185, 195
  binding, xxxi, 1–3, 10–11, 13–4, 33, 45, 65–6, 81–2, 91, 95–6, 120, 128–9, 141, 159, 169, 190–1, 193, 204, 218, 220, 222–3, 226, 237
  bioanalyzer, xxx, 8–10, 12, 14–7, 20–4, 26, 40–1, 187, 209, 237
  bio-defense, 65
bioinformatics, 150–4, 159
  promoter identification, 83
    tumor suppressor genes, 148
biomarkers, 179–200
  definitions, 180
  microarrays for discovering, 116, 179–200
    methods and approaches, 180–94
    troubleshooting, 194–6
biopharmaceutical, 179
biotin, xxv, 86–7, 94–6, 101, 193, 212, 214
biotinylation, transcriptional start site microarrays, 87, 94–5
bisulfite, 148, 152, 160
  conversion and sequencing of methylated DNA, 150, 154–6, 159
  blocking, 45, 63, 78, 80, 128, 131, 141
  blood, xxx, xxxi, 33, 143, 181–8, 199, 201–7, 209, 216, 222–4
  see also peripheral blood; RNA
  blueprint, 5
  boric acid, 165, 167, 170
  bovine serum albumin (BSA), 45, 78, 212–4, 232–3
breast cancer recurrence after tamoxifen adjuvant therapy, 183
1-bromo-3-chloropropane, 39, 206
bromophenol blue, 208
buffers
  hybridization, BAC microarrays, 128–9
  post-hybridization, BAC microarrays, 131
  printing, 119

Caenorhabditis elegans, 33
CAGE, *see* cap analysis of gene expression
Calbiochem, 138, 238
calf thymus, 232, 233
cancer, xxviii, 32, 47–8, 66, 70, 73, 80, 83, 116–8, 131, 134, 143–5, 153, 160–1, 171, 181, 198–200
  biomarkers, 182–3
  cell immortalization in pathogenesis of, 147

CGH applications, 174–6
DNA copy number changes as hallmarks of, 115
genetic changes associated with, 161
canonical, xxiv, 82
cap, xxv, 16, 18, 25–6, 40, 82, 86–7, 95, 101, 104, 139
cap analysis of gene expression (CAGE), 82–3, 84, 85, 101, 104, 109, 110, 111
cerebellum gene library, 110
cap trapping of full-length cDNAs, 86, 87, 95–7
capillary, 122, 233, 234
carcinogenesis, cell immortalization in, 147
cardiovascular disease, 32, 47
CCD-based scanners, BAC microarrays, 131–2, 141
CCSP-2, 182
CEL Associates, 44
cell cycle, 4, 30
cell division, 151
cell immortalization, epigenetic analysis, 147–60
methods and approaches, 149–58
troubleshooting, 158–9
cell line, xxiii, xxviii, 52, 70, 73–4, 134, 152, 174–5, 181
centrifuge, 10–11, 13–4, 16, 18, 22, 25, 27–9, 40–1, 43, 45–6, 56–63, 78, 93–4, 96–9, 135–7, 155, 163–4, 166–70, 172–3, 190–6, 207, 212–3, 218–21, 230, 234
centrifuged pellet of precipitated nucleic acid, pitfalls, 100
cerebellum, 83
-specific genes and gene library, 89, 90, 91, 110, 111
cerebral cortex genes, 89, 91
cervical scrapes, HPV microarray test, xxiii, 73–5
CGH, *see* comparative genomic hybridization
chaperones, 33
charge-coupled device-based scanners, BAC microarrays, 131–2
chemiluminescent, 2, 216–20
chemistry, surface, of slides/substrates for BAC microarrays, 120, 139
chemogenomics, 179
chemotherapy, 183
chimpanzee, 204
ChIP, 237
chloroform, 40, 57, 93, 95–9, 140–1, 163–4, 166, 207, 211
CHP, 215
chromatin immunoprecipitation, CpG island microarrays for, 237
chromosome, xxvi, xxviii, 3, 115–6, 134, 143–4, 147, 161–2, 168, 174–6
clean-room, 5, 70
clinical endpoints, 180, 181
clone, xxi, xxvi, xxviii, 33, 67, 70, 73, 75–6, 109, 116–9, 122, 133–5, 144, 156, 168, 239
Clontech, 60–1, 211
cluster, vi, xxiv, 35, 37, 38, 46, 49, 52, 84, 85, 147
cocktail, 8–9, 11–4, 24, 189–92
cognate, xxiii, 84, 226
co-hybridization, 231
colon cancer-secreted protein gene, 182
color palette, xix
colorimetric, 2
column, xxi, 1, 7, 10–11, 13–4, 16, 18, 22–3, 25, 27, 43, 63, 88, 95–9, 137, 158, 174, 188, 190–1, 193–4, 196, 207, 211–2, 218–21, 229–31
commercial products, *see* suppliers
comparative genomic hybridization (CGH), xxvii, xxviii, xxix, 115–8, 125–8, 131–4, 143, 144, 161
microarray/matrix, xxix, 115, 161–77
bacterial artificial chromosome, *see* bacterial artificial chromosome
methods and approaches, 162–72
troubleshooting, 177
competitive hybridization, xxviii, 108, 125–6, 134
complementary, 52, 54–5, 59, 67, 75, 84–5, 87, 127, 129, 226
complementary DNA and RNA, *see* DNA; RNA
complex, xx, 4–5, 32–3, 66, 81, 126–7, 131–2, 179, 201–2, 226
computer, 2–3, 38, 46, 82, 112–3, 153, 157, 174
concordance, 89
conjugate, xxv, 86
contact printing, 1, 6
control, xix, xxiii, xxix, 1, 5–6, 30, 35, 36, 39, 47, 68, 70–1, 73–4, 76–7, 81–2, 88, 105–6, 109–13, 119, 121, 123, 128–31, 139–41, 149, 153, 159, 162, 174–5, 177, 184, 195–7, 199, 202–3, 205, 212–3, 216–7, 228, 230, 236
control DNA for amplified differential gene expression microarrays, 52–5, 58, 60, 62
reassociation, 59
copy number, xxviii, 115–6, 125–6, 132–4, 162–3, 168, 174–7
core promoter sequence, xxiv, 82
correlation, 4–5, 55, 88, 91, 106, 110–11, 185–6, 202
Cot-1 DNA (for repetitive sequence blocking in BAC microarray hybridizations), 45, 127–8, 131, 138, 141, 142, 168, 169, 171, 172
cover slips, slide, BAC microarray hybridizations under, 129
CpG islands, 147
methylation
detection, 149–59
status, 147, 148, 148
microarrays for chromatin immunoprecipitation, 237
in promoters, *see* promoters

CPT tube, 184
cross-hybridization, 72, 126, 225
cross-link, 44, 77, 79, 177
cross-linking in HPV DNA microarray fabrication, 70
CTAB, 91–2
CT adapters (for amplified gene expression microarrays), 52–5
  ligation to *Taq*I-digested cDNA templates, 52, 53, 55, 58–9
CT primers (for amplified gene expression microarrays), 52–5
cyanine dyes
  Cy3, xix, xxvii, 15–6, 26–9, 42–3, 45, 53, 55, 60–3, 72, 87–8, 99, 105–6, 108, 126, 132, 137, 167–71, 174, 230–1
  Cy5, xix, xxvii, 15–6, 26–9, 42–3, 45, 53, 55, 60–3, 71–3, 77, 87–8, 99, 105–6, 108, 126, 132, 137, 167–71, 174, 230–1
  see also fluorescent dye
cynomolgus, *see* macaque
cytochrome P450, 33
cytogenetic, 161, 163
cytosol, xx, 32, 33

data, xix, xxviii, 2, 4–5, 29, 34–5, 37–9, 46, 52–3, 72, 75, 79, 82–3, 88–91, 101, 103–14, 118, 125, 127, 131–4, 142, 148, 150–2, 157–9, 170, 173–4, 177, 181–6, 194–8, 201, 205, 209–10, 214–7, 221–2, 224
  BAC microarrays, processing and analysis, xxviii, 132–3
  CGH, acquisition and analysis, 173–4
  differential gene expression detection, analysis, 37–8
  increasing utility, 103–14
    methods and approaches, 104–13
    troubleshooting, 113
  macaque gene expression studies, analysis, xxx, 214–16, 221–2
database, xvi, 2, 4, 5, 7, 35, 38, 46, 67, 89-91, 104, 105, 108-111, 114, 117, 118, 159, 160
degenerate-oligonucleotide-primer PCR, 119
degrade, 8, 11, 24, 29, 43–4, 79, 95, 125, 140–1, 177
denature, xxvii, 9, 12–3, 17, 19, 24, 42, 45, 52, 56, 59, 63, 75, 98, 137–8, 141, 154, 169, 170–2, 190, 209–10, 213, 228, 234
denaturing gel electrophoresis of macaque RNA, xxx, 208–9
Denhardt's, 62
DEPC, 57, 184, 189, 193–4, 196, 207–8, 210–5
detection, xxii, 2, 4, 51, 56, 65–8, 70–6, 78, 80, 88, 105, 108, 110, 118, 127, 132, 143, 148, 150, 161, 177, 185, 202–3, 205, 215, 222, 227
development, 48, 81, 85, 100, 133, 147, 151, 161, 179–83, 198–9, 202
  aromatic hydrocarbon receptor in, 33–4
  disorders of, DNA copy number changes as hallmarks of, 115
diagnostic, 1, 4, 45, 66–7, 71, 75, 115, 189, 239, 240
diethylpyrocarbonate, 184
differential gene expression, data analysis in detection of, 37–8
differential gene expression microarrays, amplified (ADGE), xxi, 51–64
  methods and approaches, 52–63
  troubleshooting, 63–4
digest, xxi, xxvi, xxix, 14, 39, 52–3, 55, 58–9, 93, 118–9, 127, 135, 156, 162, 165–8,
dimethylsulfoxide, 75
diol, 93–5, 199
dioxin (TCCD) toxicogenomics, xvii, xx, 31–49
  methods and approaches, 34–46
    three phase approach, 34–9
  troubleshooting, 47
discriminant analysis, 106–7, 112, 114
  candidate parameters, 106–7
  multivariate, 104, 105–6, 108
disease, 32, 34, 47, 65, 83, 115, 125, 201–5
  BAC microarrays for, 116
  biomarkers, 180, 181, 182, 183, 198
displace, 127
distal, xxiv, 82
dithiothreitol (DTT), 17, 42, 45, 46, 56, 98, 193, 210, 212, 228
divergence, 89
DMSO, 61, 75–6, 230
DNA
  for BAC microarray probes
    preparation, xxvi, 124–8, 138
    reference samples, 125, 126
    troubleshooting, 140–1, 142
  for CGH microarrays
    extraction from cells/tissues, 163–4
    labelling, 167–8, 170–1
    restriction digests, *see* restriction digests
    troubleshooting, 177
  copy number changes
    as hallmarks of cancer and developmental disorders, 115
    surveys, 115–16
  herring sperm, in BAC microarray hybridizations, 128–9
  ligase, 57–9, 75, 135, 210
  methylated, *see* methylated DNA
  precipitated and centrifuged pellet, pitfalls, 100
cDNA (complementary DNA), xix, xxiv, xxv, 1, 3, 23–29, 31, 35, 52, 53, 65, 67, 70, 72, 73, 81, 82, 107, 109, 114, 115, 162, 167-169, 176, 177, 184, 196, 225, 227-234, 237
  for amplified differential gene expression microarrays
    1st-strand synthesis, 56–7

2nd-strand synthesis, 57–8
double-stranded cDNA, *TaqI* digestion, *see TaqI*
for biomarker discovery from blood
1st-strand synthesis for 1st amplification of small-quantity RNA, 189
1st-strand synthesis for 2nd amplification of small-quantity RNA, 192
2nd-strand synthesis for 1st amplification of small-quantity RNA, 190
2nd-strand synthesis for 2nd amplification of small-quantity RNA, 192–3
for dioxin toxicogenomics
aminoallyl-modified, *see* aminoallyl-modified cDNA
synthesis, 41–2
for human genome microarray
1st-strand purification, 18–19
1st-strand synthesis for 2nd round amplification, 12
1st-strand synthesis from small-quantity RNA for 1st round amplification, 8–9
1st-strand synthesis from total human RNA, 17–18
1st-strand-tailed, conversion to T7 promoter-containing double-stranded cDNA, 20
1st-strand tailed, preparation, 19–20
2nd-strand synthesis for 1st round amplification, 9
2nd-strand synthesis for 2nd round amplification, 12–13
aminoallyl-modified, *see* aminoallyl-modified cDNA
double-stranded cDNA, purification, 10
transcription, *see* transcription
for human genome microarray studies of macaque gene expression
1st-strand synthesis, 210
2nd-strand synthesis, 210, 217–18
double-stranded cDNA purification, 211, 218–19
labelled, *see* fluorescent dye
for transcriptional start site microarrays
1st-strand 5′ end priming-site addition, 97–8
1st-strand amplification, 85–7
1st-strand purification, 93
1st-strand synthesis, 85–7, 92–3
2nd-strand synthesis, 98–9
double-stranded cDNA, cRNA amplification from, 99–100
full-length, cap trapping and purification, 86, 87, 95–7
see also reverse transcription
cDNA microarrays
for CGH analysis, 162, 176
DNA labelling, 167
surface acoustic wave mixing in hybridization to, 233–4
post-hybridization washes, 234–5
dsDNA (double-stranded DNA)
cDNA, *see* cDNA *(main entry above)*
for HPV detection/genotyping, production, 67, 75–6
DNA methyltransferase inhibitors (incl. 5-aza-dC), 147, 148, 149, 150, 152–3
DNase, 11, 14, 39–41, 163, 165, 167, 170–1, 188, 191
downregulate, 54, 56, 148, 150, 157
drug discovery and development, xxviii, 1, 4–5, 134, 198–9
biomarkers in, 180, 181, 182, 183
genomics initially focusing on, 179
dual-channel, 36, 37
dual-color, 108, 171, 198
duplex, 52, 54, 69, 75, 78
dwell time, printing pins, 121–2
dye, xix, xxvii, xxix, 15–6, 26–9, 36–7, 42–4, 51, 55, 61–3, 72, 79, 127, 131–2, 137, 141, 162–3, 165, 167–8, 170–1, 174, 226, 230–1
fluorescent, *see* fluorescent dye
DyeSaver, 28, 29
Dye-swap, 171, 174

E17.5 (EST library), 110, 111, 190
E. coli, 57, 75, 210
EDTA, 17, 23–4, 26, 41, 57, 59, 92–3, 95, 97–9, 163, 165–8, 170–1, 184, 208, 213
electrophoresis, xxii, xxx, 71, 76, 79, 118, 145, 165–8, 170–1, 195, 208–9
elute, xxx, 10–11, 14, 16, 18, 22, 25, 27, 43, 59–62, 77, 96, 156, 166, 184, 187–8, 191, 193–4, 207, 209, 219, 221, 230–1
embryo, 32–3, 83, 109
endocrine, 202
endpoints, clinical and surrogate, 180, 181
enhancer, xxiv, 8, 11–2, 14, 33, 82
enzo, 193, 211
enzymatic tailing, 19
enzyme, xxvii, xxix, 8–9, 11–4, 17, 19–21, 52, 55–8, 60, 79, 86–7, 93, 95, 97, 119, 126, 135, 137, 140–1, 155, 162, 165, 182, 189–92, 195, 216–20, 228
epidemiology, 47, 75
epigenetic, 147–50, 152, 154, 156–8, 160
epigenetic analysis of cell immortalization, *see* cell immortalization
EPPS, 59
Erie Scientific, 28, 225, 227, 232–3, 236
erythrocytes, RNA isolation from, 184
eukaryotic, xxiv, 81–2, 212–3
gene regulatory regions, xxiv, 81–2
evaluation, 103–4, 133, 181–3
'ever changed' genes, identification, 151, 157
Evergreen Scientific, 234
exon, xxiv, 83, 85

exon–intron, xxiv, 85
expressed sequence tags (ESTs), 7, 104
  cluster definition, 84
  data analysis, 109, 110, 111
expression, gene, *see* gene expression
extracellular matrix, xxxi, 223

false-negative/-positive, 73
FANTOM transcripts, 84, 109
FDA, 1, 30, 199
Fermentas, 165, 167, 170
filtering, data, BAC microarrays, 132
first-round, xxvi, 8–9, 11, 189–91
first-strand cDNA, 8–9, 12–3, 17–9, 56–7, 85, 87, 92–3, 96–7, 189, 192, 218
Fisher Scientific, 213–4, 239
fluorescence, xix, xxix, 3, 29, 37, 88–91, 119–20, 127, 131, 162–3, 166, 173–4, 177, 203, 205
fluorescent dye (predominantly Cy3 and Cy5)-labelled probes
  DNA for BAC microarrays
    preparation, xxvii, 126, 137
    troubleshooting, 127, 141
  DNA for CGH microarrays, preparation, 167–8, 170–1
  cDNA for amplified differential gene expression microarrays, preparation, 55, 60–2
  cDNA for dioxin toxicogenomics (aminoallyl-modified cDNA), preparation, 43–4
  cDNA for human genome microarrays (aminoallyl-modified cDNA)
    hybridization, 28–9
    preparation, 26
    purification, 27
  cDNA for surface acoustic wave mixing studies, 230
    dye preparation, 227
    preparation, 230–1
    purification, 231–2
  dsDNA of HPV
    dye quality problems, 79–80
    preparation, 72–3, 77
  one vs two-color approach, 108–9
  aRNA (for human genome microarrays)
    preparation, 15–16
    purification on columns, 16
  cRNA (for transcriptional start site microarrays), xxv, 87–8
  signals, *see* signals
fluorophore, xxvii, 126, 132, 229
formalin, 124–5
formamide, xxx, 45, 128, 131, 171–2, 208–9, 232–3
Fraction V, 232–3
freezing–thawing of blood RNA samples, effects, 185
functional genomics, 31, 34–5, 198
  definition, 31
  non-human primate studies, 202

gain, xxix, 112, 126, 142, 161–2, 174–7
GAPDH, xxii, 69–71, 73
Gaussian, 37
GE Healthcare, 60, 95, 97–8, 230
gel electrophoresis of macaque RNA, denaturing, xxx, 208–9
gel filtration, 211, 212
gelsolin gene promoter, 83
GenBank, 7, 67, 84
gene content, 7
gene expression, xix–xi, xxiv, xxx, 1, 3–7, 29–34, 36–8, 48, 51–2, 54–6, 58, 60, 62, 64, 81–4, 86, 88–92, 94, 96, 98, 103–6, 108–11, 113–5, 131, 133, 147–9, 157, 161, 174, 180, 182–4, 196, 198–9, 201–4, 206, 208, 210, 212, 214–6, 218, 220, 222, 224, 237
  absolute expression values, *see* absolute expression values
  on ArrayIt H25K human genome microarray, xix
  differential, *see* differential gene expression
  primate (non-human), *see* primate gene expression
  serial analysis (SAGE), 109, 110, 111
  see also Riken Expression Array Database
gene expression profiling
  cancers, 182–3
  transcriptional start sites, *see* transcription
gene ontology analysis, 38, 148, 150, 157–60
gene product, 1, 33, 38, 81
GENESPRING, 215, 222, 239
Genetix BAC microarray resources, 75, 76, 117
  printing pins, 123
Genisphere, xix, 17, 19–24, 28, 29
genome, xix, xxiv, xxvii, 1, 4–8, 10, 12, 14, 16–24, 26, 28–31, 48–9, 67, 80–1, 84–5, 88, 100–1, 109, 113–4, 116–9, 125–7, 143–4, 147–8, 151, 153, 156, 162–3, 174, 177, 179, 201, 203–4, 215, 217, 221–2, 236
genome representation/sampling, xxvii, 126–7
genome-wide BAC microarrays, 116
genomics, xx, 34–6, 38, 40, 42, 44, 46, 48, 64, 89–91, 101, 114, 116–7, 143–4, 163, 198–200
  definition, 31, 179
  dioxin toxicology, *see* dioxin
  focus/applications of, 179–80
  functional, *see* functional genomics
  hybridization, xxix, 115, 161–2, 164, 166, 168, 170, 172, 174, 176, 182
  instability, 147
  sequence, xxiv, 4, 84–5, 174
  see also comparative genomic hybridization

Genotech, 77
genotype, 31, 67, 72–3
genotyping, xxii, xxiii, 3–5, 30, 65–8, 70–8, 80, 118, 162
  of HPV, *see* human papillomavirus
Gentra, 154
GFX, 229–31
GIS, xxiv, 84–5
global, 31, 63, 88, 179, 185, 215, 222, 239
  normalization of BAC microarray data, 133
glyceraldehyde 3-phosphate dehydrogenase cDNA (PCR control target), 70
glycogen, 57, 61, 184, 211
GOMINER, 151, 157–8
granulocytes, RNA isolation from, 184
GSC, xxiv, 84–5, 88

H25K, xix, 6–7, 17, 19–24, 26, 28–9
haplotype, 162
healthcare, xiv, 60, 95, 97–8, 230
heart, 33, 83
HeLa, xxiii, 73, 74
hemoglobin, 184, 185
HEPA, 121
hepatic fibrosis, 33
herring sperm DNA in BAC microarray hybridizations, 128–9
heterodimer, 33
heterogeneous, 79, 125, 143, 170, 183
high
  background, *see* background
  density, 6, 56, 121–3, 162, 177
  resolution, 115
  throughput, 28, 34, 36, 56, 116, 119, 121–2, 124, 131, 137, 139, 147–8, 150, 172, 183, 235
  hippocampus, 88–91
hippocampal genes, 89, 91
HIV, 65
homeostasis, 33, 81
HPV, *see* human papillomavirus
HSP90, xx, 32, 33
human gene, 6, 30, 147, 202, 216
human genome microarrays, whole, xix, 1–30, 116, 118, 144, 147, 151, 177, 179, 203–4
  commercial products, 6, 203
  methods and approaches, 4–29
    recommended protocols, 8–29
  troubleshooting, 29
human papillomavirus (HPV) detection and genotyping, xxii–xxiii, 65-80
  clinical applications, 73–5
  methods and approaches, 66–77
    probe labeling/hybridization/detection, 72–3, 78
    sensitivity and specificity of microarrays, 71–2
    target and microarray construction, 66–70, 76–7
  troubleshooting, 78–9
humidity, 44, 77, 79, 119, 121, 123–4, 136, 138–9, 142, 173, 228
humidity control, BAC microarrays, 121, 139
hybridization, xxii, xxiii, xxv, xxvi, xxviii, xxxix, 5–6, 28, 35–6, 43, 45–7, 51, 56–7, 60, 62–5, 70–4, 77–80, 84, 86, 88–9, 96, 99, 108–9, 115, 118–9, 122, 124–32, 134, 136, 138, 141–2, 161–4, 166, 168–74, 176, 182, 185, 188, 194, 205, 213–5, 222, 225–8, 230–7
hybridization, comparative genomic, *see* comparative genomic hybridization
hybridization to microarrays
  amplified differential gene expression microarrays, 62–3
    high background problems, 63–4
    weak signals, 63
  BAC microarrays, 128–31, 138–9
    in CGH analysis, 171–2
    repetitive sequence blocking, 127–8, 129, 141, 142
    troubleshooting, 141–2
  in CGH analysis
    BAC microarrays, 171–2
    cDNA microarrays, 168–70
  comparative genomic, *see* comparative genomic hybridization
  dioxin toxicogenomics, fluorescent cDNA mixtures, 45–6
  HPV detection and genotyping, 70, 72–3, 78
    high background problems, 80
    weak signals, 79-80
  mixing, *see* mixing
  whole human genome microarrays
    fluorescent cDNA mixtures, 28–9
    macaque gene expression studies, 213–14
hydrazine, 94
hydrolysis, xxv, 86
hydrophobicity of slides/substrates for BAC microarrays, 120, 139
hydroquinone, 154
hydroxylamine, 26, 23–1
hyperplasia, 33

IgG, 214
Illumina Sentrix Human-6 Expression Beadchip, 6
ImageJ, xix
image processing, BAC microarrays, 132
immortalization, *see* cell immortalization
immune system, 33
immunotoxicity, 32
in vitro transcription, 11, 99, 184, 211
incorporate, xxvii, 44, 55, 61, 67, 72, 88, 107, 127, 132, 166, 180, 202, 226
indirect, xxvii, xxix, 37, 41–2, 55, 60–1, 83, 127, 162–3, 227
infectious, 65

inhibitor, 24, 147–9, 193, 212
input, xxi, 54–6, 113, 116, 124–5, 158, 184–5, 195–6
insert, xxvi, 78, 116, 118, 144, 208, 218–20
Intel, 3, 104, 112–13
intensity, xxviii, 55–6, 90–1, 105–6, 131–4, 141–2, 173–4, 226
interphase, 161
intron, xxiv, 84–5, 115
invertebrate, 33–4
Invitrogen, 17, 23, 42, 57–8, 75, 77–8, 92, 117, 137–8, 155, 165–8, 170–1, 194, 210, 212, 214, 228, 230, 232–3, 239
isoamyl alcohol, 57, 163–4, 211
isopropanol, 39–41, 45, 78, 95–6, 100, 163–4, 171–2, 206
IVT, 11, 14, 20–1, 184, 190–1, 193–6, 211–2, 216–20

Junction, xxiv, xxxi, 85, 223

KEGG, 148, 150
Kinase, xxxi, 51, 223
Klenow, xxvii, 20, 127, 137, 141, 167, 171

L1 HPV gene, 70
Labeling, xix, xxv, xxvii, xxix, xxx, 2, 4, 6–7, 15–6, 23, 26, 29, 35, 41–2, 44, 55–6, 60–1, 72–3, 77, 79–80, 85–7, 91, 100–1, 108–9, 118, 124, 126–7, 137, 141, 162, 167–8, 170–1, 174, 177, 183–5, 187–8, 196–8, 205, 209, 211–2, 216–20, 227, 229–30
labelling, fluorescent dye, *see* fluorescent dye
lacZ, 70
laser, 2, 73, 125, 131–2, 183, 185
laser-based scanners, BAC microarrays, 131, 132
LB, 75, 155–6
least-squares, 133
leukocytes, RNA isolation from, 184
Life Technologies, 194, 239
LifterSlip, 28–9, 225, 227, 232–4, 236
ligand, 33–4
ligate, xxi, xxvi, 52–3, 75, 87, 98, 119, 135
ligation methods
  amplified gene expression microarrays, 52, 53, 55, 58–9
  transcriptional start site microarrays, 87–8
linear, xxvii, 44, 54, 107, 119, 127, 131–2, 185
linker, xxv, xxvi, 67, 86–7, 97–8, 119, 135
linker-mediated PCR, BAC clone generation via, xxvi, 119, 135–6
liver, 32–3, 83
LM-PCR, xxvi, 119, 127, 135–6, 144
locally-weighted scatter plot smoothing, 133
loci, 116, 125
loss, xxix, 33, 48, 86–7, 109, 116, 123, 126, 147, 161–2, 174–7
low abundance, 226
low-risk, xxii, 72
lumen, 225–6
luria, 75, 155
lymphocyte, 183–4
lyophilization, 23
lysate, 40–1, 185, 187–8

macaque (incl. cynomolgus), xxxi, 216, 222–3
  alcoholism, 201–2
  gene expression, *see* primate gene expression
MAGE-OM and MAGE-ML (Microarray Gene Expression Object Model and the XML-based format), 38, 39
magnetic, xxv, 86–7, 91, 95–6, 196–7
magnification, xxi, 51–2, 55–6
Mahalanobis distances, 106, 107, 108
malignancy, *see* cancer
marker, 115–6, 118, 165, 167, 180, 183, 202, 205
marker-based microarrays, *see* biomarkers
MAS, xxx, 6, 203, 215
master mix, 8, 11–4, 17, 19, 42, 45, 56, 156, 190–3
matrix CGH, *see* comparative genomic hybridization
MaxD system, 38–9
MDS Analytical Technologies, 8–12, 14–6, 29
mechanical mixing systems, 226
medium, 40, 88, 101, 152, 155–6, 163
MEDLINE, 2
melanoma, uveal, chromosomal alterations, 174–5
membrane, xx, xxxi, 3–4, 18, 22, 25, 27, 32, 43, 166, 219, 223, 229–30
MES, 212–4
metabolism, 33–4, 202
methylate, 147, 149–50, 159
methylated DNA
  CpG islands, *see* CpG islands
  extraction, 154
methylation, 147–50, 152, 156, 159–60
methyltransferase, 147–9
MIAME (Minimum Information About a Microarray Experiment), 38, 104, 223–4
MicroAmp, 216–7, 219–20
Micro Spotting Plus, 119
microarray
  biomarker, 179, 181, 183, 196, 198
  biomarker discovery, 179, 181, 198
  data, xxviii, 2, 38–9, 89, 103–6, 108–10, 112–4, 134, 148, 150, 152, 157, 159, 181–3, 185, 195–6, 205
  fabrication, 66, 77, 80, 83
  hybridization, xxii, 45, 57, 62, 70–3, 99, 128, 130, 169, 171–2, 188, 194, 213, 225–6, 228, 230, 232–4, 236
  printing, xxvi, 119–20, 122, 144, 229
  scanning, 131

Microarray Gene Expression Object Model (MAGE-OM) and the XML-based format (MAGE-ML), 38, 39
microbial, 204
Microcon, 23, 25, 137, 168
microdissection, 125, 142, 183, 185
microplate, 75–6, 79, 117, 121, 123, 135–6, 140, 166, 197, 234
microplates for BAC microarrays, 123, 140
  handling, 120
Millipore, 23, 137, 168, 239
Minimum Information About a Microarray Experiment (MIAME), 38, 104, 223–4
miniaturization, 3
mismatch, 67, 125, 203, 215, 226
mixing in microarray hybridization
  limited, 225–6
  surface acoustic wave-enhanced, *see also* surface acoustic wave mixing
model, 1–5, 33, 37–8, 44, 48, 75, 81, 107, 133, 180–2, 201–4, 224
modification, 66–8, 91, 120, 161, 207, 225, 235
Molecular Devices, xix
Molecular Probes, 214
Molecular Research, xxx, 205, 209, 239
monkey, macaque, *see* primate gene expression
mononuclear cells, peripheral blood, RNA isolation from, 184
mono-reactive, 26
MOPS, 208
morphology, 120–1, 123–4, 132
motion control, 1, 121
mouse, 3, 33, 36, 44, 48, 82–4, 88, 101, 109–10
MPG, 95–6
multivariate discriminant analysis, 104, 105–6, 108
mutation, 147
MY09 and MY11 primers for PCR, HPV detection and genotyping, 69, 70, 72, 73, 77
MYH, xxii, 68, 71, 77, 80

N-terminal modifications, 67
Nanodrop, xxx, 8–10, 12, 14–7, 20–4, 26, 29, 40–1, 167–8, 207, 209, 237
nascent, xxiv, xxv, 24, 43, 82, 86
National Cancer Institute, 153
National Center for Biotechnology Information, 2
neonatal, 32
neoplasms, malignant, *see* cancer
neuroblastoma cell line, CGH profile, 174, 175
neuroblastoma, 174–5, 177
New England Biolabs, 40, 135, 163, 165, 239
NHP, xxx, 201–4, 209, 214, 224
NIH, xix, 2, 64, 151, 157–8
NimbleGen Human Genome Expression Microarray, 6
Nippongene, 99
nitrocellulose, 4
nonhuman, primate, xxx, 201–2, 204, 206, 208, 210, 212, 214, 216, 218, 220, 222, 224
nonlinear, 105, 107
nonparametric, 37
normalize, xxi, xxviii, xxx, 37, 46, 105, 134, 174, 215, 222
normalization of BAC microarray data, xxviii, 132–3, 134
nuclease, xxx, 8, 19, 28–9, 39–41, 140, 189, 196, 205–6, 209, 216–8, 220, 228–33, 235–6
nucleic acid, xxxi, 45, 48–9, 52, 64–5, 71, 80, 94–5, 100–1, 114, 125, 144–5, 160, 168, 177, 186–8, 223, 237
  precipitated and centrifuged pellet, pitfalls, 100
  nucleotide, xxiv, xxvii, 5, 21, 55, 59, 68–9, 79, 82, 115, 126–7, 141, 149, 162, 166, 193
  nucleus, xx, 32–3
  Nunc, 163
  nylon, 4, 203

oligo(dT), xxv, 42, 44, 56, 92–3, 210, 216–7
  priming in cDNA synthesis for transcriptional start site microarrays, 85–6
  oligonucleotides, xxiii, 3–7, 20, 31, 36, 44, 53, 55, 65–9, 73–6, 79, 83, 86–91, 98, 115–6, 119, 126, 162, 176, 203–5, 222, 225–9, 232–4, 236
oligonucleotide microarrays, 5, 36, 66, 88, 176, 225–9, 232–4, 236
  CGH, 162
  dioxin toxicogenomics, 36–7
  HPV detection and genotyping, 65–6
    construction, 68–70, 76–7
  hybridization to
    problems, 226
    surface acoustic wave mixing in, 232–3
    washes, 234–5
  primate (nonhuman) gene expression, 205
  transcriptional start site microarrays, detection sensitivity, 88–91
oligonucleotide sequences for transcriptional start site microarrays, 86
oligonucleotide target, xxiii, 66, 68–9, 74–6, 88, 204–5
oncogene, 115, 143, 147, 160–1, 177, 199–200, 224
oncogenesis, cell immortalization in, 147
ontology (gene) analysis, 150, 157–8, 159
operon, 44
otx2 promoter, 83
overstroke, 124, 139
oxidation of mRNA diol groups, 94–5, 226
ozone, 127, 131, 139, 141

PANTHER, xxxi, 215, 222, 223
paraffin, 118, 124, 183
parallel, 3, 20, 66, 195
parameter, xxviii, 93, 104–8, 112–3, 120, 123–4, 132, 134–5, 153, 233
passive hybridization, 225
pathogen, 65–6
pathway, xx, 32, 34, 147–8, 150, 157, 159, 201, 215, 222
patient amplicon, 3
PBMC, xvi, 184
PBS, xvi, 136, 163, 187, 188
pellet (precipitated and centrifuged) of nucleic acid, pitfalls, 100
peripheral blood, 202
  preparing and storing, 205–6
  RNA isolation from, *see* RNA
Perkin Elmer, 99, 117, 137, 174
pGEM-T Easy vector, double-stranded HPV DNA sequences, 67, 70
pharmaceutical industry and genomics, 179–80
pharmacodynamic, 183
pharmacogenetics, 48, 179
pharmacogenomics, 179
phenol, 57, 93, 95–9, 140–1, 163–4, 166, 208, 211
phenotype, 31, 82
phospholipidosis, drug-induced, 181
phosphoramidite, 69
photomask, 6
photomultiplier tube, 2, 131–2
phycoerythrin, 214, 227
piezoelectric, 227
pins (printing) for BAC microarrays, 44, 70, 79, 117, 120, 122–3, 139–40, 144, 229
  dwell time, 121–2
planar, 1, 4
plant, 1, 3–4, 204, 239
plasma membrane, xx, 32
platen levelness, BAC microarrays, 121–2
plate stacker, 121, 123
platform, 7, 36, 48, 70, 78, 88–90, 103–4, 115–6, 124–6, 133, 162, 168, 173, 176, 196, 203–5, 210, 215, 217, 225, 227
  choice, vii, 118,
  plot, xxviii, xxx, 2, 89, 111, 133–4, 153, 162, 174, 185–6, 197, 215
  poly(A), 23, 39, 45, 62–3, 85, 87, 169, 212, 217
  polyadenylate, xxiv, 82
  polyethylene glycol, 128
  polyinosinic acid, 8
  polylysine, 4
polymerase chain reaction (PCR), xxi–xxiii, xxvi, 18, 27, 39, 42, 51–5, 57, 59–62, 80, 88–93, 101, 103–8, 112–3, 116, 120, 127, 141, 144, 148–50, 154–6, 165–7, 170, 177, 181–3, 185, 188, 202, 205, 223–4
  degenerate-oligonucleotide-primer, 119
  in HPV detection/genotyping, xxii, 66–7, 70, 71–2, 75–6, 77
  linker-mediated (LM), BAC clone generation via, xxvi, 119, 135–6
  methylation-specific, 159
  see also reverse transcription–PCR (RT–PCR)
polypropylene, 163–4
polysaccharide removal in total RNA preparation, 91, 92
pool, xxvi, 108, 119, 125, 207, 219
population, 31, 47, 124–5, 161, 183–4
positive control, xxiii, 73–4
precipitated and centrifuged pellet of nucleic acid, pitfalls, 100
precision, 5, 173–4
primate, xxx, 201–2, 204, 206, 208, 210, 212, 214, 216, 218, 220, 222, 224
primate gene expression (macaque), xxx–xxxi, 201–24
  methods and approaches, 203–23
  troubleshooting, 223–4
primers, xxi, xxiv, xxv, xxvii, 8–9, 12–3, 17, 23–4, 52–5, 59–61, 66–70, 72–3, 75, 77, 85–8, 92–3, 98, 105, 119–20, 127, 135, 148, 153, 155–6, 159, 166–7, 170, 189–90, 192, 202, 210, 216–7, 228
  for amplified gene expression microarrays, 52, 52–5
  for cDNA synthesis for transcriptional start site microarrays, 85–6
  for PCR, for HPV DNA microarrays, 67, 69
  see also random primer DNA labelling
priming site on 5′ end of 1st-strand cDNA, addition, 97–8
printed microarrays
  BAC
    preparation of BAC clones, xxvi, 118–19, 135–6
    preparation and processing, 120–4
    printing pins, *see* pins
  HPV, 67, 69, 70, 76, 77
    troubleshooting, 79
  viral DNA (in general), 65
  see also slide; substrate
printing, xxvi, 1, 5–6, 66–7, 69–70, 76–7, 79, 113, 118–124, 136, 139–41, 144, 147, 229
probes, xix, xxi, xxiv, xxvii, 2, 28, 31, 39, 44–46, 52–3, 55, 60–3, 66, 72–3, 77–8, 80, 84–5, 108, 113, 115–6, 124, 126–31, 137–8, 141–2, 167–9, 171–3, 195, 204, 212–5, 222, 225–7, 232, 236–7
  for BAC microarrays, *see* DNA
  hybridization, *see* hybridization
  preparation/purification
    for cDNA microarray CGH analysis, 168–9
    for macaque gene expression studies, 212–13

processing, 5–6, 109, 120, 124, 129, 132, 136, 142, 184, 194, 233, 236
profiling, xxiv, 1, 3, 34, 51, 81–2, 84, 86, 88, 90, 92, 94, 96, 98, 100, 124, 148, 177, 179–80, 182–4, 186, 188, 190, 192, 194, 196, 198–200, 204
Promega, 58, 75, 95, 135, 137, 154, 163, 165, 167, 170–1, 212
promoters, xxiv, xxv, 20–1, 32, 44, 67, 81–8, 98, 147–8, 150, 152–3
  CpG islands and their methylation
    identification of CpG islands, 153–4
    tumor suppressor gene, 147, 148
  identification, 82–3
    bioinformatic, *see* bioinformatics
protein, xx, xxiv, xxxi, 1, 3, 32–3, 39–41, 51, 57, 81–3, 93, 95, 97–9, 125, 140–1, 163–4, 177, 182, 223, 235, 237
proteinase K, 39–41, 93, 95, 97–9, 163–4
  digestion of 1st-strand cDNA (for transcriptional start site microarrays), 93
pSP6-containing double-stranded HPV DNA sequences, 67, 69
pT7, *see* T7 promoter
PubMed, 2
purification, xxx, 6, 10–11, 13–14, 16, 18, 22, 24, 27, 42–3, 59–63, 75–7, 91, 93, 95, 154, 166, 168, 184, 187, 190–1, 193–5, 207, 209, 211, 218, 220, 229–31

Qiagen, xxx, 18, 22, 27, 42–3, 59–61, 75, 77, 99, 152, 154–5, 165–6, 194, 207, 209
QRT-PCR, 88, 103–8, 112–3
quadratic, 51–2, 54–5
quantitative, 4, 5, 51, 72, 88–9, 101, 103, 109, 150, 177, 182, 198
QuantumDot, 227
quartz, 221
quench, 26–7, 95, 231

random primer DNA labelling, xxvii, 17, 127, 167, 170
rate of transcription, xxiv, 82
reactive group, 120, 124, 129
READ (Riken Expression Array Database), 108, 109, 110
readout, xix, 4–5
real-time, 88–9, 101, 224
reanneal, xxvii, 21, 75, 128
reassociation of control and tester DNA (for amplified differential gene expression microarrays), 59
receptor signaling pathway, 32
red blood cells, RNA isolation from, 184
reducing agent, 124, 142
redundant, 5
reference DNA for BAC microarray, 125, 126
regression, 37, 133
relative expression level, 108
repetitive sequence blocking in BAC microarray hybridization, 127–8, 129, 141, 142
reporter molecules, fluorescent, *see* fluorescent dye
representation, genome, xxvii, 126–7
representative transcript set (RTS), 109, 111
restriction digests of DNA for CGH microarrays, 165
  purification of products, 166
restriction enzyme, xxix, 55, 119, 126, 135, 155, 162, 165
retinoic acid, 33
retrospective, 125
reverse transcriptase, 17, 23–4, 42, 51, 56, 66, 86–7, 103, 181, 210, 216, 228
reverse transcription, 24, 87, 212, 219,
  dioxin toxicogenomics, 42–3
  macaque gene expression studies, 216–17
  surface acoustic wave mixing studies, 228–9
  see also cDNA
reverse transcription-PCR (RT-PCR), 51, 181
  quantitative (QRT-PCR), 103, 104–6, 113
  transcriptional start site microarray production, 88–91
RiboAmp (Arcturus), 8–12, 14, 185, 189–92
ribonuclease, 140, 196
Riken Expression Array Database (READ), 108, 109, 110
RNA
  for biomarker discovery, small-quantity (from blood), 179–200
    1st round amplification, 189–91
    2nd round amplification, 192–3
    isolation, 186–8
    troubleshooting, 194–6
  for dioxin toxicogenomics, total, isolation, 39–41
  for human genome microarrays, small-quantity
    1st round amplification, 8–11
    2nd round amplification, 12–15
    cDNA synthesis from, *see* DNA
  for human genome microarrays, total, cDNA synthesis from, *see* DNA
    precipitated and centrifuged pellet, pitfalls, 100
  for primate (non-human) gene expression, 204–24
    denaturing gel electrophoresis, xxx, 208–9
    extraction, 206–7
    purification, 207
    for biomarker discovery, 196
    for dioxin toxicogenomics, 47
    for human genome microarray use, 29
    for transcriptional start site microarrays, 100
  for transcriptional start site microarrays, total purification, 91
    integrity number, xxx, 209

RNA – *continued*
profiling, 179–80, 182–4, 186, 188, 190, 192, 194, 196, 198, 200
quality considerations (and possible degradation)
sample, xxx, xxxi, 8, 11–2, 14–6, 22–4, 39, 41–2, 44, 92, 95, 104, 109–10, 179–80, 182, 184, 186, 188, 190, 192, 194, 196, 198, 200, 207–9, 213, 215, 217, 221–3
aRNA (amplified RNA) for human genome microarrays xvi, 12, 14–6, 29, 44, 191, 192
fluorescent labelling, *see* fluorescent dye-labelled probes
purification, 11
sense-strand aRNA, 22–3
sense-strand aRNA
aminoallyl-modified cDNA synthesis from, 23–4
purification, 22–3
synthesis, *see* transcription
cRNA (complementary RNA), xxv, 85–8, 99, 195–6, 212–3, 220–2, 236
for macaque gene expression studies
fragmentation, 212–13
purification of labelled material, 219–20
quantity and quality assessment, 221
for transcriptional start site microarray production, 86, 87
amplification from double-stranded cDNA, 99
fluorescent labelling, xxv, 87–8
mRNA, oxidation and biotinylation of diol groups, 94–5
tRNA, 45, 95–6, 128–9, 138, 168–9, 171–2, 232–4
tRNA, yeast, in BAC microarray hybridizations, 128–9
RNase, 8–29, 39–42, 57, 95, 100, 140, 163, 165, 167, 170–1, 188–9, 192–4, 208, 210–2
RNase A treatment of DNA for BAC microarrays, 140
RNase H, 57, 210
RNase I, 24, 95, 193, 212
robot, 1, 5, 44, 70, 77, 113, 120, 197
Roche, 45, 117, 128, 138, 168, 189, 232, 233
rodent, 32, 181
rotavirus, 65
RTS, 109, 111
RT-PCR, *see* reverse transcription–PCR

Saccharomyces cerevisiae, 4
SAGE, 51, 82, 109–11, 113
saline sodium citrate-based printing buffers, 119
salt concentration in DNA for BAC microarrays, 140–1
sample preparation, 118, 179–80, 182, 184, 186, 188, 190, 192, 194, 196, 198, 200
'sandwich' hybridizations, BAC microarrays, 129
SARS, 65
SAW, 227–8, 236
scaling factor, 184, 195
scaling normalization, 133
scanner, xix, 28–9, 46, 62–3, 73, 78, 117, 131–2, 141–2, 169, 173–4, 205, 210, 235
scanning (and scanners for probe detection)
amplified differential gene expression microarrays, 62–3
weak signals, 63
BAC microarray, 131–2, 139
CGH analysis, 173–4
HPV detection and genotyping, 72–3, 78
macaque gene expression studies, 214–16
scatter plot (of fluorescent signals), xxx, 89, 111, 186
RNA-generated microarrays datasets
fresh vs frozen blood RNA, 185
macaques, xxx, 215
smoothing, locally-weighted, 133
SDS, 28, 45–6, 62, 78, 94–5, 97–9, 128, 131, 136, 139, 169, 172, 232–5
second-round, xxvi, 12, 14, 192–4
SeeGH software, 117, 133
segment, xxvi, 125
sense, xix, xxv, 17, 19–24, 29, 54, 67, 69, 86, 88, 113
SenseAmp, xix, 17, 19–22, 29
sensitivity, xxii, 37, 51, 56, 71–2, 88, 106, 110, 116, 126–7, 129, 131–2, 141–3, 170, 176, 185
Sentrix, 6
Sephadex, 137, 211–2
sequence motif, 5
serial analysis of gene expression (SAGE), 51, 82, 109, 110, 111
serum, 3, 45, 78, 212
shareware, 38
Sigma, 8, 42, 45, 57, 61, 152, 163, 165–7, 170–1, 212–4, 228, 232–3, 239–40
signal-to-noise ratio, xxviii, 115, 120, 132, 134, 222, 228
signals (fluorescent)
intensity
with BAC microarrays, gradients on slide, 141–2
with blood RNA samples, 185, 186
spot intensity values, 105
with transcriptional start site microarrays, 90, 91
scatter plot, *see* scatter plot
weak, *see* weak signals
see also background, high
silane, 4, 44, 68, 70
-treated slides for BAC microarrays, 120
silicon, 2, 239
single-nucleotide polymorphism (SNP) microarrays, 115, 162

slide (microarray), 6, 28, 31, 35–6, 44–7, 63, 65, 67–8, 70, 77–9, 117, 119–22, 124–5, 128–32, 135–6, 138–42, 163, 169–70, 173, 226–9, 232–7
BAC microarrays
hybridizations under cover slips, 129
intensity gradients, 141–2
number, 121
processing, 124, 136–7
scanning after hybridization, 131–2
surface chemistry, 120, 139
dioxin toxicogenomics, production, 44
SlideBooster, 226, 227, 228
small-quantity
RNA, 8, 9, 179–80, 182–6, 188–90, 192, 194, 196, 198, 200
RNA profiling, 179, 180, 182–4, 186, 188, 190, 192, 194, 196, 198, 200
smooth muscle, 34, 36
SMRT microarray, 118, 119, 120, 124, 125
SNP, 115–6, 124
microarrays, 162
software, xix, xxviii, xxx, xxxi, 38, 45–6, 108, 117, 131, 133–4, 150, 157, 163, 173–4, 196, 203, 210, 215, 222, 223
Solgent, 75, 77
sorbitol, 87, 92–3
SP6 promoter (pSP6) primer-containing double-stranded HPV DNA sequences, 67, 69
specificity, xxii, xxiv, 37, 71–3, 84–5, 89–91, 110, 127, 147, 233
Spectral Genomics, 117
Spectral Chip 2600, 116
spectrophotometer, 15, 23, 127, 137, 154–5, 164, 166–7, 207, 221, 237
spin column, 22, 194, 212, 219
spleen, 197
spot, xxi, xxvi, xxviii, 28–9, 37, 44, 47, 63, 65, 76, 79, 103–10, 112–3, 119–24, 129, 132–4, 139–40, 174, 215, 222, 225–6
spot reliability evaluation score for DNA microarrays (SRED), 103, 104–8
calculation method, 112
theoretical background, 106–8
Spotfire, 215, 222, 240
SpotLight, 28–9
spotting solutions, 119
SRC promoter, 83
SRED, 103–6, 108, 112–4
SSC, 44–6, 62, 78, 119, 128, 131, 135–6, 139, 142, 169, 172–3, 232–5
SSC-based printing buffers, 119
staining, macaque gene expression studies using human genome microarrays, 214–16
standard deviation, 105
static, 120, 129, 225, 235
statistics, 35, 49, 159
statistical analysis of data, 105, 108
BAC microarrays, 133
differential gene expression detection, 37–8
storage, whole blood, 205–6
streptavidin, xxv, 86–7, 91, 95, 214
submegabase-resolution tiling set (SMRT) microarray, 118, 119, 120, 124, 125
substrate, 1, 4, 6, 65, 67, 119–21, 127, 139, 169, 203, 225, 227
substrates for BAC microarrays
numbers, 121
surface chemistry, 120
supernatant, 40–1, 43, 46, 58, 62, 78, 80, 92–4, 96–100, 136, 155, 164, 172, 206–7
superscript, 17, 23–4, 42, 56–7, 87, 92–3, 210, 228
suppliers of commercial products
human genome microarrays, 6, 203
website addresses, 238–40
SurePrint, 6
surface, 4, 5, 28–9, 41, 44–6, 63, 65–7, 77–8, 113, 119–20, 122, 124, 129, 131, 136, 142, 173, 208, 225–7
surface acoustic wave mixing to enhance hybridization, 225–37
methods and approaches, 227–37
troubleshooting, 237
surface chemistry of slides/substrates for BAC microarrays, 120, 139
surrogate endpoints, 180, 181
SymAtlas database, 91

t.p.m., 109–10
T7, xxv, 20–1, 44, 67, 69, 70, 75, 85–8, 98–9, 185, 193, 210, 212, 216, 217
T7 promoter, xxv, 44, 86–7, 98
T7 promoter (pT7)-containing double-stranded cDNA
for human genome microarrays
preparation, 20
transcription, 21
for transcriptional start site microarrays, preparation, 87
T7 promoter (pT7)-containing double-stranded HPV DNA sequences, 67, 69
tags, xxiv, 83–5, 104, 109–11, 113, 127
expressed sequence, *see* expressed sequence tags
in expression profiling of transcriptional start sites, 84
cluster definition, 84
TaKaRa, 92, 97–8
tamoxifen adjuvant therapy, breast cancer recurrence after, 183
*Taq*I-digested double-stranded cDNA (for amplified gene expression microarrays)
ligation of adapters to, 52, 53, 55, 58–9
preparation, 52

target, xxiii, 2, 4, 6, 65–78, 109, 113, 115–6, 119, 124–5, 127–30, 133, 139–41, 144, 159, 161, 163, 169, 176, 179–81, 183, 201–2, 204–5, 225–7, 236
  construction, vi, 66, 67
  design, xxiv, 84, 85, 225
target sequence (in microarrays), 88, 89, 204, 248
  HPV, preparation, 66–70, 76–7
  transcriptional start sites
    design, xxiv, 83, 84
    preparation, 84
TATA box, xxiv, 82
TBE, 165, 167, 170–1
TCCD, *see* dioxin
technology, 1–6, 30, 38, 83, 87, 113–4, 131, 147, 161, 176, 182, 185, 195, 198, 216, 225–6, 228–9, 235–6, 239
TeleChem International BAC microarray resources, 6, 17, 19–24, 26, 28, 29, 45, 62, 76, 117, 119, 138–9, 173, 229, 240
  microplates, 123
  printing pins, 123
template, xxv, xxvii, 20–1, 42–4, 52–3, 58–61, 71–3, 77, 86, 98–9, 125, 127, 141, 228
Tempus blood collection, 184, 186–8
terminal transferase, 23
tester DNA
  for amplified differential gene expression microarrays, 52–5, 60, 61, 62
    reassociation, 59
  for BAC microarrays, 125, 126
2,3,7,8-tetrachlorodibenzo-*p*-dioxin, *see* dioxin
TGF, 34
TheOnyx system, 196, 197
theoretical background, 106
therapeutic, 115, 180, 198, 224
thermocycler, 60–1, 67, 135, 154–5, 196–7
thiol linkers, N-terminal modifications, 67
threshold cycle (Ct) values, transcriptional start site expression profiling, 88
TIFF, xix, 29
toxicant, 31–3
toxicogenomics, xx, 31–2, 34, 36, 38, 40, 42, 44, 46, 48, 179
  of dioxin, *see* dioxin
  toxin, xx, 32
  transactivation, xx, 32–3
transcription, xxiv, xxxi, 10, 24, 32–3, 35, 42, 54, 109–11, 147, 149, 153, 158, 184, 212, 216, 223, 228, 237
  cDNA to produce amplified RNA (for biomarker discovery from blood)
    1st round, 191
    2nd round, 193
    2nd round sample purification, 194
  cDNA to produce amplified RNA (for human genome microarray)
    1st round, 11
    2nd round, 14–15
    of T7 promoter-containing cDNA templates, 21
  cDNA to produce amplified RNA (for human genome microarray for macaque gene expression analysis), 211, 219–20
  gene expression profiling of start sites/regulatory regions, xxiv–xxv, 81–101
    methods and approaches, 83–99
    recommended protocols, 91–9
    troubleshooting, 100
  initiation, 82–3
  start site, xxiv, 82–5, 149, 153, 158
  unit, 109, 111
transcription factor genes, 88–91
transcriptional units (TUs), 109–10
transcriptome, 82, 101, 204, 215, 222
transcripts, xxiv, 6, 44, 83–7, 103, 108–10, 204, 214, 228
translocate, 33
treatment, 31, 35, 37–8, 41, 95, 115, 129, 140–2, 147–50, 152–4, 156, 159–60, 163, 188, 196, 205
trehalose, 87, 92, 93
TRI reagent, xxx, 39–41, 205–6, 209
triplicate, xxviii, 134
Tris, 17, 23–6, 41, 92–3, 95–7, 163, 165–8, 170, 211–2, 228–9
trypsin, 163
TSS, xxiv, xxv, 82–8, 100–1
TT adapters (for amplified gene expression microarrays), 52–5
  ligation to *Taq*I-digested cDNA templates, 52, 53, 55, 58–9
TT primers (for amplified gene expression microarrays), 52–5
tumor(s), malignant, *see* cancer
tumor necrosis factor-α response in HUVEC cells, 195
tumor promoter, 32
tumor suppressor genes, 147
  promoters, CpG islands and their methylation, 147, 148
turbo labeling, 15–6
Tween, 163, 213–4
two-color, 4, 28–9, 31, 174, 177, 231

U95A chips, Affymetrix, 204
U133A 2.0 Human Genome Microarray, *see* Affymetrix
UCSC, 84, 117, 153
unbound, xx, 32, 46, 124, 142, 222
upregulate, 149–50
UV, 44, 77, 165, 167, 170–1, 221
uveal melanoma, chromosomal alterations, 174–5

vacuum, 23, 43–4, 57–62, 93, 100, 136, 155, 166, 169, 187, 197, 211, 219, 229–32

validation, 7, 73, 111, 177, 180–3, 196, 202
variance, 105
  analysis of (ANOVA), differential gene expression detection, 37–8
  coefficient of, double-round RNA amplification and, 195–6
vascular, 32–3, 47
vector, xxvi, 66–70, 75, 108, 118–9, 156, 214, 240
Vector Laboratories, 214, 240
vertebrate, 33–4
viscous, 139
visual representation of BAC microarray data, 133
VizX Labs, 215, 222
vortex, 16, 21, 28, 40, 43, 45–6, 56–62, 79, 93, 96, 99, 164, 167, 170, 172, 193–4, 206–7, 211, 217–8, 220, 234–6

washing, 47, 62–4, 78, 100, 128, 130–1, 139–42, 169, 190–1, 193, 214, 216, 229, 234–6
  post-hybridization, *see* hybridization
  wave, 79, 129, 208, 225–7, 235
weak signals
  amplified differential gene expression, 63
  BAC microarrays, 141
  HPV DNA microarrays, 79–80
wetting, 139
Whatman, 234–5
white blood cells, RNA isolation from, 184
whole blood, xxx, xxxi, 183–4, 187–8, 203–7, 209, 222–3
whole genome, 4, 7, 81
  labelling, xxvii, 126–7
  whole human genome, xix, 1–2, 4–8, 10, 12, 14, 16–24, 26, 28–30, 236
  microarrays, *see* human genome microarray
  wild-type, 34, 36

X chromosome, 147
xenobiotic, xx, 32–4
X-gal, xvii, 75

yeast, 3–4, 30, 45, 95–6, 128–9, 138, 168–9, 171–2, 232–4
yeast tRNA in BAC microarray hybridizations, 128–9